T0350082

FROM CATEGORIES TO HOMOTOPY THEORY

Category theory provides structure for the mathematical world and is seen everywhere in modern mathematics. With this book, the author bridges the gap between pure category theory and its numerous applications in homotopy theory, providing the necessary background information to make the subject accessible to graduate students or researchers with a background in algebraic topology and algebra.

The reader is first introduced to category theory, starting with basic definitions and concepts, before progressing to more advanced themes. Concrete examples and exercises illustrate the topics, ranging from colimits to constructions such as the Day convolution product.

Part II covers important applications of category theory, giving a thorough introduction to simplicial objects including an account of quasi-categories and Segal sets. Diagram categories play a central role throughout the book, giving rise to models of iterated loop spaces, and feature prominently in functor homology and homology of small categories.

Birgit Richter is Professor of Mathematics at the University of Hamburg. She is the co-editor of *Structured Ring Spectra* (2004) and *New Topological Contexts for Galois Theory and Algebraic Geometry* (2009).

From Categories to
Homotopy Theory

BIRGIT RICHTER
University of Hamburg

CAMBRIDGE
UNIVERSITY PRESS

University Printing House, Cambridge CB2 8BS, United Kingdom

One Liberty Plaza, 20th Floor, New York, NY 10006, USA

477 Williamstown Road, Port Melbourne, VIC 3207, Australia

314–321, 3rd Floor, Plot 3, Splendor Forum, Jasola District Centre, New Delhi – 110025, India

79 Anson Road, #06-04/06, Singapore 079906

Cambridge University Press is part of the University of Cambridge.

It furthers the University's mission by disseminating knowledge in the pursuit of education, learning, and research at the highest international levels of excellence.

www.cambridge.org
Information on this title: www.cambridge.org/9781108479622
DOI: 10.1017/9781108855891

First published 2020

A catalogue record for this publication is available from the British Library.

ISBN 978-1-108-47962-2 Hardback

To my father

Contents

Introduction

Category theory has at least two important features. The first one is that it allows us to structure our mathematical world. Many constructions that you encounter in your daily life look structurally very similar, like products of sets, products of topological spaces, and products of modules, and then you might be delighted to learn that there is a notion of a product of objects in a category and all the above examples are actually just instances of such products, here in the category of sets, topological spaces, and modules, so you don't have to reprove all the properties products have, because they hold for every such construction. So category theory helps you to recognize things as what they are.

It also allows you to express objects in a category by something that looks apparently way larger. For instance, the Yoneda lemma describes a set of the form $F(C)$ (where C is an object of some category and F is a functor from that category to the category of sets) as the set of natural transformations between another nice functor and F. This might look like a bad deal, but in this set of natural transformations you can manipulate things and this reinterpretation for instance gives you cohomology operations as morphisms between the representing objects.

Another feature is that you can actually use category theory in order to build topological spaces and to do homotopy theory. A central example is the nerve of a (small) category: You view the objects of your category as points, every morphism gives a 1-simplex, a pair of composable morphisms gives a 2-simplex, and so on. Then you build a topological space out of this by associating a topological n-simplex to an n-simplex in the nerve, but you do some nontrivial gluing, for instance, identity morphisms don't really give you any information so you shrink the associated edges. In the end you get a CW complex BC for every small category C. Properties of categories and functors translate into properties of this space and continuous maps between such spaces. For instance, a natural transformation between two functors gives rise to a homotopy between the induced maps, and an equivalence of categories gives a homotopy equivalence of the corresponding classifying spaces.

Classifying spaces of categories give rise to classifying spaces of groups but you can also use them and related constructions to build the spaces of the algebraic K-theory spectrum of a ring, you can give models for iterated based loop spaces and you can construct explicit models of homotopy colimits and much more.

This book has two parts. The first one gives an introduction to category theory describing its basic definitions and constructions, so this part focuses on the first feature of category theory. The second part presents applications to homotopy theory. Here, "homotopy theory" does not mean any precisely confined area of algebraic topology but rather some collection of topics that is heavily influenced by my research interests and my personal taste. An emphasis is on simplicial methods, on functor categories, on some concepts that are crucial for algebraic K-theory, on models for iterated based loop spaces, and on applications to homological algebra. The book also contains an account of functor homology and of homology of small categories. These are two concepts that in my opinion deserve to be wider known by working mathematicians, in particular, by working algebraic topologists. Many prominent examples of homology theories can be expressed that way, and even if you are not interested in homology theories per se, you might stumble across a spectral sequence whose E^2 term happens to consist of such homology groups – and then it might be helpful to recognize these groups, because then you have other means of understanding and calculating them.

One thing that you might realize is that I love diagrams. If I see a proof that uses a lengthy reformulation for showing that one thing (functor, natural transformation, etc.) is the same as a second thing, then I usually don't understand such a proof before I "translate" it into a diagram that has to commute, so I usually end up drawing the corresponding diagram. My hope is that this approach isn't just helpful for me. So in a lot of places in this book you will find proofs that more or less just consist of showing that a certain diagram commutes by displaying the diagram and dissecting its parts. I also love examples and therefore there are plenty of examples in the book. There are also some exercises. These are not meant to be challenging, but they want to nudge you to actually learn how to work with the concepts that are introduced and how to deal with examples. One danger in category theory is that one learns the abstract theory, and then, if confronted with an example, one doesn't really know what to do. I hope that the examples and exercises help to avoid this.

My hope is that this book will bridge a gap: There are several very good accounts on category theory, for instance, [**Rie16, Bo94-1, Bo94-2, ML98, Sch70**], and there are many excellent sources on applications of categorical methods to topics such as algebraic K-theory [**Q73, Gr76, DGM13, W13**], the theory of iterated loop spaces, models for categories of spectra and many more, but in the latter texts it is assumed that the reader is familiar with the relevant concepts from category theory and sometimes it can be difficult to collect the necessary background. This is not a book on ∞-categories, but I cover quasi-categories, joins, slices, cocartesian fibrations, and the category Θ_n and some other things related to ∞-categories, and these might help you with digesting Jacob Lurie's books [**Lu09, Lu∞**] and other sources. For an overview on quasi-categories and some of their applications, I recommend Moritz Groth's survey [**Gro20**]. For a comparison between different models of ∞-categories, Julie Bergner's book [**B18**] is an excellent source. My book is also not an introduction to model categories, but if you want to dive deeper into some of the applications and you decide to read the papers that I mention, then you will need them. There are very good sources for learning about model categories, and in increasing level of complexity I recommend [**DwSp95, Ho99, Hi03**], but of course also the original account [**Q67**].

In my opinion (many people would disagree), you should not learn category theory before you have seen enough examples of categories in your mathematical life, so before you feel the *need* for category theory.

I assume that you have some background in algebra and algebraic topology. In several places, I will use concepts from homological algebra, and I recommend Chuck Weibel's book [**W94**] if you need background on that.

How you read this book heavily depends on your background and I therefore refrain from giving a *Leitfaden*. If you know some basic category theory, you might jump ahead to the applications, and if you then realize that for a specific topic you need to look things up, then you can go back to the corresponding spot in the first part of the book. Similarly, some of you might know what a simplicial object is. Then of course you should feel free to skip that section.

I assume that the axiom of choice holds. I will *not* give an introduction to set theory in this book. For diagram categories, I will assume smallness in order to avoid technical problems.

Whenever you see a $A \subset B$ that means $a \in A \Rightarrow a \in B$, so that's what other people might denote by $A \subseteq B$. By $\square$, I denote the end of a proof.

Acknowledgments: This book had a long gestation period. It all started with a lecture course that I gave in Hamburg in the academic term of 2010/11.

I thank Clemens Heine for encouraging me to write the book and Inna Zakharevich who persuaded me to actually finish it. I would like to thank the Isaac Newton Institute for Mathematical Sciences for support and hospitality during the program *Homotopy Harnessing Higher Structures*, supported by EPSRC grant number EP/R014604/1. I finished a first draft when I was in Cambridge in 2018. I was an associate member of the SCR of Churchill College Cambridge, and I am grateful for the hospitality and the welcoming atmosphere in Churchill College. This really helped me to transform a rough draft of the book into a final version.

Part I

Category Theory

1

Basic Notions in Category Theory

1.1 Definition of a Category and Examples

If you want to define a category, it is not enough to specify the objects that you want to consider; you always have to say what kind of morphisms you want to allow.

Definition 1.1.1 A *category* $\mathcal{C}$ consists of

(1) A class of objects, $\text{Ob}\mathcal{C}$.
(2) For each pair of objects C_1 and C_2 of $\mathcal{C}$, there is a set $\mathcal{C}(C_1, C_2)$. We call the elements of $\mathcal{C}(C_1, C_2)$ the *morphisms from C_1 to C_2 in $\mathcal{C}$*.
(3) For each triple C_1, C_2, and C_3 of objects of $\mathcal{C}$, there is a composition law

$$\mathcal{C}(C_1, C_2) \times \mathcal{C}(C_2, C_3) \to \mathcal{C}(C_1, C_3).$$

We denote the composition of a pair (f, g) of morphisms by $g \circ f$.
(4) For every object C of $\mathcal{C}$, there is a morphism 1_C, called the *identity morphism on C*.

The composition of morphisms is associative, that is, for morphisms $f \in \mathcal{C}(C_1, C_2)$, $g \in \mathcal{C}(C_2, C_3)$, and $h \in \mathcal{C}(C_3, C_4)$, we have

$$h \circ (g \circ f) = (h \circ g) \circ f,$$

and identity morphisms do not change morphisms under composition, that is, for all $f \in \mathcal{C}(C_1, C_2)$,

$$1_{C_2} \circ f = f = f \circ 1_{C_1}.$$

We will soon see plenty of examples of categories. Despite the fact that for some categories this notation is utterly misleading, it is common to denote morphisms as arrows. If $f \in \mathcal{C}(C_1, C_2)$, then we represent f as $f : C_1 \to C_2$.

As for functions, we call $C_1 = s(f)$ the *source (or domain) of f* and $C_2 = t(f)$ the *target (or codomain) of f* for all $f \in C(C_1, C_2)$.

The identity morphism 1_C is uniquely determined by the object C: if both 1_C and $1'_C$ are identity morphisms on C, then

$$1_C = 1_C \circ 1'_C = 1'_C.$$

One can visualize the unit and associativity conditions geometrically. Omitting the objects from the notation, the rule $f \circ 1 = f = 1 \circ f$ can be expressed as two (glued) triangles,

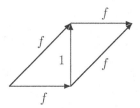

and the associativity constraint corresponds to a tetrahedron.

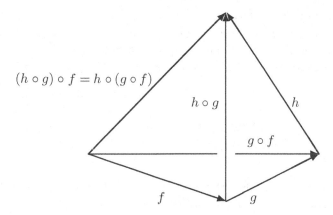

These pictures are more than mere illustrations; this will become clear when we discuss nerves and classifying spaces of small categories in 11.1 and 11.2.

Remark 1.1.2 Sometimes one does not require that the morphisms constitute a set, but one would allow classes of morphisms as well. In such contexts, our definition would be called a *locally small category*, that is, one in which for every pair of objects $C_1, C_2, C(C_1, C_2)$ is a set and not a class.

Remark 1.1.3 Some authors require that the sets of morphisms in a category are pairwise disjoint. If (C_1, C_2) is different from (C'_1, C'_2), then

$\mathcal{C}(C_1, C_2) \cap \mathcal{C}(C_1', C_2') = \varnothing$. This is related to the question of how you define a function: If X and Y are sets, then a function can be viewed as a relation $f \subset X \times Y$ with the property that, for all x in X, there is a unique $y \in Y$ with $(x, y) \in f$, or you could say that a function is a triple (X, Y, f) with $f \subset X \times Y$ with the same uniqueness assumption. In the latter definition, the domain and target are part of the data. In the first definition, f could also be a function for some other X' and Y'.

Definition 1.1.4

- A category $\mathcal{C}$ is *small* if the objects of $\mathcal{C}$ are a set (and not a proper class).
- A small category is *finite* if its set of objects is a finite set and every set of morphism is finite.
- A category is *discrete* if the only morphisms that occur in it are identity morphisms.

In particular, you can take any class X and form the *discrete category associated with X*, by declaring the elements of X to be the objects and by allowing only the identity morphisms as morphisms. If X is a set, then this category is small.

You are probably already familiar with several examples of categories.

Examples 1.1.5

- Sets: The category of sets and functions of sets. Here, the objects form a proper class.
- Gr: The category of groups and group homomorphisms.
- Ab: The category of abelian groups and group homomorphisms.
- K-vect: Here, K is a field and K-vect is the category of K-vector spaces and K-linear maps.
- R-mod: Here, R is an associative ring with unit and R-mod is the category of (left) R-modules and R-linear maps.
- Top: The category of topological spaces and continuous maps.
- Top_*: The category of topological spaces with a chosen basepoint and continuous maps preserving the basepoint.
- CW: The category of CW complexes and cellular maps.
- Ch: The category of (unbounded) chain complexes of abelian groups together with chain maps. Here, objects are families of abelian groups $(X_i)_{i \in \mathbb{Z}}$ with boundary operators $d_i \colon X_i \to X_{i-1}$

$$\cdots \xrightarrow{d_{n+2}} X_{n+1} \xrightarrow{d_{n+1}} X_n \xrightarrow{d_n} X_{n-1} \xrightarrow{d_{n-1}} \cdots.$$

The d_i are linear maps and satisfy $d_i \circ d_{i+1} = 0$ for all $i \in \mathbb{Z}$. We denote such a chain complex by (X_*, d). A chain map from (X_*, d) to (Y_*, d') is a family of linear maps $(f_i \colon X_i \to Y_i)_{i \in \mathbb{Z}}$ such that $f_i \circ d_{i+1} = d_i' \circ f_{i+1}$ for all $i \in \mathbb{Z}$, so the following diagram commutes:

$$
\begin{array}{ccccccccc}
\cdots \xrightarrow{d_{n+2}} & X_{n+1} & \xrightarrow{d_{n+1}} & X_n & \xrightarrow{d_n} & X_{n-1} & \xrightarrow{d_{n-1}} & \cdots \\
& \downarrow{f_{n+1}} & & \downarrow{f_n} & & \downarrow{f_{n-1}} & & \\
\cdots \xrightarrow{d_{n+2}'} & Y_{n+1} & \xrightarrow{d_{n+1}'} & Y_n & \xrightarrow{d_n'} & Y_{n-1} & \xrightarrow{d_{n-1}'} & \cdots .
\end{array}
$$

We also consider the variant of the category of nonnegatively graded chain complexes, $\mathsf{Ch}_{\geq 0}$, where $X_i = 0$ for all negative indices i. An important variant is to allow different ground rings than the integers, so we might consider chain complexes of R-modules for some associative and unital ring R, and then, the boundary operators and chain maps are required to be R-linear.

There are other examples of categories where you might find the morphisms slightly nonstandard.

Examples 1.1.6

(1) Let Corr be the category of correspondences. Objects of this category are sets, and the morphisms $\mathrm{Corr}(S, T)$ between two sets S and T are the subsets of the product $S \times T$. If you have $U \subset R \times S$ and $V \subset S \times T$, then $U \times V$ is a subset of $R \times S \times S \times T$. You can take the preimage of $U \times V$ under the map $j \colon R \times S \times T \to R \times S \times S \times T$ that takes the identity on R and T and the diagonal map on S and then project with $p \colon R \times S \times T \to R \times T$. This gives the composition. The identity morphism on the set S is the diagonal subset

$$\Delta_S = \{(s, s) \mid s \in S\} \subset S \times S.$$

(2) Let X be a partially ordered set (poset, for short), that is, a nonempty set X together with a binary relation $\leq$ on X that satisfies that $x \leq x$ for all $x \in X$ (reflexivity), that $x \leq y$ and $y \leq z$ implies $x \leq z$ (transitivity), and if $x \leq y$ and $y \leq x$, then $x = y$ (antisymmetry).

We consider such a poset as a category, and by abuse of notation, we call this category X. Its objects are the elements of X, and the set of morphisms $X(x, y)$ consists of exactly one element if $x \leq y$. Otherwise, this set is empty.

(3) Quite often, we will view categories as diagrams. For instance, let [0] be the category with one object and one morphism, the identity on that object.

Similarly, let $[1] = \{0, 1\}$ be the category with two objects 0 and 1, coming with their identity morphisms and one other morphism from 0 to 1. This corresponds to the poset $0 < 1$ viewed as a category: $0 \to 1$.

When we draw diagrams like that, we usually omit the identity morphisms and we don't draw composites in posets. For every poset $[n] = 0 < 1 < \cdots < n$, we get the corresponding category $0 \to 1 \to \cdots \to n$.

(4) Let X be a topological space and let $\mathfrak{U}(X)$ denote its family of open subsets of X. We can define a partial order on $\mathfrak{U}(X)$ by declaring that $U \leq V$ if and only if $U \subset V$.

(5) If $\mathcal{C}$ is an arbitrary category and if C is an object of $\mathcal{C}$, then the endomorphisms of C, $\mathcal{C}(C, C)$ form a monoid, that is, a set with a composition that is associative and possesses a unit. Thus, every category can be thought of as a *monoid with many objects*.

Conversely, if $(M, \cdot, 1)$ is a monoid with composition $\cdot$ and unit 1, then we can form the category that has one object $*$ and has M as its set of endomorphisms. We denote this category by $\mathcal{C}_M$.

There are several constructions that build new categories from old ones.

Definition 1.1.7

- We will need the *empty category*. It has no object and hence no morphism.
- If we have two categories $\mathcal{C}$ and $\mathcal{D}$, then we can build a third one by forming their *product* $\mathcal{C} \times \mathcal{D}$. As the notation suggests, the objects of $\mathcal{C} \times \mathcal{D}$ are pairs of objects (C, D) with C an object of $\mathcal{C}$ and D an object of $\mathcal{D}$. Morphisms are pairs of morphisms:

$$\mathcal{C} \times \mathcal{D}((C_1, D_1), (C_2, D_2)) = \mathcal{C}(C_1, C_2) \times \mathcal{D}(D_1, D_2),$$

and composition and identity morphisms are formed componentwise:

$$(f_2, g_2) \circ (f_1, g_1) = (f_2 \circ f_1, g_2 \circ g_1), \quad 1_{(C,D)} = (1_C, 1_D).$$

This is indeed a category.

- Given two categories $\mathcal{C}$ and $\mathcal{D}$, we can also form their disjoint union, $\mathcal{C} \sqcup \mathcal{D}$. Its objects consist of the disjoint union of the objects of $\mathcal{C}$ and $\mathcal{D}$. One defines

$$(\mathcal{C} \sqcup \mathcal{D})(X, Y) := \begin{cases} \mathcal{C}(X, Y), & \text{if } X, Y \text{ are objects of } \mathcal{C}, \\ \mathcal{D}(X, Y), & \text{if } X, Y \text{ are objects of } \mathcal{D}, \\ \varnothing, & \text{otherwise.} \end{cases}$$

- If we want a limited amount of interaction between $\mathcal{C}$ and $\mathcal{D}$, we can form the *join of $\mathcal{C}$ and $\mathcal{D}$*, denoted by $\mathcal{C} * \mathcal{D}$. The objects of $\mathcal{C} * \mathcal{D}$ are the disjoint union of the objects of $\mathcal{C}$ and the objects of $\mathcal{D}$, and as morphism, we have

$$(\mathcal{C}*\mathcal{D})(X, Y) = \begin{cases} \mathcal{C}(X, Y), & \text{if } X \text{ and } Y \text{ are objects of } \mathcal{C}, \\ \mathcal{D}(X, Y), & \text{if } X \text{ and } Y \text{ are objects of } \mathcal{D}, \\ \{*\}, & \text{if } X \text{ is an object of } \mathcal{C} \text{ and } Y \text{ is an object of } \mathcal{D}, \\ \varnothing, & \text{otherwise.} \end{cases}$$

So the join is not symmetric: There are morphisms from $\mathcal{C}$ to $\mathcal{D}$ but not from $\mathcal{D}$ to $\mathcal{C}$.

- Let $\mathcal{C}$ be an arbitrary category. Let $\mathcal{C}^o$ be the category whose objects are the same as the ones of $\mathcal{C}$ but where

$$\mathcal{C}^o(C, C') = \mathcal{C}(C', C).$$

We denote by f^o the morphism in $\mathcal{C}^o(C, C')$ corresponding to $f \in \mathcal{C}(C', C)$.

The composition of $f^o \in \mathcal{C}^o(C, C')$ and $g^o \in \mathcal{C}^o(C', C'')$ is defined as $g^o \circ f^o := (f \circ g)^o$. The category $\mathcal{C}^o$ is called the *dual category of* $\mathcal{C}$ or the *opposite category of* $\mathcal{C}$.

If you consider the preceding example of the category $\mathcal{C}_M$ from above, then the dual $(\mathcal{C}_M)^o$ is the category associated with the opposite of the monoid M, M^o. Here, M^o has the same underlying set as M, but the multiplication is reversed:

$$m \cdot^o m' := m' \cdot m.$$

1.2 EI Categories and Groupoids

Definition 1.2.1 We call a morphism $f \in \mathcal{C}(C, C')$ in a category $\mathcal{C}$ an *isomorphism* if there is a $g \in \mathcal{C}(C', C)$, such that $g \circ f = 1_C$ and $f \circ g = 1_{C'}$.

We denote g by f^{-1}, because g is uniquely determined by f.

Definition 1.2.2
- A category $\mathcal{C}$ is an *EI category* if every endomorphism of $\mathcal{C}$ is an isomorphism.
- A category $\mathcal{C}$ is a *groupoid* if every morphism in $\mathcal{C}$ is an isomorphism.

Of course, every groupoid is an EI category. In any EI category, the endomorphisms of an object form a group.

Examples 1.2.3
- Consider the category $\mathcal{I}$ whose objects are the finite sets $\mathbf{n} = \{1, \dots, n\}$ with $n \geq 0$ and $\mathbf{0} = \varnothing$. The morphisms in $\mathcal{I}$ are injective functions. Hence, the

endomorphisms of an object **n** constitute the symmetric group on n letters, Σ_n, and $\mathcal{I}$ is an EI category.

- Dually, let Ω be the category of finite sets and surjections, that is, Ω has the same objects as $\mathcal{I}$, but $\Omega(\mathbf{n}, \mathbf{m})$ is the set of surjective functions from the set **n** to the set **m**. Again, the endomorphisms of **n** consist of the permutations in Σ_n, and Ω is an EI category.

- Let $\mathcal{C}$ be any category. Then one can build the associated *category of isomorphisms of* $\mathcal{C}$, $\mathrm{Iso}(\mathcal{C})$. This has the same objects as $\mathcal{C}$, but we take

$$\mathrm{Iso}(\mathcal{C})(C_1, C_2) = \{f \in \mathcal{C}(C_1, C_2) \mid f \text{ is an isomorphism}\}.$$

Hence, for all categories $\mathcal{C}$, the category $\mathrm{Iso}(\mathcal{C})$ is a groupoid. We call the category $\mathrm{Iso}(\mathcal{I}) = \mathrm{Iso}(\Omega)$ the *category of finite sets and bijections*, Σ, that is, Σ has the same objects as Ω but

$$\Sigma(\mathbf{n}, \mathbf{m}) = \begin{cases} \Sigma_n, & \text{if } n = m, \\ \varnothing, & \text{otherwise.} \end{cases}$$

- If G is a group, then we denote by $\mathcal{C}_G$ the category with one object $*$ and $\mathcal{C}_G(*, *) = G$ with group multiplication as composition of maps. Then, $\mathcal{C}_G$ is a groupoid. Hence, every group gives rise to a groupoid. Vice versa, a groupoid can be thought of as a *group with many objects*.

- Let X be a topological space. The *fundamental groupoid of X*, $\Pi(X)$, is the category whose objects are the points of X, and $\Pi(X)(x, y)$ is the set of homotopy classes of paths from x to y:

$$\Pi(X)(x, y) = [[0, 1], 0, 1; X, x, y].$$

The endomorphisms $\Pi(x, x)$ of $x \in X$ constitute the fundamental group of X with respect to the basepoint x, $\pi_1(X, x)$.

- Another important example of a groupoid is the *translation category of a group*. If G is a discrete group, then we denote by $\mathcal{E}_G$ the category whose objects are the elements of the group and

$$\mathcal{E}_G(g, h) = \{hg^{-1}\}, \quad g \xrightarrow{\quad hg^{-1} \quad} h.$$

This category has the important feature that there is precisely one morphism from one object to any other object, so every object has equal rights.

For the symmetric group on three letters, Σ_3, the diagram of objects and (nonidentity) morphisms looks as follows:

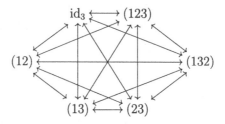

Here, we use cycle notation for permutations. Note that the upper-right triangle depicts the translation category of the cyclic group of order three.

1.3 Epi- and Monomorphisms

Often, we will need morphisms with special properties. In the category of sets, one can use elements in order to test whether a function is surjective or injective. In a general category, we do not have a notion of elements, but we always have sets of morphisms. Epimorphisms and monomorphisms are defined using morphisms as test objects. For the category of sets, this is straightforward: if a function $f : S \to T$ is injective, then $f(s_1) = f(s_2)$ implies that $s_1 = s_2$. So, this is also true for morphisms $h_1, h_2 : U \to S$. If $f \circ h_1 = f \circ h_2$, then $h_1 = h_2$. A similar consideration applies to surjective functions.

Definition 1.3.1 Let $\mathcal{C}$ be a category and $f \in \mathcal{C}(C_1, C_2)$. Then, f is an *epimorphism* if for all objects D in $\mathcal{C}$ and all pairs of morphisms $h_1, h_2 : C_2 \to D$, the equation $h_1 \circ f = h_2 \circ f$ implies that $h_1 = h_2$.

Epimorphisms are therefore right-cancellable.

Exercise 1.3.2 Beware, epimorphisms might not be what you think they are. Consider the category of commutative rings with unit and show that the unique morphism from the integers into the rational numbers is an epimorphism.

Remark 1.3.3 Of course, every identity morphism is an epimorphism and the composition of two epimorphisms is an epimorphism. If $g \circ f$ is an epimorphism, then so is g, because $h_1 \circ g = h_2 \circ g$ implies that $h_1 \circ g \circ f = h_2 \circ g \circ f$, and by assumption, this yields $h_1 = h_2$. Note that every isomorphism is an epimorphism.

Proposition 1.3.4 *Epimorphisms in the category of sets and functions are precisely the surjective functions.*

Proof   Let $f: X \to Y$ be a surjective function of sets and let $h_1, h_2: Y \to Z$ be functions with $h_1 \circ f = h_2 \circ f$. Let $y \in Y$ be an arbitrary element. Then, there is an $x \in X$ with $f(x) = y$. Hence

$$h_1(y) = h_1(f(x)) = h_2(f(x)) = h_2(y),$$

and thus, $h_1 = h_2$. Conversely, let $f: X \to Y$ be an epimorphism. Let $Z = \{z_1, z_2\}$ with $z_1 \neq z_2$ and define $h_1: Y \to Z$ as $h_1(y) = z_1$ for all $y \in Y$ and let h_2 be the test function of the image of f, that is, $h_2(y) = z_1$ if y is in the image of f and $h_2(y) = z_2$ if y lies outside of the image of f. Then, $h_1 \circ f = h_2 \circ f$ is the constant function with value z_1, and hence, $h_1 = h_2$, which implies that f is surjective.   □

Similarly, one can prove that epimorphisms of topological spaces are surjective continuous maps. It is more involved to show that epimorphisms of groups are surjective group homomorphisms. See, for example, [**Bo94-1**, 1.8.5.d].

Dual to the notion of an epimorphism is the one of a monomorphism.

Definition 1.3.5 Let $\mathcal{C}$ be a category. A morphism $f \in \mathcal{C}(C_1, C_2)$ is a *monomorphism* if $f^o \in \mathcal{C}^o(C_2, C_1)$ is an epimorphism.

So, monomorphisms are left-cancellable. For all objects D in $\mathcal{C}$ and all $h_1, h_2: D \to C_1$, the equation $f \circ h_1 = f \circ h_2$ implies that $h_1 = h_2$.

One can use morphism sets as test objects because the very definition of mono- and epimorphisms gives the following criteria.

Proposition 1.3.6 *A morphism $f \in \mathcal{C}(C_1, C_2)$ is a monomorphism if and only if for all objects D in $\mathcal{C}$, the induced map $\mathcal{C}(D, f): \mathcal{C}(D, C_1) \to \mathcal{C}(D, C_2)$ is an injective function of sets.*

Dually, a morphism $f \in \mathcal{C}(C_1, C_2)$ is an epimorphism if and only if for all objects D in $\mathcal{C}$, the induced map $\mathcal{C}(f, D): \mathcal{C}(C_2, D) \to \mathcal{C}(C_1, D)$ is an injective function of sets.

Monomorphisms of sets are injective functions. Monomorphisms of commutative rings with unit are injective ring homomorphisms: Let $f: R_1 \to R_2$ be a monomorphism of commutative rings with unit. Consider the polynomial ring in one variable over the integers, $\mathbb{Z}[X]$. A morphism $h: \mathbb{Z}[X] \to R_1$ determines and is determined by the element $r = h(X)$ of R_1. For all $r, s \in R_1$, we define $h_1: \mathbb{Z}[X] \to R_1$ via $h_1(X) = r$ and $h_2: \mathbb{Z}[X] \to R_1$ via $h_2(X) = s$. If $f(r) = f(s)$, then $f \circ h_1 = f \circ h_2$, and thus, $h_1 = h_2$, which implies $r = s$, so f is injective. The converse is easy to see.

There are categories where monomorphisms should be handled with care, that is, where monomorphisms do not behave like injective maps. A standard example is the category of divisible abelian groups (see Exercise 1.3.13).

A topologically minded example is the category of connected Hausdorff topological groups [**HM13**, A3.10].

Exercise 1.3.7 Show that, in the category of monoids, the inclusion of the additive monoid of natural numbers into the integers is a monomorphism and an epimorphism.

Definition 1.3.8 A morphism $r \in \mathcal{C}(C_1, C_2)$ is called a *retraction* if there is an $s \in \mathcal{C}(C_2, C_1)$, such that $r \circ s = 1_{C_2}$. In this situation, s is called a *section* and C_2 is a *retract of C_1*.

Proposition 1.3.9 *Retractions are epimorphisms, and sections are monomorphisms.*

Proof   We only prove the first claim; the second is dual. Let r be a retraction with section s. If $h_1 \circ r = h_2 \circ r$, then $h_1 = h_1 \circ r \circ s = h_2 \circ r \circ s = h_2$.   □

Remark 1.3.10 Be careful: the converse of the preceding statement is often wrong. For instance, let $\mathcal{C}$ be the category of groups. Then, a surjective group homomorphism f does not have a section in general. There is a section of the underlying function on sets, but this section does not have to be a group homomorphism in general.

The example of the category of commutative rings with units shows that there are categories where morphisms f that are epimorphisms and monomorphisms do not have to be isomorphisms: take $f \colon \mathbb{Z} \to \mathbb{Q}$. It is a monomorphism because it is injective, and it is an epimorphism but certainly not an isomorphism.

You might know the notions of projective and injective modules from homological algebra. The following is the categorical analog of these properties.

Definition 1.3.11
- An object P in a category $\mathcal{C}$ is called *projective* if for every epimorphism $f \colon M \to Q$ in $\mathcal{C}$ and every $p \colon P \to Q$, there is a $\xi \in \mathcal{C}(P, M)$ with $f \circ \xi = p$:

- Dually, an object I in a category $\mathcal{C}$ is called *injective* if for every monomorphism $f \colon U \to M$ in $\mathcal{C}$ and every $j \colon U \to I$, there is a $\zeta \in \mathcal{C}(M, I)$ with $\zeta \circ f = j$:

Remark 1.3.12 We think of the morphism ξ as a *lift* of p to M and of the morphism ζ as an *extension* of j to M. Note that uniqueness of the morphisms ξ and ζ is *not* required.

In the category of sets, every object is injective and projective, assuming the axiom of choice for projectivity. In the category of left R-modules for R an associative ring with unit, projectivity and injectivity are precisely defined as in homological algebra. Examples of projective modules are free modules or R-modules of the form Re, where e is an idempotent element of R, that is, $e^2 = e$. Injective $\mathbb{Z}$-modules, that is, injective abelian groups, are divisible abelian groups. These are abelian groups A, such that $nA = A$ for all natural numbers $n \neq 0$. Thus, $\mathbb{Q}$ and the discrete circle $\mathbb{Q}/\mathbb{Z}$ are injective abelian groups.

Exercise 1.3.13 Show that in the category of divisible abelian groups, the canonical projection map $\mathbb{Q} \to \mathbb{Q}/\mathbb{Z}$ is a monomorphism.

Projectivity and injectivity are preserved by passing to retracts.

Proposition 1.3.14 *If P is a projective object of a category $\mathcal{C}$ and if $i \colon U \to P$ is a monomorphism in $\mathcal{C}$ with a retraction $r \colon P \to U$, then U is projective. Similarly, if $i \colon J \to I$ is a monomorphism with retraction $r \colon I \to J$ and I is injective, then J is injective.*

Proof Let $f \colon M \to Q$ be an epimorphism. If U maps to Q via g, then P maps to Q via $g \circ r$. Thus, there is a morphism $\xi \colon P \to M$ with $f \circ \xi = g \circ r$, and therefore, $\xi \circ i$ satisfies $f \circ \xi \circ i = g \circ r \circ i = g$.

In the second case, if U maps to M via the monomorphism f and $j \colon U \to J$, then $i \circ j \colon U \to I$ has an extension $\zeta \colon M \to I$ with $\zeta \circ f = i \circ j$, and hence, $r \circ \zeta$ is the required extension of j to M. $\qquad\square$

We also get certain splitting properties for injective and projective objects.

Proposition 1.3.15 *If $q \colon Q \to P$ is an epimorphism and if P is projective, then q has a section. Dually, if $j \colon I \to J$ is a monomorphism and I is injective, then j has a retraction.*

Proof   We show the second claim and leave the first claim as an exercise. Consider the diagram

$$
\begin{array}{ccc}
 & & I \\
 & {}^{1_I}\nearrow & \Big\uparrow{\zeta} \\
I & \xrightarrow{\ j\ } & J
\end{array}
$$

By the injectivity of I, we get an extension of 1_I to J, ζ, satisfying $\zeta \circ j = 1_I$. Thus, ζ is a retraction for j. $\square$

1.4 Subcategories and Functors

Definition 1.4.1 Let C be a category. A *subcategory* D *of* C consists of a sub-collection of objects and morphisms of C, called the objects and morphisms of D, such that

- for all objects D_1, D_2 of D, there is a set of morphisms $D(D_1, D_2) \subset C(D_1, D_2)$;
- if $f \in D(D_1, D_2)$, then D_1, D_2 are objects of D;
- for all objects D of D, the identity morphism 1_D is an element of $D(D, D)$; and
- if $f \in D(D_1, D_2)$, $g \in D(D_2, D_3)$, then the composition of f and g in C satisfies $g \circ f \in D(D_1, D_3)$.

Hence, a subcategory of a category is a subcollection of objects and morphisms of the category that is closed under composition, identity morphisms, and source and target. Note that a subcategory again forms a category.

Definition 1.4.2 A subcategory D of C is called *full* if for all objects D, D' of D

$$D(D, D') = C(D, D').$$

The category of abelian groups is a full subcategory of the category of groups. However, the category I of finite sets and injections is *not* a full subcategory of the category of finite sets. We restricted the morphisms.

Definition 1.4.3 A *functor* F from a category C to a category D

- Assigns to every object C of C an object $F(C)$ of D.
- For each pair of objects C, C' of C, there is a function of sets

$$F \colon C(C, C') \to D(F(C), F(C')), \quad f \mapsto F(f).$$

- The following two axioms hold:

$$F(g \circ f) = F(g) \circ F(f) \quad \text{for all } f \in \mathcal{C}(C, C'), g \in \mathcal{C}(C', C''),$$

$$F(1_C) = 1_{F(C)}$$

for all objects C of $\mathcal{C}$.

Like for morphisms, we use the arrow notation $F : \mathcal{C} \to \mathcal{C}'$ to indicate a functor.

Examples 1.4.4

(1) The inclusion of a subcategory into its ambient category defines a functor.

(2) The identity map on objects and morphisms of a category $\mathcal{C}$ define the *identity functor*

$$\text{Id}_{\mathcal{C}} : \mathcal{C} \to \mathcal{C}.$$

(3) Let $(-)_{ab} : \mathsf{Gr} \to \mathsf{Ab}$ be the functor that assigns to a group G the factor group of G with respect to its commutator subgroup: $G/[G, G]$. The resulting group is abelian, and the functor is called the *abelianization*.

(4) Often, we will consider functors that forget part of some structure. These are called *forgetful functors*. For instance, we can consider the underlying set $U(X)$ of a topological space X, and this gives rise to the forgetful functor

$$U : \mathsf{Top} \to \mathsf{Sets}.$$

Similarly, if K is a field, then every K-vector space V has an underlying abelian group $U(V)$, and this gives rise to a forgetful functor

$$U : K\text{-vect} \to \mathsf{Ab}.$$

Here, we used that continuous maps are in particular functions of sets and that K-linear maps are morphisms of abelian groups.

You should come up with at least five more examples of such forgetful functors.

(5) To a pair of topological spaces (X, A) and to a fixed $n \in \mathbb{N}_0$, you can assign the nth singular homology group of (X, A), $H_n(X, A)$. Then, this defines a functor from the category of pairs of topological spaces to abelian groups.

(6) If you consider topological spaces with a chosen basepoint and if you assign to such a space (X, x) its fundamental group with respect to the basepoint x, $\pi_1(X, x)$, then this defines a functor from Top_* to the category of groups

$$\pi_1 : \mathsf{Top}_* \to \mathsf{Gr}.$$

(7) A functor $F\colon [0] \to \mathcal{C}$ corresponds to a choice of an object in $\mathcal{C}$, namely $F(0)$.

(8) A functor $F\colon [1] \to \mathcal{C}$ corresponds to the choice of two objects in $\mathcal{C}$, $F(0)$ and $F(1)$, and a morphism between them, $F(0 < 1)$:

$$F(0) \xrightarrow{\ F(0<1)\ } F(1).$$

(9) Let E be the category with two objects 0 and $0'$ and an isomorphism between them, that is, a morphism $f \in E(0, 0')$ and a morphism $g \in E(0', 0)$, such that $g \circ f = 1_0$ and $f \circ g = 1_{0'}$. Then, a functor F from E to any category $\mathcal{C}$ picks an isomorphism in $\mathcal{C}$ between $F(0)$ and $F(0')$. The category E is therefore often called the *wandering isomorphism*.

(10) A functor $F\colon [2] \to \mathcal{C}$ corresponds to the choice of a composable pair of morphisms in $\mathcal{C}$, so $F(0 < 2) = F(1 < 2) \circ F(0 < 1)$.

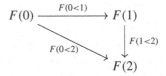

(11) We can assign to a set S the free group generated by S, $\mathrm{Fr}(S)$. A function of sets $f\colon S \to T$ induces a group homomorphism

$$\mathrm{Fr}(f)\colon \mathrm{Fr}(S) \to \mathrm{Fr}(T)$$

and hence Fr is a functor from the category of Sets to the category of groups.

(12) Similarly, we can send a set S to the free abelian group generated by S, $\mathrm{Fra}(S)$. This assignment is a functor as well.

(13) An innocent-looking but very important example of a functor is the (covariant) *morphism functor*: For an arbitrary category $\mathcal{C}$ and any object C_0 of $\mathcal{C}$, we can consider the map

$$C \mapsto \mathcal{C}(C_0, C)$$

that sends an object C of $\mathcal{C}$ to the set of morphisms from C_0 to C in $\mathcal{C}$. This defines a functor

$$\mathcal{C}(C_0, -)\colon \mathcal{C} \to \mathsf{Sets}.$$

(14) Another important functor that is crucial for the discussion of limits and colimits later is the constant functor. Consider two arbitrary nonempty

categories $\mathcal{C}$ and $\mathcal{D}$ and choose an object D of $\mathcal{D}$. The *constant functor from $\mathcal{C}$ to $\mathcal{D}$ with value D* is

$$\Delta_D \colon \mathcal{C} \to \mathcal{D}, \quad \Delta_D(C) = D, \quad \Delta_D(f) = 1_D$$

for all objects C in $\mathcal{C}$ and all $f \in \mathcal{C}(C_1, C_2)$.

(15) Let M be a smooth manifold and let $C(M)$ be the real vector space of all smooth real-valued functions on M. We denote by Sm the category of smooth manifolds and smooth maps. The assignment $M \mapsto C(M)$ defines a functor from the dual category of Sm to the category of real vector spaces

$$C \colon \mathsf{Sm}^o \to \mathbb{R}\text{-vect.}$$

(16) Let X be a topological space, and let $\mathcal{C}$ be an arbitrary category. As discussed earlier, $\mathfrak{U}(X)$ denotes the category of open subsets of X. Its objects are the open subsets of X, and if U and V are objects of $\mathfrak{U}(X)$ with $U \subset V$, then there is a morphism $i_U^V \colon U \to V$.

A *presheaf F on X* is a functor $F \colon \mathfrak{U}(X)^{op} \to \mathcal{C}$.

Often, the morphisms $F(i_U^V) \colon F(V) \to F(U)$ are called *restriction maps* and are denoted by $\mathrm{res}_{V,U}$. The property of F to being a functor is then equivalent to requiring that $\mathrm{res}_{U,U} = \mathrm{id}_U$ for all objects U of $\mathfrak{U}(X)$, and for open subsets $U \subset V \subset W$ in X, it doesn't matter whether you restrict from W to V and then from V to U or you restrict directly from W to U:

$$\mathrm{res}_{W,U} = \mathrm{res}_{V,U} \circ \mathrm{res}_{W,V}.$$

Typical examples of presheaves are sets of functions on a topological space X, such as the continuous functions from X to the reals. If $p \colon E \to M$ is a smooth vector bundle on a smooth manifold M, then setting $F(U)$ to be the set of smooth sections of p on the open subset $U \subset M$ defines a presheaf.

If $\mathcal{C}$ is a concrete category (see 5.1.12), then the elements of $F(U)$ are called *sections of F on U*, and $F(X)$ are the *global sections*. Sometimes, these notions are also used for general $\mathcal{C}$.

Remark 1.4.5 If $F \colon \mathcal{C} \to \mathcal{D}$ is a functor, then of course you can collect all objects of the form $F(C)$ for objects C of $\mathcal{C}$ and all morphisms $F(f)$ for f a morphism in $\mathcal{C}$, but beware that the image of a functor is not a subcategory of $\mathcal{D}$ in general.

Take, for instance, $\mathcal{C}$ as the category that consists of two disjoint copies of the poset [1]:

$$\mathcal{C}: \quad C_0 \xrightarrow{\ f\ } C_1 \quad C_0' \xrightarrow{\ f'\ } C_1'$$

and let $\mathcal{D}$ be the category [2]. Then, we can define a functor from $\mathcal{C}$ to $\mathcal{D}$ by declaring that $F(f) = (0 < 1)$, $F(f') = (1 < 2)$. As f cannot be composed with f' in $\mathcal{C}$, the composition $(0 < 2)$ is not in the image of $\mathcal{C}$ under F [**Sch70**, I.4.1.4].

Functors $F\colon \mathcal{C} \to \mathcal{D}$ are often called *covariant functors*, whereas functors $F\colon \mathcal{C}^o \to \mathcal{D}$ are called *contravariant functors from $\mathcal{C}$ to $\mathcal{D}$*. Thus, these are assignments from the class of objects in $\mathcal{C}$ to the class of objects in $\mathcal{D}$, so that on the level of morphism sets, we get

$$F\colon \mathcal{C}(C, C') \to \mathcal{D}(F(C'), F(C))$$

with $F(g \circ f) = F(f) \circ F(g)$ and $F(1_C) = 1_{F(C)}$.

Singular cochains (or singular cohomology groups) define a contravariant functor from the category of topological spaces to the category of cochain complexes (or graded abelian groups).

If you assign to a vector space its dual vector space, then for every K-linear map $f\colon V \to W$, you get a K-linear map $f^*\colon W^* \to V^*$, which is defined as $\varphi \mapsto \varphi \circ f$

$$
\begin{array}{ccc}
V & \xrightarrow{\ f\ } & W \\
 & \searrow{\scriptstyle \varphi \circ f} & \downarrow{\scriptstyle \varphi} \\
 & & K
\end{array}
$$

This turns the process of building the dual of a vector space into a contravariant functor from the category of K-vector spaces to itself.

Example 1.4.6 Contravariant functors from the fundamental groupoid of a space X, $\Pi(X)$, to the category of abelian groups, $\mathcal{G}\colon \Pi(X)^o \to \mathsf{Ab}$, are called (abelian) bundles of groups on X or a system of local coefficients on X. This can be used to define homology with local coefficients (see, for instance, [**Ste43**], [**Wh78**, Chapter VI], or [**DK01**, Chapter 5]).

For every point $x \in X$, we get an abelian group $\mathcal{G}(x)$, and for every homotopy class $[w]$ of a path from x to y, there is a group homomorphism $\mathcal{G}([w])\colon \mathcal{G}(y) \to \mathcal{G}(x)$. Note that the $\mathcal{G}([w])$s are automatically isomorphisms with inverse $\mathcal{G}([\bar{w}])$, where $\bar{w}$ is the time-reversed path of w.

Exercise 1.4.7 Consider two groups G, G' and the corresponding categories $\mathcal{C}_G$ and $\mathcal{C}_{G'}$ with one object and morphisms G and G'. Show that functors $F : \mathcal{C}_G \to \mathcal{C}_{G'}$ correspond to group homomorphisms $f : G \to G'$.

Exercise 1.4.8 Let $(S, \leq)$ and $(T, \leq)$ be two posets. A morphism of posets is an order-preserving function $f : S \to T$, that is, if $s_1 \leq s_2$ in S, then $f(s_1) \leq f(s_2)$ in T. Show that functors from the category S to the category T are precisely morphisms of posets.

Let $\mathcal{C}$ be a category, and let C be an object in $\mathcal{C}$. We can use the functors $\mathcal{C}(C, -)$ and $\mathcal{C}(-, C)$ for testing whether C is projective or injective. The following criterion is a direct consequence of the definitions, bearing in mind that epimorphisms in the category Sets are precisely surjective functions.

Proposition 1.4.9
- *The object C is projective if and only if $\mathcal{C}(C, -) : \mathcal{C} \to$ Sets preserves epimorphisms.*
- *The object C is injective if and only if $\mathcal{C}(-, C) : \mathcal{C}^o \to$ Sets sends monomorphisms to epimorphisms.*

Functors can be used to compare categories.

Definition 1.4.10
- A functor $F : \mathcal{C} \to \mathcal{D}$ is an *isomorphism of categories* if there exists a functor $G : \mathcal{D} \to \mathcal{C}$ with the properties $F \circ G = \mathrm{Id}_{\mathcal{D}}$ and $G \circ F = \mathrm{Id}_{\mathcal{C}}$.

  In particular, F induces a bijection between the classes of objects of $\mathcal{C}$ and $\mathcal{D}$ and on the morphism sets.
- A functor is *full* if the assignment

$$F : \mathcal{C}(C, C') \to \mathcal{D}(F(C), F(C')), \ f \mapsto F(f) \tag{1.4.1}$$

  is surjective for all pairs of objects C, C' of $\mathcal{C}$.
- A functor is *faithful* if the assignment in (1.4.1) is injective for all pairs of objects C, C' of $\mathcal{C}$.
- A functor is *fully faithful* if the assignment in (1.4.1) is a bijection for all pairs of objects C, C' of $\mathcal{C}$.
- A functor $F : \mathcal{C} \to \mathcal{D}$ is *essentially surjective* if for all objects D of $\mathcal{D}$, there is an object C of $\mathcal{C}$, such that $F(C)$ is isomorphic to D.

Exercise 1.4.11 Let R be an associative ring with unit and denote by R^{op} the ring that has the same underlying abelian group as R but whose multiplication is reversed. Show that the categories of left R-modules and of right R^{op}-modules are isomorphic.

Exercise 1.4.12 Recall the join of categories from Definition 1.1.7. Show that there is an isomorphism of categories between $(\mathcal{C} * \mathcal{D})^o$ and $\mathcal{D}^o * \mathcal{C}^o$.

Prove that there is an isomorphism between the categories $[0] * [0]$ and $[1]$. Show that $[i] * [j]$ is isomorphic to $[i + j + 1]$.

Exercise 1.4.13 Let $F: \mathcal{C} \to \mathcal{D}$ be a faithful functor. Show that F detects monomorphisms, that is, if $F(f)$ is a monomorphism, then f is a monomorphism. Do full functors detect epimorphisms?

Remark 1.4.14
- Faithful functors can forget structure. For instance, the forgetful functor from groups to sets is faithful.
- There are examples of functors that are fully faithful but not isomorphisms of categories. The important point is that full faithfulness does not imply that the functor is essentially surjective, and it does not rule out that the functor maps different objects to the same image.
- If $\mathcal{D}$ is a subcategory of $\mathcal{C}$, then there is an inclusion functor $I: \mathcal{D} \to \mathcal{C}$. The category $\mathcal{D}$ is a full subcategory of $\mathcal{C}$ if the inclusion functor is a full functor.

We can compose functors, and we have identity functors; thus, categories behave like objects in a category. This can be made precise.

Definition 1.4.15 We denote by cat the category whose objects are all small categories and whose morphisms between a category $\mathcal{C}$ and a category $\mathcal{D}$ are all functors from $\mathcal{C}$ to $\mathcal{D}$.

Why do we restrict to small categories? We insisted on the morphisms between two objects forming a set. Take, for instance, the category of sets. The functors from Sets to itself contain the constant functors, so for each set, there is a constant functor, with that set as its value. This would already be a proper class of functors.

For a small category, we can define a suitable notion of connectedness.

Definition 1.4.16 Let $\mathcal{C}$ be a small category. Two objects C_1 and C_2 are said to be equivalent if there is a morphism in $\mathcal{C}$ between C_1 and C_2. We consider the equivalence relation generated by this relation. Thus, two objects $C, \tilde{C}$ are equivalent if there is a finite zigzag of morphisms of $\mathcal{C}$ connecting C and $\tilde{C}$:

If every object of $\mathcal{C}$ is connected to any other object in $\mathcal{C}$, that is, if there is just one such equivalence class, then we call the category $\mathcal{C}$ *connected*.

1.5 Terminal and Initial Objects

Some categories possess special objects.

Definition 1.5.1
- An object t of a category $\mathcal{C}$ is called *terminal* if there exists a unique morphism $f_C\colon C \to t$ in $\mathcal{C}$ from every object C of $\mathcal{C}$ to t.
- Dually, an object s is called *initial* if there exists a unique morphism $f^C\colon s \to C$ in $\mathcal{C}$ from s to every object C of $\mathcal{C}$.
- An object 0 is a *zero object* in $\mathcal{C}$ if it is terminal and initial.

Remark 1.5.2 If t is terminal, then the endomorphisms of t consist only of the identity map of t, and dually, the set of endomorphisms of an initial object is $\{1_s\}$.

A small category that possesses an initial or a terminal object is connected.

Terminal and initial objects are unique up to isomorphism. If $\mathcal{C}$ has a zero object 0, then for all pairs of objects C, C' in $\mathcal{C}$, there is the unique morphism

$$C \xrightarrow{\ f_C\ } 0 \xrightarrow{\ f^{C'}\ } C'.$$

This is often called the *zero morphism* and is denoted by $0\colon C \to C'$.

Exercise 1.5.3
- Show that any composite of a morphism with the zero morphism is zero.
- If $f\colon C \to C'$ is a monomorphism and if the composition $f \circ g$ is the zero morphism, then $g = 0$.

Examples 1.5.4
- In the category of sets Sets, the empty set is the initial object and any set with one element is terminal. There is no zero object in Sets. This changes if we consider the category of pointed sets, Sets$_*$. The objects of Sets$_*$ are sets with a chosen basepoint, and morphisms are functions of sets that map the basepoint in the source to the basepoint in the target. In this example, every set with one element is a zero object.
- Let R be an associative ring with unit. The category of left R-modules has a zero object, and this is the zero module 0. For $R = \mathbb{Z}$, we obtain that the trivial group is a zero object in the category of abelian groups Ab.

- This also applies to the category of groups: the trivial group is an initial object, and it is also terminal.
- Let G be a group. In the translation category $\mathcal{E}_G$, every object is initial and terminal.

Exercise 1.5.5 Let X be a partially ordered set. What does it say about the partial order relation on X if X has a terminal or initial object? When does X possess a zero object?

Exercise 1.5.6 Let $\mathcal{C}$ be an arbitrary category. Show that the join of $\mathcal{C}$ with $[0]$, $\mathcal{C} * [0]$, has 0 as a terminal object and that $[0] * \mathcal{C}$ has 0 as an initial object.

Definition 1.5.7 The category $\mathcal{C} * [0]$ is the *inductive cone with base* $\mathcal{C}$, and $[0] * \mathcal{C}$ is the *projective cone with base* $\mathcal{C}$.

2

Natural Transformations and the Yoneda Lemma

2.1 Natural Transformations

A typical example of a natural transformation is the connecting homomorphism for singular homology. Consider, for instance, the nth singular homology group of a pair of spaces (X, A). There is a connecting homomorphism

$$\delta \colon H_n(X, A) \to H_{n-1}(A).$$

This morphism is natural for morphisms of pairs of topological spaces: if $f \colon (X, A) \to (Y, B)$ is a continuous map $f \colon X \to Y$ with $f(A) \subset B$, then the following diagram commutes:

$$
\begin{array}{ccc}
H_n(X, A) & \xrightarrow{\ \delta\ } & H_{n-1}(A) \\
{\scriptstyle H_n(f)}\downarrow & & \downarrow{\scriptstyle H_{n-1}(f|_A)} \\
H_n(Y, B) & \xrightarrow{\ \delta\ } & H_{n-1}(B)
\end{array}
$$

Thus, δ is a morphism between functors from the category of pairs of topological spaces to the category of abelian groups, the functor $(X, A) \mapsto H_n(X, A)$ and the functor $(X, A) \mapsto H_{n-1}(A)$. The general definition is as follows:

Definition 2.1.1

- Let F, G be two functors from $\mathcal{C}$ to $\mathcal{C}'$. A *natural transformation* η from F to G consists of a class of morphisms $\eta_C \in \mathcal{C}'(F(C), G(C))$, the *components of η*, such that for every morphism $f \in \mathcal{C}(C_1, C_2)$,

$$\eta_{C_2} \circ F(f) = G(f) \circ \eta_{C_1},$$

that is, the diagram

$$F(C_1) \xrightarrow{\eta_{C_1}} G(C_1)$$

with vertical maps $F(f)$ and $G(f)$, and bottom

$$F(C_2) \xrightarrow{\eta_{C_2}} G(C_2)$$

commutes.

- Two functors $F, G : C \to C'$ are *naturally isomorphic* if there is a natural transformation η from F to G, such that the components $\eta_C \in C'(F(C), G(C))$ are isomorphisms for all objects C of C.

Remark 2.1.2 If η is a natural isomorphism, then the morphisms $(\eta_C)^{-1}$ assemble to a natural transformation η^{-1}, such that η^{-1} is again a natural isomorphism.

Example 2.1.3 Two local coefficient systems $G, G' : \Pi(X)^o \to$ Ab are isomorphic if there is a natural isomorphism between them. If G is isomorphic to a constant local coefficient system, then G is called *simple*.

Exercise 2.1.4 Let G and G' be groups and let $F, F' : C_G \to C_{G'}$ be two functors. When do we have a natural transformation $\eta : F \Rightarrow F'$?

We denote morphisms and functors by arrows. For natural transformations, it is common to use $\Rightarrow$ as a symbol. If η is a natural transformation from F to G, then we write $\eta : F \Rightarrow G$ for that fact. If we want to encorporate the functors F and G into the picture, then we use a 2-cell

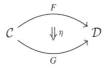

as the corresponding diagram. You can compose natural transformations. Let F, G, and H be functors from C to D and let $\eta : F \Rightarrow G, \nu : G \Rightarrow H$. Then, the composition $\nu \circ_1 \eta$ is a natural transformation from F to H with components $(\nu \circ_1 \eta)_C = \nu_C \circ \eta_C$. This is often called the *vertical composition of two natural transformations*:

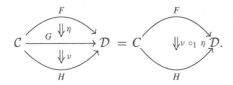

We fix some notation for the composition of natural transformations and functors. If $\eta\colon F \Rightarrow G$ is a natural transformation between the functors $F, G\colon \mathcal{C} \rightarrow \mathcal{C}'$ and if $H\colon \mathcal{C}' \rightarrow \mathcal{C}''$ is a functor, then we denote by $H(\eta)\colon H \circ F \Rightarrow H \circ G$ the natural transformation with components

$$H(\eta)_C := H(\eta_C)\colon H(F(C)) \rightarrow H(G(C)).$$

Similarly, for a functor $K\colon \tilde{\mathcal{C}} \rightarrow \mathcal{C}$, the natural transformation $\eta_K\colon F \circ K \Rightarrow G \circ K$ is given by

$$(\eta_K)_C := \eta_{K(C)}\colon F(K(C)) \rightarrow G(K(C)).$$

There is also the *horizontal composition*. Let $\mathcal{C}$, $\mathcal{D}$, and $\mathcal{E}$ be three categories and $F, G\colon \mathcal{C} \rightarrow \mathcal{D}$ and $F', G'\colon \mathcal{D} \rightarrow \mathcal{E}$ be functors with natural transformations $\eta\colon F \Rightarrow G$ and $\psi\colon F' \Rightarrow G'$. Then, η and ψ can be used to construct a natural transformation between the composite functors $F' \circ F$ and $G' \circ G$. To this end, consider an arbitrary object C of $\mathcal{C}$ and the diagram

$$
\begin{array}{ccc}
F'(F(C)) & \xrightarrow{\ \psi_{F(C)}\ } & G'(F(C)) \\
{\scriptstyle F'(\eta_C)}\big\downarrow & & \big\downarrow{\scriptstyle G'(\eta_C)} \\
F'(G(C)) & \xrightarrow{\ \psi_{G(C)}\ } & G'(G(C)).
\end{array}
$$

This diagram commutes because of the naturality of ψ applied to the morphism η_C. Define the composition $\psi \circ_0 \eta$ as the composite. This is depicted as

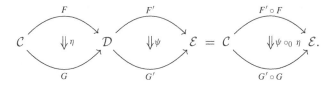

We always have the identity natural transformation $\mathrm{ID}\colon F \Rightarrow F$, whose components are the identity on the objects $F(C)$:

$$\mathrm{ID}_C = 1_{F(C)}.$$

There is an *interchange law* for the vertical and horizontal compositions. If there are three categories involved and two triples of functors, then in the diagram

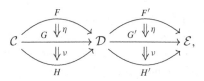

the horizontal composition of $\nu' \circ_1 \eta'$ and $\nu \circ_1 \eta$ agrees with the vertical composition of $\nu' \circ_0 \nu$ and $\eta' \circ_0 \eta$:

$$(\nu' \circ_1 \eta') \circ_0 (\nu \circ_1 \eta) = (\nu' \circ_0 \nu) \circ_1 (\eta' \circ_0 \eta), \qquad (2.1.1)$$

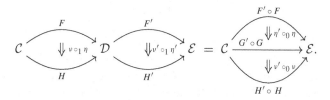

This should remind you of the interchange law for the multiplication in a double-loop space, ensuring, for instance, that the homotopy groups $\pi_n(X, x_0)$ of a pointed topological space (X, x_0) are abelian for all $n \geq 2$. In the following, we will not use the notation $\circ_0$ and $\circ_1$, and we hope that it is clear from the context what kind of composition is used.

Notation 2.1.5 Let $\mathcal{C}$ be a small category and let $\mathcal{C}'$ be an arbitrary category. Then, the functors from $\mathcal{C}$ to $\mathcal{C}'$ form a category whose morphisms are the natural transformations. We denote this category by $\mathsf{Fun}(\mathcal{C}, \mathcal{C}')$.

Remark 2.1.6 We can jazz up the example cat to something that is called a 2-category: the objects of cat are all small categories, the morphisms between a small category $\mathcal{C}$ and a small category $\mathcal{D}$ themselves form a category where the objects are the functors from $\mathcal{C}$ to $\mathcal{D}$ and the morphisms are the natural transformations between such functors. We will say more about higher categories in 9.5.

Examples 2.1.7

- Let U denote the forgetful functor from the category of abelian groups to the category of groups. There is a natural transformation between the functors Fr and $U \circ$ Fra from the category Sets to the category Gr, η. This is given by the abelianization $(-)_{\mathrm{ab}}$. The free abelian group generated by a set S, $\mathrm{Fra}(S)$ is isomorphic to the abelianization of the free group generated by S, $\mathrm{Fr}(S)/[\mathrm{Fr}(S), \mathrm{Fr}(S)]$, and any function $f: S \to T$ of sets renders the following diagram commutative:

$$\begin{array}{ccc} \mathrm{Fr}(S) & \xrightarrow{\eta_S} & \mathrm{Fra}(S) \\ {\scriptstyle \mathrm{Fr}(f)} \downarrow & & \downarrow {\scriptstyle \mathrm{Fra}(f)} \\ \mathrm{Fr}(T) & \xrightarrow{\eta_T} & \mathrm{Fra}(T) \end{array}$$

- Let V be a K-vector space. There is a canonical map from V to its double dual

$$\iota_V : V \to V^{**}, v \mapsto (\varphi \mapsto \varphi(v)), v \in V, \varphi \in V^*.$$

For any K-linear map $f : V \to W$, we have $\iota_W \circ f = f^{**} \circ \iota_V$, and therefore, ι is a natural transformation between the identity functor and the functor that sends a vector space to its double dual and sends a K-linear f to f^{**}.

- The category of presheaves on a topological space X with values in a category $\mathcal{C}$ forms a functor category. Its objects are the presheaves on X and its morphisms are the natural transformations, that is, for two presheaves F and G on X a morphism from F to G is a family of morphisms in $\mathcal{C}$ $(f_U : F(U) \to G(U))_{U \in U(X)}$, which is compatible with the restriction maps; that is,

$$\begin{array}{ccc} F(V) & \xrightarrow{f_V} & G(V) \\ {\scriptstyle \mathrm{res}_V^U} \downarrow & & \downarrow {\scriptstyle \mathrm{res}_V^U} \\ F(U) & \xrightarrow{f_U} & G(U) \end{array}$$

commutes for all open subsets $U \subset V$ of X. For instance, if X is a smooth manifold, then the inclusion of the smooth real-valued functions into the continuous real-valued functions is a morphism of presheaves.

For groups, we know the notion of the center of a group. It consists of all elements in the group that commute with all other elements. In category theory, we can express centrality as follows:

Definition 2.1.8 Let $\mathcal{C}$ be a small category. The *center of* $\mathcal{C}$, $Z(\mathcal{C})$, is the set of natural transformations from the identity functor on $\mathcal{C}$ to itself.

Example 2.1.9 Consider the case of a category $\mathcal{C}_G$ for a group G. A natural transformation η from $\mathrm{Id}_{\mathcal{C}_G}$ to itself consists of a morphism η_* from the only object of $\mathcal{C}_G$ to itself; hence, η_* is an element of the group G, and we call this element g_η. The naturality condition then means that for all $h \in G$, the composition $h \circ \eta_* = h \cdot g_\eta$ is equal to the composition $\eta_* \circ h = g_\eta \cdot h$. Hence, the center of the category $\mathcal{C}_G$ is the center of the group.

Exercise 2.1.10 Show that the center of a small category is always an abelian monoid.

Exercise 2.1.11 Prove that the center of the category of left R-modules for an associative ring with unit is the center of the ring R. It helps to study η_R first.

2.2 The Yoneda Lemma

The Yoneda lemma controls natural transformations from a representable functor to another functor.

Definition 2.2.1 A functor $F: \mathcal{C} \to$ Sets is *representable* if it is isomorphic to a morphism functor; that is, there is an object C of $\mathcal{C}$ and a natural isomorphism

$$\eta: \mathcal{C}(C, -) \Rightarrow F.$$

In this case, the pair (C, η) is called a *representation* of F.

Of course, morphism functors $\mathcal{C}(C, -)$ are representable with the pair (C, ID) as a representation.

Theorem 2.2.2 (Yoneda lemma) *Let $\mathcal{C}$ be a category and let F be a functor from the category $\mathcal{C}$ to the category* Sets.

(1) For each object C of $\mathcal{C}$, there is a bijection $Y(F, C)$ between the set of all natural transformations from $\mathcal{C}(C, -)$ to F, $\text{Nat}(\mathcal{C}(C, -), F)$, and the set $F(C)$.

(2) The bijections $(Y_F)_C := Y(F, C)$ are the components of a natural transformation $Y_F: \text{Nat}(\mathcal{C}(-, -), F) \Rightarrow F$.

(3) If $\mathcal{C}$ is small, then the bijections $(Y_C)_F := Y(F, C)$ are the components of a natural transformation from the functor

$$\text{Nat}(\mathcal{C}(C, -), -): \text{Fun}(\mathcal{C}, \text{Sets}) \to \text{Sets}$$

to the functor $\varepsilon_C: \text{Fun}(\mathcal{C}, \text{Sets}) \to$ Sets that sends F to the set $F(C)$.

Proof

(1) Let η be such a natural transformation, then, in particular, for all objects C' of $\mathcal{C}$, there is a function

$$\eta_{C'}: \mathcal{C}(C, C') \to F(C').$$

One defines $Y(F, C)$ as

$$Y(F, C)(\eta) := \eta_C(1_C).$$

We have an evaluation map

$$\mathcal{C}(C, C') \times F(C) \to F(C')$$

that sends a pair (f, x) to $F(f)(x)$. Fixing $x \in F(C)$, this yields for $f \in \mathcal{C}(C, C')$

$$\tau(F, C)_{x, C'}(f) := F(f)(x),$$

and if we vary f, we obtain a function

$$\tau(F, C)_{x, C'} : \mathcal{C}(C, C') \to F(C'),$$

and we claim that this yields a natural transformation $\tau(F, C)_x$: $\mathcal{C}(C, -) \Rightarrow F$. For a morphism $g \in \mathcal{C}(C', C'')$, we get for all $f \in \mathcal{C}(C, C')$

$$\begin{aligned} F(g) \circ \tau(F, C)_{x, C'}(f) = F(g)(F(f)(x)) &= F(g \circ f)(x) \\ &= \tau(F, C)_{x, C''} \circ \mathcal{C}(C, g \circ f) \\ &= \tau(F, C)_{x, C''} \circ \mathcal{C}(C, g)(f) \end{aligned}$$

and thus

$$F(g) \circ \tau(F, C)_{x, C'} = \tau(F, C)_{x, C''} \circ \mathcal{C}(C, g).$$

The maps $Y(F, C)$ and $\tau(F, C)$ are inverses of each other. For any $x \in F(C)$, we have

$$Y(F, C)(\tau(F, C)_x) = \tau(F, C)_{x, C}(1_C) = F(1_C)(x) = 1_{F(C)}(x) = x.$$

On the other hand, for every natural transformation $\eta : \mathcal{C}(C, -) \Rightarrow F$ and for every $f : C \to C'$ in $\mathcal{C}$, we get that

$$\begin{aligned} \tau(F, C)_{Y(F, C)(\eta), C'}(f) = \tau(F, C)_{\eta_C(1_C), C'}(f) &= F(f)(\eta_C(1_C)) \\ &= \eta_{C'}(\mathcal{C}(C, f)(1_C)) = \eta_{C'}(f \circ 1_C) = \eta_{C'}(f). \end{aligned}$$

(2) For the claim about the naturality in $\mathcal{C}$, we consider the functor $N : \mathcal{C} \to$ Sets, which is defined on objects as

$$N(C) := \mathrm{Nat}(\mathcal{C}(C, -), F).$$

For $f : C \to C'$, we set $N(f) : \mathrm{Nat}(\mathcal{C}(C, -), F) \to \mathrm{Nat}(\mathcal{C}(C', -), F)$ as

$$N(f)(\eta) := \eta \circ \mathcal{C}(f, -).$$

The natural transformation $Y_F \colon N \Rightarrow F$ is defined as $(Y_F)_C :=$ $Y(F, C)$, and the following straightforward calculations

$$(Y(F, C') \circ N(f))(\eta) = Y(F, C')(\eta \circ C(f, -)) = (\eta \circ C(f, -))(1_{C'})$$
$$= \eta_{C'}(f)$$

and

$$(F(f) \circ Y(F, C))(\eta) = F(f)(\eta_C(1_C)) = \eta_{C'} \circ C(f, -)(1_C) = \eta_{C'}(f)$$

yield the claim. We also get $(Y_F)_{C'} \circ N(f) = F(f) \circ (Y_F)_C$.

(3) If C is a small category, then the functors from C to the category of sets form a category $\mathsf{Fun}(C, \mathsf{Sets})$, with the natural transformations as morphisms. For a fixed object C of C, we define a functor

$$\Upsilon \colon \mathsf{Fun}(C, \mathsf{Sets}) \to \mathsf{Sets}, \quad \Upsilon(F) := \mathsf{Nat}(C(C, -), F).$$

Let G be an object of $\mathsf{Fun}(C, \mathsf{Sets})$, that is, a functor from C to the category of sets, and let $\gamma \colon F \Rightarrow G$ be a natural transformation. We define

$$\Upsilon(\gamma) \colon \mathsf{Nat}(C(C, -), F) \to \mathsf{Nat}(C(C, -), G)$$

as $\Upsilon(\gamma)(\eta) := \gamma \circ \eta$.

We consider the evaluation functor at an object C in C, $\varepsilon_C \colon \mathsf{Fun}(C, \mathsf{Sets}) \to \mathsf{Sets}$, which sends a functor $F \colon C \to \mathsf{Sets}$ to

$$\varepsilon_C(F) := F(C)$$

and $\varepsilon_C(\gamma) = \gamma_C$.

We define $Y_C \colon \Upsilon \Rightarrow \varepsilon_C$ as

$$(Y_C)_F := Y(F, C).$$

Then, a calculation shows that $(Y_C)_G \circ \Upsilon(\gamma) = \varepsilon_C(\gamma) \circ (Y_C)_F$ for any G as above. $\qquad \square$

Remark 2.2.3 The Yoneda lemma is one of the most important tools in category theory and its applications. You can express the set $F(C)$ as a set of natural transformations, and often, this is a means to control the functor F. You can also use it to control the objects in C as follows:

For instance, a representation of a functor F, that is, a natural isomorphism $\eta \colon C(C, -) \Rightarrow F$, is determined by $F(C)$.

Corollary 2.2.4 *Let C be an arbitrary category and let C, C' be two objects of C. Then, there is a bijection between the set of natural transformations between the representable functors $C(C, -)$ and $C(C', -)$ and the set of morphisms $C(C', C)$.*

This innocent-looking corollary has many applications. For instance, cohomology operations of singular cohomology can be described using this result. For that application, we actually need a contravariant version of the Yoneda lemma and its consequences.

First, we describe how different representations of a representable functor can be.

Lemma 2.2.5 *If a functor* $F : C \to D$ *has two representations,* $\eta : C(C, -) \Rightarrow F$ *and* $\tau : C(C', -) \Rightarrow F$, *then* C *and* C' *are isomorphic in* C.

Proof　We define $f : C \to C'$ as $\eta_{C'}^{-1}(\tau_{C'}(1_{C'}))$ and $g \in C(C', C)$ as $\tau_C^{-1}(\eta_C(1_C))$. We have to check that $f \circ g = 1_{C'}$ and $g \circ f = 1_C$. We prove the first equality; the proof of the second one is similar. The diagram

$$
\begin{array}{ccccc}
1_{C'} \in C(C', C') & \xrightarrow{\tau_{C'}} & F(C') & \xleftarrow{\eta_{C'}} & C(C, C') \\
{\scriptstyle (-)\circ g}\downarrow & & {\scriptstyle F(g)}\downarrow & & \downarrow{\scriptstyle (-)\circ g} \\
g \in C(C', C) & \xrightarrow{\tau_C} & F(C) & \xleftarrow{\eta_C} & C(C, C) \ni 1_C
\end{array}
$$

commutes. As $\tau_C(g) = \eta_C(1_C)$, we get that $f \circ g = \eta_C^{-1}\tau_C(g) = 1_C$. □

Exercise 2.2.6 Dualize the Yoneda lemma to the context of contravariant functors from the category C to the category of sets.

The nth singular cohomology group of a space X with coefficients in an abelian group A is isomorphic to the homotopy classes of maps from X to an Eilenberg–Mac Lane space of type (A, n). The latter is a space denoted by $K(A, n)$ of the homotopy type of a CW space, such that

$$
\pi_i K(A, n) = \begin{cases} A, & \text{if } i = n, \\ 0, & \text{otherwise.} \end{cases}
$$

The $K(A, n)$ are infinite loop spaces, and hence, the set of homotopy classes of maps $[X, K(A, n)]$ is actually an abelian group for all $n \geq 0$ and

$$
H^n(X; A) \cong [X, K(A, n)]
$$

is an isomorphism of abelian groups that is natural in the space X. Thus, the functor $X \mapsto H^n(X; A)$ is representable. A cohomology operation $\varphi_{(A,n),(B,m)} : H^n(X; A) \to H^m(X; B)$, which is natural in X, can hence be identified with a natural transformation between the functors $X \mapsto [X, K(A, n)]$ and $X \mapsto [X, K(B, m)]$, and these in turn are in bijection with

$[K(A, n), K(B, m)] \cong H^m(K(A, n); B)$. Here, we actually get an isomorphism of abelian groups: we will later see versions of the Yoneda lemma in enriched settings (see 9.3 and Proposition 2.4.5). As $K(A, n)$ doesn't have nontrivial cohomology groups below degree n (due to the Hurewicz theorem), these operations are trivial for $m < n$. For $A = B = \mathbb{F}_p$, a prime field, the collection of all such cohomology operations constitutes the *Steenrod algebra*.

More generally, Brown's representability theorem states that every generalized cohomology theory can be represented by an Omega spectrum ([**Ad74**], [**Sw75**, Chapter 9]).

By changing our point of view, we can use the Yoneda lemma to say something about the assignment that sends an object C of a small category $\mathcal{C}$ to the representable functor $\mathcal{C}(C, -)$. A morphism $f \in \mathcal{C}(C, C')$ induces $\mathcal{C}(f, -): \mathcal{C}(C', -) \to \mathcal{C}(C, -)$, and hence, we get a functor

$$Y: \mathcal{C}^o \to \mathsf{Fun}(\mathcal{C}, \mathsf{Sets}).$$

Theorem 2.2.7 (Yoneda embedding) *The functor Y is fully faithful, and two morphism functors $\mathcal{C}(C', -)$ and $\mathcal{C}(C, -)$ are isomorphic if and only if the objects C and C' are isomorphic.*

Proof   Corollary 2.2.4 ensures that the functor Y is fully faithful.

If C and C' are isomorphic, then there are morphisms $f: C \to C'$ and $g: C' \to C$, such that the precomposition with f and g induces a natural isomorphism between the functors $\mathcal{C}(C', -)$ and $\mathcal{C}(C, -)$.

If we assume that the functors $\mathcal{C}(C', -)$ and $\mathcal{C}(C, -)$ are isomorphic, then there is a natural isomorphism $\eta: \mathcal{C}(C, -) \to \mathcal{C}(C', -)$, which corresponds to an element $f_\eta \in \mathcal{C}(C', C)$ by Corollary 2.2.4. The inverse of η, say τ, corresponds to an $f_\tau \in \mathcal{C}(C, C')$, and the Yoneda lemma implies that f_τ is inverse to f_η.   □

Morphisms in a category induce natural transformations of morphism functors. The Yoneda lemma 2.2.2 relates properties of morphisms to the properties of the induced natural transformation.

Lemma 2.2.8 *Let $f \in \mathcal{C}(C_1, C_2)$ and let $\mathcal{C}(f, -)$ denote the induced natural transformation with*

$$\mathcal{C}(f, -)_{C_3} = \mathcal{C}(f, C_3): \mathcal{C}(C_2, C_3) \to \mathcal{C}(C_1, C_3).$$

Then, $\mathcal{C}(f, C_3)$ is a monomorphism for all objects C_3 of $\mathcal{C}$ if and only if f is an epimorphism. Dually, $\mathcal{C}(f, C_3)$ is an epimorphism for all objects C_3 of $\mathcal{C}$ if and only if f has a left inverse.

Note that the natural transformations from $\mathcal{C}(C_2, -)$ to $\mathcal{C}(C_1, -)$ are in bijection with $\mathcal{C}(C_1, C_2)$ and under this bijection $\mathcal{C}(f, -)$ corresponds to f.

Proof For every $g \in \mathcal{C}(C_2, C_3)$, $\mathcal{C}(f, -)_{C_3}(g) = g \circ f$. This shows the first claim. If $\mathcal{C}(f, -)_{C_3}$ is an epimorphism, then it is surjective, and thus, there is an $h \in \mathcal{C}(C_2, C_1)$ with $h \circ f = 1_{C_1}$ and thus f has a left inverse. Conversely, if f has a left inverse h and if $g \in \mathcal{C}(C_1, C_3)$ is any morphism, then g is equal to $\mathcal{C}(f, -)(gh)$. □

2.3 Equivalences of Categories

Often, we would like to compare different categories. In topology, you have notions of homeomorphisms and homotopy equivalences. Homeomorphisms correspond to actual isomorphisms in the category of topological spaces, whereas homotopy equivalences are a weaker but nonetheless very important method to compare spaces. The corresponding notions in category theory are isomorphisms of categories and equivalences of categories. We will see a direct connection to properties of topological spaces in Theorem 11.2.4.

Definition 2.3.1 A functor F from $\mathcal{C}$ to $\mathcal{C}'$ is called an *equivalence of categories* if there is a functor G from $\mathcal{C}'$ to $\mathcal{C}$, such that there is a natural isomorphism between the identity functor on $\mathcal{C}'$ and $F \circ G$ and one between $G \circ F$ and the identity functor on $\mathcal{C}$.

Example 2.3.2 Let $\mathcal{E}_G$ be the translation category of a discrete group G, as in Example 1.2.3. Then, $\mathcal{E}_G$ is equivalent to the category $[0]$ with one object and one identity morphism. There is a unique functor $P\colon \mathcal{E}_G \to [0]$, sending every object to 0 and every morphism to the identity morphism on 0. We define $F\colon [0] \to \mathcal{E}_G$ via $F(0) = e$, where e denotes the neutral element in G, and we set $F(1_0) = e$. The composite $P \circ F$ is the identity functor on the category $[0]$, whereas the composite $F \circ P$ sends any morphism $h\colon g \to hg$ in $\mathcal{E}_G$ to $e\colon e \to e$.

We define $\eta\colon F \circ P \Rightarrow \mathrm{Id}_{\mathcal{E}_G}$ by setting

$$\eta_g\colon F \circ P(g) = e \to g = \mathrm{Id}_{\mathcal{E}_G}(g)$$

to be the morphism $g: e \to g$ in the translation category. As the diagram

$$
\begin{array}{ccc}
\eta_g : F \circ P(g) = e & \xrightarrow{\quad g \quad} & g \\
{\scriptstyle F \circ P(h) = e} \downarrow & & \downarrow {\scriptstyle \mathrm{Id}(h) = h} \\
\eta_{hg} : F \circ P(hg) = e & \xrightarrow{\quad hg \quad} & hg
\end{array}
$$

commutes for all $h, g \in G$ and as $\mathcal{E}_G$ is a groupoid, this defines a natural isomorphism.

Exercise 2.3.3 Show that the composition of two equivalences of categories is an equivalence.

Prove that two auto-equivalences of a category $\mathcal{C}$, F and F', do not have to be isomorphic [**Sch70**, 16.2.3].

We will later see in Theorem 2.5.1 that functors that are essentially surjective, full, and faithful are actually equivalences of categories.

2.4 Adjoint Pairs of Functors

The definition of adjoint functors goes back to Kan [**K58a**] and is one of the fundamental concepts of category theory.

Definition 2.4.1 Let $\mathcal{C}$ and $\mathcal{C}'$ be categories. An *adjunction between $\mathcal{C}$ and $\mathcal{C}'$* is a pair of functors $L: \mathcal{C} \to \mathcal{C}'$, $R: \mathcal{C}' \to \mathcal{C}$, such that for each pair of objects C of $\mathcal{C}$ and C' of $\mathcal{C}'$, there is a bijection of sets

$$\varphi_{C,C'}: \mathcal{C}'(L(C), C') \cong \mathcal{C}(C, R(C')), \tag{2.4.1}$$

which is natural in C and C'.

The functor L is then *left adjoint to R*, and R is *right adjoint to L*. We call (L, R) an *adjoint pair of functors*.

The naturality condition on the bijections $\varphi_{C,C'}$ can be spelled out explicitly as follows: For all morphisms $f: C \to D$ in $\mathcal{C}$ and $g: C' \to D'$ in $\mathcal{C}'$, the diagram

$$
\begin{array}{ccccc}
\mathcal{C}'(L(D), C') & \xrightarrow{\mathcal{C}'(Lf, C')} & \mathcal{C}'(L(C), C') & \xrightarrow{\mathcal{C}'(L(C), g)} & \mathcal{C}'(L(C), D') \\
{\scriptstyle \varphi_{D,C'}} \downarrow & & \downarrow {\scriptstyle \varphi_{C,C'}} & & \downarrow {\scriptstyle \varphi_{C,D'}} \\
\mathcal{C}(D, R(C')) & \xrightarrow{\mathcal{C}(f, R(C'))} & \mathcal{C}(C, R(C')) & \xrightarrow{\mathcal{C}(C, R(g))} & \mathcal{C}(C, R(D'))
\end{array}
$$

commutes.

One often denotes adjunctions as $\mathcal{C} \underset{R}{\overset{L}{\rightleftarrows}} \mathcal{C}'$ or as $L: \mathcal{C} \rightleftarrows \mathcal{C}': R$.

Examples 2.4.2 A prototypical example of an adjunction is a forgetful functor and a "free" functor: if $R = U$ is a forgetful functor, and if a left adjoint of U exists, then the defining property from (2.4.1) means that for each morphism from C to $U(C')$ in the underlying category, there is a unique corresponding morphism from $L(C)$ to C', so, in this sense, $L(C)$ is the free object associated with C.

- For instance, let U be the functor that maps an abelian group A to the underlying set of A, and Fra is the functor that sends a set S to the free abelian group with basis S, $\bigoplus_S \mathbb{Z}$. Then, Fra is left adjoint to U: for any function f from S to the underlying set of an abelian group A, there is a unique morphism of abelian groups from Fra(S) to A extending f that is determined by sending the basis element of the copy of $\mathbb{Z}$ in component $s \in S$ to $f(s)$.
- Similarly, we can consider the forgetful functor from the category of Lie algebras over a field K to the category Sets. For simplicity, assume that the characteristic of K is not equal to 2 or 3. This forgetful functor has as a left adjoint functor that sends a set S to the free Lie algebra, $L_K(S)$, over K generated by the set S. An ad hoc way to describe $L_K(S)$ is to consider the K-vector space generated by all iterated formal Lie brackets on pairwise different elements of S, that is, expressions of the form $[s_1, [[s_2, s_3], s_4]]$, and then to divide out by the antisymmetry and Jacobi relation. If you want to learn more about this, then Reutenauer's book [**Re93**] is an excellent source.

Exercise 2.4.3 Let U denote the forgetful functor from the category of topological spaces to the category of sets. Does U have a left adjoint? What about a right adjoint?

Exercise 2.4.4 Fix an arbitrary commutative ring with unit, k, and a small category $\mathcal{C}$. Functors from $\mathcal{C}$ to the category of k-modules are called $\mathcal{C}$-modules. Note that the set of natural transformations between two $\mathcal{C}$-modules carries the structure of a k-module.

For any object C of $\mathcal{C}$, the functor $C' \mapsto k\{\mathcal{C}(C, C')\}$ is a $\mathcal{C}$-module. Prove the following k-linear version of the Yoneda lemma by using the free-forgetful adjunction between the category of Sets and the category of k-modules:

Proposition 2.4.5 *For each object C of $\mathcal{C}$ and for every $\mathcal{C}$-module F, there is an isomorphism of k-modules $Y_{F,C}$ between the k-module of all natural transformations from $k\{\mathcal{C}(C, -)\}$ to F, $\mathsf{Nat}(k\{\mathcal{C}(C, -)\}, F)$, and the k-module $F(C)$.*

We can describe adjunctions in a different manner:

Proposition 2.4.6 *The following are equivalent:*

(1) The functor L is left adjoint to R (and hence R is right adjoint to L).
(2) There are natural transformations $\eta\colon \mathrm{Id} \Rightarrow R \circ L$ and $\varepsilon\colon L \circ R \Rightarrow \mathrm{Id}$ with the properties that

$$\varepsilon_L \circ L(\eta) = \mathrm{Id}_L \text{ and } R(\varepsilon) \circ \eta_R = \mathrm{Id}_R;$$

hence, the diagrams

$$L(C) \xrightarrow{L(\eta)} LRL(C) \text{ and } R(C') \xrightarrow{\eta_{R(C')}} RLR(C')$$

with $1_{L(C)}$, $\varepsilon_{L(C)}$ to $L(C)$; and $1_{R(C')}$, $R(\varepsilon_{LR(C')})$ to $R(C')$

commute for all objects C of $\mathcal{C}$ and C' of $\mathcal{C}'$.

The transformation η is called the unit *of the adjunction and ε is the* counit.

Proof   Given an adjunction, we obtain η_C as

$$\eta_C := \varphi_{C,LC}(1_{LC}), \ \varepsilon_{C'} := \varphi^{-1}_{RC',C'}(1_{RC'}).$$

Here, $1_{LC} \in \mathcal{C}'(LC, LC)$ and $1_{RC'} \in \mathcal{C}(RC', RC')$. The naturality of the bijections $\varphi_{C,C'}$ guarantees that η and ε are indeed natural transformations. A calculation shows that they satisfy the required properties.

Conversely, if η and ε are natural transformations satisfying

$$\varepsilon_L \circ L(\eta) = \mathrm{Id}_L \text{ and } R(\varepsilon) \circ \eta_R = \mathrm{Id}_R,$$

and if $f \in \mathcal{C}'(LC, C')$, then we define $\varphi_{C,C'}(f)\colon C \to RC'$ as

$$C \xrightarrow{\eta_C} RLC \xrightarrow{R(f)} RC'.$$

By construction, $\varphi_{C,C'}$ is natural in C and C' and is a bijection. $\square$

Remark 2.4.7 Proposition 2.4.6 allows us to make the isomorphism $\varphi_{C,C'}$ and its inverse explicit. In the proof, we saw that $\varphi_{C,C'}(f) = R(f) \circ \eta_C$. Conversely, if $g \in \mathcal{C}(C, R(C'))$, then $\varphi^{-1}_{C,C'}(g) = \varepsilon_{C'} \circ L(g)$. Often, for instance, if R is a forgetful functor, then $\varphi_{C,C'}(f)$ can be thought of as a morphism that has less information than f, and one might denote $\varphi_{C,C'}(f)$ by $f^\flat$ (thinking of the underlying category as being lower than the original one). Similarly, $\varphi^{-1}_{C,C'}(g)$ might be denoted by $g^\sharp$. Note, however, that this usage is not at all standard. Some authors actually use the reverse notation.

Exercise 2.4.8 Show that adjunctions can be composed. Let $L: \mathcal{C} \rightleftarrows \mathcal{C}': R$ and $L': \mathcal{C}' \rightleftarrows \mathcal{C}'': R'$ be two adjunctions. Prove that $L'L: \mathcal{C} \rightleftarrows \mathcal{C}'': RR'$ is an adjunction. What are the unit and the counit of this adjunction?

Proposition 2.4.9 *If (L, R) is an adjoint pair of functors, then each of the functors L and R determines the other functor uniquely up to isomorphism.*

Proof Assume that R has two left adjoints functors, L and L'. Then, for all objects C of $\mathcal{C}$ and D of $\mathcal{D}$, we have bijections

$$\mathcal{D}(LC, D) \cong \mathcal{C}(C, RD) \cong \mathcal{D}(L'C, D)$$

that are natural in C and D. By Lemma 2.2.5, we get that $LC \cong LC'$, and this isomorphism is also natural in C; hence, L and L' are isomorphic as functors. The argument for L with two right adjoints is similar. $\square$

Exercise 2.4.10 Let $F: \mathcal{C} \to \mathcal{D}$ be a functor. Express the existence of a right adjoint of F in terms of representability of functors.

The counit of an adjunction actually tells us about the properties of the right adjoint functor.

Proposition 2.4.11 *Let $L: \mathcal{C} \rightleftarrows \mathcal{C}': R$ be an adjunction and let $\varepsilon: LR \Rightarrow Id$ be its counit. Then,*

(1) The functor R is faithful if and only if the morphism $\varepsilon_{C'}: LR(C') \to C'$ is an epimorphism for all objects C' of $\mathcal{C}'$.

(2) The functor R is full if and only if for all objects C' of $\mathcal{C}'$, the morphism $\varepsilon_{C'}: LR(C') \to C'$ has a left inverse.

Hence, the functor R is fully faithful if and only if $\varepsilon_{C'}$ is an isomorphism for all objects C'. In this case, both ηR and $L\eta$ are natural isomorphisms.

Proof The functor R gives a function on sets from $\mathcal{C}'(C_1', C_2')$ to $\mathcal{C}(R(C_1'), R(C_2'))$, and we compose this map with $\varphi_{R(C_1'), C_2}^{-1}$ to end up in $\mathcal{C}'(LR(C_1'), C_2')$:

$$\mathcal{C}'(C_1', C_2') \to \mathcal{C}'(LR(C_1'), C_2').$$

This map is determined by its effect on $1_{C_1'}$, but then, we obtain $\varepsilon_{C_1'}$ as its image. The Yoneda lemma then implies that the map above is $\mathcal{C}'(\varepsilon_{C_1'}, C_2')$. As $\varphi_{R(C_1'), C_2}^{-1}$ is an isomorphism, R is full (or faithful) if and only if $\mathcal{C}'(\varepsilon_{C_1'}, -)$ induces an epimorphism (or monomorphism). We saw in Lemma 2.2.8 that this is the case if and only if $\mathcal{C}'(\varepsilon_{C_1'}, -)$ is a split monomorphism (or an epimorphism).

The identities

$$R(\varepsilon_C) \circ \eta_{RC} = 1_{RC}, \ \varepsilon_{LD} \circ L(\eta_D) = 1_{LD}$$

hold for all objects C of $\mathcal{C}$ and D of $\mathcal{D}$. If we assume (1) and (2), then ε_{LD} and ε_C are isomorphisms, and hence, η_{RC} and $L\eta_D$ are isomorphisms as well. $\square$

Exercise 2.4.12 Can one dualize Proposition 2.4.11 and obtain statements about the unit and the properties of L?

How does one actually construct adjunctions? Sometimes you might be able to guess what an adjoint should be on objects. The following concepts then allow you to construct an actual adjunction from these partial data.

Definition 2.4.13 Let $F: \mathcal{C} \to \mathcal{D}$ be an arbitrary functor and let D be an object of $\mathcal{D}$. A *reflection of D at F* is a pair (G_D, η_D), where

- G_D is an object in $\mathcal{C}$; and
- $\eta_D: D \to F(G_D)$ is a morphism in $\mathcal{D}$,

with the property that for all objects C of $\mathcal{C}$ and for all $g: D \to F(C)$, there is a unique $f: G_D \to C$ with

$$F(f) \circ \eta_D = g,$$

$$\begin{array}{ccc}
D & \xrightarrow{\eta_D} & F(G_D) \\
{\scriptstyle g} \downarrow & \swarrow{\scriptstyle F(f)} & \\
F(C). & &
\end{array}$$

As usual, the universal property ensures uniqueness. In the following, an isomorphism between two reflections (G_D, η_D) and (G'_D, η'_D) of D at F is a morphism $f \in \mathcal{C}(G_D, G'_D)$ with $F(f) \circ \eta_D = \eta'_D$.

Lemma 2.4.14 *Let $F: \mathcal{C} \to \mathcal{D}$ be an arbitrary functor and let D be an object of $\mathcal{D}$. Then, a reflection of D at F is unique up to unique isomorphism.*

Proof We assume that both (G'_D, η'_D) and (G_D, η_D) are reflections of D at F. The defining property then yields a unique $f: G_D \to G'_D$ with $F(f) \circ \eta_D = \eta'_D$ and vice versa, there is a unique $f': G'_D \to G_D$, with the property that $F(f') \circ \eta'_D = \eta_D$. However, this also implies

$$F(f \circ f') \circ \eta'_D = F(f) \circ F(f') \circ \eta'_D = \eta'_D \text{ and } F(f' \circ f) \circ \eta_D$$
$$= F(f') \circ F(f) \circ \eta_D = \eta_D.$$

The uniqueness of $f' \circ f$ and $f \circ f'$ therefore ensures that

$$f' \circ f = 1_{G_D} \text{ and } f' \circ f = 1_{G'_D}. \qquad \square$$

A reflection gives us an object G_D, together with a structure morphism $\eta_D \colon D \to F(G_D)$, and this map is optimal in the above sense. If such a reflection exists for every object, then we can actually glue these reflections together and get a functor.

Lemma 2.4.15 *Let* $F \colon \mathcal{C} \to \mathcal{D}$ *be a functor and assume that for all objects D of $\mathcal{D}$, there is a reflection of D at F, and let (G_D, η_D) denote our choice of a reflection of D at F.*

Then there is a unique functor $L \colon \mathcal{D} \to \mathcal{C}$, such that

- *for all objects D of $\mathcal{D}$: $L(D) = G_D$; and*
- *the morphisms $\eta_D \colon D \to F(G_D)$ yield a natural transformation $\eta \colon \mathrm{Id} \Rightarrow FL$.*

Proof Of course, we define $L(D) := G_D$. Let $h \colon D \to D'$ be a morphism in $\mathcal{D}$ and let $(G_D, \eta_D), (G_{D'}, \eta_{D'})$ be the reflections of D and D', respectively, at F. Consider the solid diagram

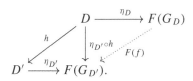

There is a unique $f \colon G_D \to G_{D'}$ with

$$F(f) \circ \eta_D = \eta_{D'} \circ h.$$

We define $L(h)$ as f and show now that this assignment actually produces a functor. Note that by construction, this ensures that the morphisms η_D are natural.

Let $k \colon D' \to D''$ be another morphism in $\mathcal{D}$ and let $g \colon G_{D'} \to G_{D''}$ be the uniquely defined morphism that satisfies $F(g) \circ \eta_{D'} = \eta_{D''} \circ k$. We get

$$F(L(k) \circ L(h)) \circ \eta_D = FL(k) \circ FL(h) \circ \eta_D,$$

because F is a functor. By definition

$$FL(h) \circ \eta_D = F(f) \circ \eta_D = \eta_{D'} \circ h,$$

so, in total, we get that the following diagram commutes:

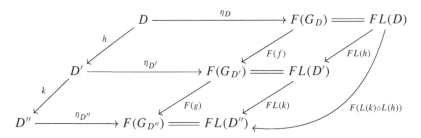

in particular, $F(L(k) \circ L(h)) \circ \eta_D = \eta_{D''} \circ k \circ h$. On the other hand, by construction, we get that $FL(k \circ h) \circ \eta_D = \eta_{D''} \circ (k \circ h)$, and thus, uniqueness implies

$$L(k \circ h) = L(k) \circ L(h).$$

We still have to show that $L(1_D) = 1_{L(D)}$ holds. Consider the diagram

$$\begin{array}{ccc}
D & \xrightarrow{\eta_D} & F(G_D) \\
\downarrow{\scriptstyle 1_D} & & \downarrow{\scriptstyle F(f)} \\
D & \xrightarrow{\eta_D} & F(G_D).
\end{array}$$

We get a unique f with the property that $F(f) \circ \eta_D = \eta_D \circ 1_D$. But the identity on G_D satisfies the same equality, and hence, $f = 1_{G_D} = 1_{L(D)}$. $\square$

We can now provide a criterion for testing when a functor is a left adjoint to a given functor F. In particular, any functor constructed by Lemma 2.4.15 actually gives rise to a left adjoint functor.

Proposition 2.4.16 *A functor $L \colon \mathcal{D} \to \mathcal{C}$ is left adjoint to a functor $F \colon \mathcal{C} \to \mathcal{D}$ if and only if there is a natural transformation $\eta \colon \mathrm{Id} \Rightarrow F \circ L$, such that for all objects D of $\mathcal{D}$, the pair $(L(D), \eta_D \colon D \to FL(D))$ is a reflection.*

Proof Assume the existence of η with the required properties. Then, we have to construct a counit $\varepsilon \colon FL \Rightarrow \mathrm{Id}$. By assumption, for every FC, we have the reflection $(LFC, \eta_{FC} \colon FC \to FLFC)$. This ensures that there is a unique $\varepsilon_C \colon LFC \to C$, with the property that the diagram

$$\begin{array}{ccc}
FC & \xrightarrow{\eta_{FC}} & FLFC \\
\downarrow{\scriptstyle 1_{FC}} & \swarrow{\scriptstyle F(\varepsilon_C)} & \\
FC & &
\end{array}$$

commutes. We have to show that the ε_C are actually natural in C. Let $h : C \to C'$ be a morphism in $\mathcal{C}$. As η is a natural transformation, the diagram

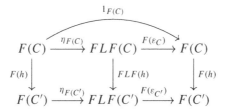

commutes and we obtain

$$F(\varepsilon_{C'} \circ LF(h)) \circ \eta_{FC} = F(\varepsilon_{C'}) \circ FLF(h) \circ \eta_{FC} = F(\varepsilon_{C'}) \circ \eta_{FC'} \circ F(h)$$
$$= 1_{FC'} \circ F(h) = F(h).$$

On the other hand, as F is a functor and thanks to the property of ε_C, we also get

$$F(h \circ \varepsilon_C) \circ \eta_{FC} = F(h) \circ F(\varepsilon_C) \circ \eta_{FC} = F(h) \circ 1_{FC} = F(h).$$

The uniqueness of the morphisms involved therefore shows $h \circ \varepsilon_C = \varepsilon_{C'} \circ LF(h)$, and thus, ε is natural. Similarly, as

$$F(\varepsilon_{LD} \circ L(\eta_D)) \circ \eta_D = F(\varepsilon_{LD}) \circ FL(\eta_D) \circ \eta_D = \eta_D = F(1_{LD}) \circ \eta_D,$$

the uniqueness of the morphisms involved shows that $\varepsilon_{LD} \circ L(\eta_D) = 1_{LD}$. As $\eta_{FC} \circ F(\varepsilon_C) = 1_{FC}$ holds for all objects C of $\mathcal{C}$ by assumption, this shows that L is left adjoint to F by Proposition 2.4.6.

Let us now assume that L is left adjoint to F and let $\varphi_{D,C} : \mathcal{C}(LD, C) \cong \mathcal{D}(D, FC)$ be the binatural bijection of the adjunction. We claim that for all objects D of $\mathcal{D}$, the pair $(LD, \varphi_{D,LD}(1_{LD}))$ is a reflection.

Let C be an arbitrary object of $\mathcal{C}$ and let $k : D \to FC$ be an arbitrary morphism in $\mathcal{D}$. As $\varphi_{D,C}$ is a bijection, we can write k uniquely as $k = \varphi_{D,C}(h)$ for a unique $h : LD \to C$. The naturality of $\varphi_{D,C}$ implies that

$$
\begin{array}{ccc}
\mathcal{C}(LD, LD) & \xrightarrow{\varphi_{D,LD}} & \mathcal{D}(D, FL(D)) \\
{\scriptstyle \mathcal{C}(LD,h)} \downarrow & & \downarrow {\scriptstyle \mathcal{D}(D,F(h))} \\
\mathcal{C}(LD, C) & \xrightarrow{\varphi_{D,C}} & \mathcal{D}(D, FC)
\end{array}
$$

commutes, so chasing 1_{LD} through that diagram gives

$$
\begin{array}{ccc}
1_{LD} & \longmapsto & \varphi_{D,LD}(1_{LD}) \\
{\scriptstyle \mathcal{C}(LD,h)} \big\uparrow & & \big\uparrow {\scriptstyle F(h)} \\
h & \longmapsto & \varphi_{LD,C}(h) = k = F(h) \circ \varphi_{D,LD}(1_{LD}).
\end{array}
$$

Let $h': LD \to C$ be another morphism with the property that

$$F(h') \circ \varphi_{D,LD}(1_{LD}) = k,$$

then we obtain with a similar diagram chase

$$\varphi_{D,C}(h') = \varphi_{D,C}(\mathcal{C}(LD, h')(1_{LD})) = \mathcal{D}(D, F(h')(\varphi_{D,LD}(1_{LD}))$$
$$= F(h')(\varphi_{D,LD}(1_{LD})) = k,$$

and as $\varphi_{D,C}$ is injective, this proves that $h = h'$. $\square$

Reflections are often used in order to understand subcategories:

Definition 2.4.17 Let $\mathcal{C}'$ be a subcategory of a category $\mathcal{C}$ and let $I: \mathcal{C}' \to \mathcal{C}$ denote the inclusion functor. Then $\mathcal{C}'$ is called a *reflective subcategory* of $\mathcal{C}$ if the functor I has a left adjoint L.

Remark 2.4.18
- The functor I is always a faithful functor. Hence, Proposition 2.4.11 implies that the counit $\varepsilon_{C'}: LI(C') \to C'$ is an epimorphism for all objects C' of $\mathcal{C}'$.
- If $\mathcal{C}'$ is a reflective subcategory of $\mathcal{C}'$, then for every object C of $\mathcal{C}$, there is an object $L(C)$ of $\mathcal{C}'$, and there is a morphism $\eta_C: C \to IL(C)$, the unit of the adjunction, which satisfies the properties of Definition 2.4.13.
- If a reflective subcategory $\mathcal{C}'$ of $\mathcal{C}$ is full, then by Proposition 2.4.11 we know that the counit $\varepsilon_{C'}: LI(C') \to C'$ is an isomorphism for all objects C' of $\mathcal{C}'$. Thus, L doesn't change the objects of $\mathcal{C}'$.

Example 2.4.19 The subcategory of abelian groups inside the category of groups is full. The *abelianization functor* sends a group G to $G/[G, G]$. As

$$\mathsf{Gr}(G, A) = \mathsf{Ab}(G/[G, G], A)$$

for all abelian groups A, the abelianization functor is left adjoint to the inclusion functor and Ab is a reflective subcategory of Gr.

Exercise 2.4.20 Let Ab^f be the subcategory of Ab of all finitely generated abelian groups. Show that Ab^f cannot be reflective.

2.5 Equivalences of Categories via Adjoint Functors

We use reflections and adjunctions in order to describe equivalences of categories. In particular, every equivalence of categories is given by an adjunction.

Theorem 2.5.1 *Let $F: \mathcal{C} \to \mathcal{D}$ be an arbitrary functor. Then the following are equivalent.*

(1) The functor F possesses a left adjoint L, and the corresponding natural transformations $\varepsilon\colon LF \Rightarrow Id$ and $\eta\colon Id \Rightarrow FL$ are natural isomorphisms.

(2) There is a functor $L\colon \mathcal{D} \to \mathcal{C}$ and two arbitrary natural isomorphisms $Id \cong FL$ and $LF \cong Id$.

(3) The functor F is fully faithful and essentially surjective.

Proof The fact that (1) implies (2) is clear. If we assume (2), then there is an isomorphism from every object D of $\mathcal{D}$ to FLD, and therefore, with $C = LD$, we get that F is essentially surjective.

The composite $F \circ L$ is naturally isomorphic to the identity. This implies that $F \circ L$ if fully faithful; this in turn implies that L is faithful and F is full. Vice versa, the natural isomorphism from $L \circ F$ to the identity functor yields that $L \circ F$ is fully faithful and hence that L is full and F is faithful. Thus (2) implies (3).

Let us now assume that (3) holds. For every object D of $\mathcal{D}$, we choose an object C of $\mathcal{C}$ with $D \cong FC$. We denote C by LD and fix an isomorphism $\eta_D\colon D \to FLD$ for every D. If C' is an arbitrary object of $\mathcal{C}$ and if $g\colon D \to FC'$ is a morphism in $\mathcal{D}$, then the composite

$$FLD \xrightarrow{\;\eta_D^{-1}\;} D \xrightarrow{\;g\;} FC'$$

is of the form $F(h)$ for a unique $h\colon LD \to C$, because F is fully faithful by assumption. This shows that the pair (LD, η_D) is a reflection of D at F and hence that L is a left adjoint to F by Proposition 2.4.16. The morphisms η_D are isomorphisms by construction, and they are natural by a similar argument, as in the proof of Lemma 2.4.15. The remaining properties follow from Proposition 2.4.11. $\qquad\square$

2.6 Skeleta of Categories

Definition 2.6.1
- A category is called *reduced* if isomorphic objects are identical.
- A subcategory $\mathcal{S}$ of a category $\mathcal{C}$ is a *skeleton* if $\mathcal{S}$ is reduced and if the inclusion $\mathcal{S} \hookrightarrow \mathcal{C}$ is an equivalence of categories.

Examples 2.6.2
- Consider the category of finite sets and functions. It contains the full subcategory whose objects are the sets of the form $\{1, \ldots, n\}$ for $n \geq 0$. Here, we use the convention that the empty set is encoded by $n = 0$. The inclusion functor is full and faithful. As every finite set is in bijection with a

standardized set of the form $\{1, \ldots, n\}$ as above, the inclusion functor is also essentially surjective. Therefore, these finite sets build a skeleton.

- A similar example is the category of finite-dimensional K-vector spaces. This has as a skeleton the full subcategory of vector spaces of the form K^n for some finite natural number n. Here, $n = 0$ encodes the zero vector space.

For reduced categories, it is easy to check whether we have an equivalence.

Lemma 2.6.3 *Assume that C' and D' are reduced categories, and let $F : C' \to D'$ be a functor. Then the following are equivalent:*

(1) F is an isomorphism of categories.
(2) F is an equivalence of categories.

Proof   Of course, (1) implies (2). Assume that F is an equivalence of categories. Then it is fully faithful and essentially surjective. Assume that there are objects C_1 and C_2 of C' with $F(C_1) = F(C_2) = D$. As F is fully faithful, there are unique morphisms $f : C_1 \to C_2$ and $g : C_2 \to C_1$ with $F(f) = 1_D = F(g)$. We also get $F(f \circ g) = F(1_{C_2}) = F(1_{C_1}) = F(g \circ f)$, and again, by F being fully faithful, this ensures that f and g are inverse to each other, and hence, $C_1 = C_2$. Thus, F is bijective on objects. As it is also bijective on morphisms, F is an isomorphism. $\square$

Assuming the axiom of choice gives the following result:

Proposition 2.6.4 *Every category C possesses a skeleton.*

Proof   We declare two objects C_1 and C_2 of C to be equivalent if and only if they are isomorphic. We choose a representative of every equivalence class and consider the full subcategory C' of C consisting of these objects. Let $I : C' \to C$ be the inclusion functor. We define $P : C \to C'$ as follows: For every object C' of C' and every object C of C that is isomorphic to C', we choose an isomorphism $a_C^{C'} : C' \to C$, such that $a_{C'}^{C'} = 1_{C'}$. Then, P sends C to C', and an $f \in C(C_1, C_2)$ is mapped to $P(f) = (a_{C_2}^{C_2'})^{-1} \circ f \circ a_{C_1}^{C_1'}$:

$$
\begin{array}{ccc}
C_1' & \xrightarrow{\;P(f)\;} & C_2' \\[2pt]
{\scriptstyle a_{C_1}^{C_1'}}\Big\downarrow & & \Big\downarrow{\scriptstyle a_{C_2}^{C_2'}} \\[2pt]
C_1 & \xrightarrow{\;f\;} & C_2 \,.
\end{array}
$$

$\square$

Lemma 2.6.3 then yields the following fact:

Corollary 2.6.5 *Two categories C and D are equivalent if and only if their skeleta are isomorphic.*

3

Colimits and Limits

Limit and colimit constructions appear in many areas of mathematics. You probably know examples of colimits and limits. If you glue two topological spaces together, that is a pushout. You might have met fiber products of spaces, for instance, in the disguise of the preimage of a continuous map. The Hawaiian earring is an inverse limit. In algebra, you have encountered products of groups or direct sums of abelian groups. Sometimes, limits or colimits do not exist. If you consider the category of fields, then this category does not possess products. The underlying set of a product of fields had to be the product of the underlying sets, but then, the multiplication would have zero divisors.

3.1 Diagrams and Their Colimits

Let $\mathcal{D}$ be a small category and let $\mathcal{C}$ be an arbitrary category. A functor $F: \mathcal{D} \to \mathcal{C}$ can be thought of as a diagram in $\mathcal{C}$ with objects $F(D) \in \mathcal{C}$ for all objects D of $\mathcal{D}$ and morphisms $F(f)$ between $F(D)$ and $F(D')$ for $f: D \to D'$ in $\mathcal{D}$. Recall that we denote by $\mathrm{Fun}(\mathcal{D}, \mathcal{C})$ the category of functors from $\mathcal{D}$ to $\mathcal{C}$ with natural transformations as morphisms.

What should be the (co)limit of F if the diagram is as follows?

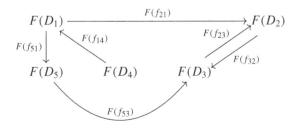

Does it exist? If yes, how can we describe it?

Definition 3.1.1 We let

$$\Delta \colon \mathcal{C} \to \mathsf{Fun}(\mathcal{D}, \mathcal{C})$$

be the functor that assigns to an object C of $\mathcal{C}$ the constant diagram, that is, $\Delta(C)(D) = C$ for all objects D of $\mathcal{D}$, and an $f \in \mathcal{D}(D, D')$ is sent to $\Delta(C)(f) = 1_C$.

If g is a morphism in $\mathcal{C}$ from C to C', then g induces a natural transformation of functors

$$\Delta_g \colon \Delta(C) \Rightarrow \Delta(C').$$

Definition 3.1.2 A *colimit of a functor* $F \colon \mathcal{D} \to \mathcal{C}$ consists of

- an object of $\mathcal{C}$, denoted by $\mathrm{colim}_{\mathcal{D}} F$; and
- a natural transformation

$$\tau \colon F \Rightarrow \Delta(\mathrm{colim}_{\mathcal{D}} F). \tag{3.1.1}$$

- The pair $(\mathrm{colim}_{\mathcal{D}} F, \tau)$ is universal with respect to all natural transformations $\tau' \colon F \Rightarrow \Delta(C)$; that is, for all such τ', there is a unique morphism $\xi \colon \mathrm{colim}_{\mathcal{D}} F \to C$, such that $\tau' = \Delta(\xi) \circ \tau$:

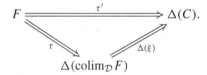

Locally, this looks like the following picture:

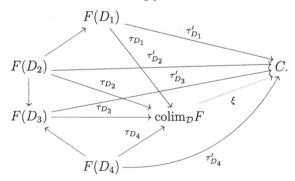

Thus, the colimit, $\mathrm{colim}_{\mathcal{D}} F$, is an object of $\mathcal{C}$ that is "as close to the diagram F as possible" with respect to morphisms *out* of the diagram.

A natural transformation η from F to $\Delta(C)$ is often called a *cone from F over C*. A colimit is then a cone, as in (3.1.1), and is sometimes called a *limiting cone* or a *universal cone* [**ML98**, p. 67].

The usual abstract nonsense argument tells you that colimits are unique up to canonical isomorphism, if they exist. If you have two colimits of the same diagram, then the universal property ensures the existence of a unique morphism from one to the other and vice versa. And again, by uniqueness, the two compositions of those two morphisms have to be identity morphisms because these also fit the bill.

The colimit over the empty category is defined to be the initial object of $\mathcal{C}$, if this exists.

Exercise 3.1.3 Show that if $G: \mathcal{D} \rightarrow \mathcal{C}$ is another functor and if $\alpha: F \Rightarrow G$ is a natural transformation, then α induces a morphism $\mathrm{colim}_{\mathcal{D}}\alpha \in \mathcal{C}(\mathrm{colim}_{\mathcal{D}}F, \mathrm{colim}_{\mathcal{D}}G)$. Prove that this turns $\mathrm{colim}_{\mathcal{D}}$ into a functor from $\mathrm{Fun}(\mathcal{D}, \mathcal{C})$ to $\mathcal{C}$.

The universal property of the colimit can equivalently be described in terms of adjunctions.

Proposition 3.1.4 *If the colimit $(\mathrm{colim}_{\mathcal{D}}F, \tau)$ exists for all functors $F: \mathcal{D} \rightarrow \mathcal{C}$, then the functor $\mathrm{colim}_{\mathcal{D}}: \mathrm{Fun}(\mathcal{D}, \mathcal{C}) \rightarrow \mathcal{C}$ is left adjoint to the diagonal functor $\Delta: \mathcal{C} \rightarrow \mathrm{Fun}(\mathcal{D}, \mathcal{C})$; that is, there are natural isomorphisms*

$$\mathcal{C}(\mathrm{colim}_{\mathcal{D}}F, C) \cong \mathrm{Fun}(\mathcal{D}, \mathcal{C})(F, \Delta(C))$$

for all functors F and all object C of $\mathcal{C}$.

Exercise 3.1.5 Let $\mathcal{D}$ be a small category that possesses a terminal object and let $F: \mathcal{D} \rightarrow \mathcal{C}$ be a functor. What is the colimit of F over $\mathcal{D}$?

Example 3.1.6 Consider the representable functor $\mathcal{D}(D, -): \mathcal{D} \rightarrow \mathrm{Sets}$ for some object D of $\mathcal{D}$. A useful fact is that

$$\mathrm{colim}_{\mathcal{D}}\mathcal{D}(D, -) \cong \{*\}.$$

This is easy to see using Proposition 3.1.4. Colimits exist in the category of Sets:

$$\mathrm{colim}_{\mathcal{D}}F = \bigsqcup_{D \text{ object of } \mathcal{D}} F(D)/\sim,$$

where we declare that an $x \in F(D)$ is equivalent to a $y \in F(D')$ if there is a morphism $f \in \mathcal{D}(D, D')$, such that $F(f)(x) = y$. This relation is not symmetric, so one has to consider the equivalence relation generated by this relation. As colimits exist, we can use the adjunction described above and get a natural bijection

$$\mathsf{Fun}(\mathcal{D}, \mathsf{Sets})(\mathcal{D}(D, -), \Delta(X)) \cong \mathsf{Sets}(\mathrm{colim}_{\mathcal{D}}\mathcal{D}(D, -), X)$$

for all sets X. The Yoneda lemma implies that

$$\mathsf{Fun}(\mathcal{D}, \mathsf{Sets})(\mathcal{D}(D, -), \Delta(X)) \cong \Delta(X)(D) = X.$$

But X is also determined by all functions from a one-point set $\{*\}$ to X, thus

$$\mathsf{Sets}(\mathrm{colim}_{\mathcal{D}}\mathcal{D}(D, -), X) \cong \Delta(X)(D) \cong X \cong \mathsf{Sets}(\{*\}, X).$$

The uniqueness of representable functors (see Lemma 2.2.5) then gives the claim.

In the everyday life of a working mathematician, the following special forms of colimits frequently show up.

3.1.1 Sequential Colimits

If $\mathcal{D}$ is the category given by the natural numbers considered as a poset

$$\mathcal{D} = (0 \to 1 \to 2 \to \cdots),$$

then a colimit of any $F : \mathcal{D} \to \mathcal{C}$ is a *sequential colimit*.

Examples 3.1.7
- If all structure maps $F(i < j)$ are monomorphisms, then we might interpret the colimit $\mathrm{colim}_{\mathcal{D}} F$ as the union of the $F(i)$s. Typical examples are increasing sequences of sets or topological spaces

$$X_0 \subset X_1 \subset X_2 \subset \cdots$$

or increasing sequences of abelian groups, vector spaces, and other algebraic objects.
- An important class of examples is CW complexes. These are the colimits of their skeleta.
- In stable homotopy theory, the stable homotopy groups of spheres are a central object of study. Let $\mathbb{S}^n$ denote the unit sphere in $\mathbb{R}^{n+1}$. As the smash product of spheres satisfies $\mathbb{S}^1 \wedge \mathbb{S}^n \cong \mathbb{S}^{n+1}$, we have stabilization maps

$$\pi_n(\mathbb{S}^m) = [\mathbb{S}^n, \mathbb{S}^{m+1}]_* \to [\mathbb{S}^{n+1}, \mathbb{S}^{m+1}]_* = \pi_{n+1}(\mathbb{S}^m)$$

that send a homotopy class $[f]$ to the homotopy class of $\mathbb{S}^1 \wedge f$. Therefore, for every m, we get a sequential colimit, and as $\pi_n(\mathbb{S}^m) = 0$ for $n < m$, we can express $\pi_n(\mathbb{S}^m)$ as $\pi_{k+m}(\mathbb{S}^m)$ in the nontrivial cases with $k \geq 0$, and get the kth *stable homotopy group of spheres* as

$$\pi_k^s = \mathrm{colim}(\pi_{k+m}(\mathbb{S}^m) \to \pi_{k+m+1}(\mathbb{S}^{m+1}) \to \pi_{k+m+2}(\mathbb{S}^{m+2}) \to \cdots).$$

The first groups are $\pi_0^s = \mathbb{Z}$, $\pi_1^s = \mathbb{Z}/2\mathbb{Z}$ generated by the stabilization of the Hopf map $\eta \colon \mathbb{S}^3 \to \mathbb{S}^2$, $\pi_2^s = \mathbb{Z}/2\mathbb{Z}$, $\pi_3^s = \mathbb{Z}/24\mathbb{Z}$, and so on.

3.1.2 Coproducts

If you build the colimit over a discrete diagram category, that is, a small category $\mathcal{D}$ that has only identity morphisms, then the colimit of a functor $F \colon \mathcal{D} \to \mathcal{C}$ is called the *(categorical) sum* or the *(categorical) coproduct* of the $F(D)$ for D an object of $\mathcal{D}$, denoted by

$$\bigsqcup_{\mathcal{D}} F(D).$$

If you take as the category $\mathcal{C}$ the category Sets and if $F \colon \mathcal{D} \to$ Sets is any functor from a discrete category $\mathcal{D}$ to Sets, then the colimit is

$$\mathrm{colim}_{\mathcal{D}} F = \bigsqcup_D F(D).$$

Similarly, the categorical sum in the category of topological spaces is given by the disjoint union of spaces with the corresponding topology.

Every coproduct comes with canonical structure maps. By assumption, we have morphisms $\tau_D \colon F(D) \to \bigsqcup_{\mathcal{D}} F(D)$.

Definition 3.1.8 We denote these morphisms by i_D and call i_D the *inclusion of $F(D)$ into $\bigsqcup_{\mathcal{D}} F(D)$*.

Definition 3.1.9 Let C be an object of $\mathcal{C}$. For the binary coproduct of C with itself, $C \sqcup C$, the universal property of the coproduct ensures the existence of a morphism

$$\nabla \colon C \sqcup C \to C$$

induced by the identity morphism on C. We call this map the *fold map*.

Note that for $F(D_1) = F(D_2) = C$, the composition of $i_{D_1} \colon F(D_1) \to F(D_1) \sqcup F(D_2) = C \sqcup C$ with the fold map ∇ is the identity morphism on $C = F(D_1)$.

Example 3.1.10 In the category of pointed topological spaces, the pointed sum (also known as the bouquet of spaces) is the coproduct; that is, for two pointed spaces (X, x_0) and (Y_0, y_0), their coproduct is given by

$$X \vee Y = X \times \{y_0\} \sqcup \{x_0\} \times Y/\sim,$$

where the point (x_0, y_0) in both summands is identified with the basepoint $[(x_0, y_0)]$ of $X \vee Y$. For $X = Y = \mathbb{S}^1$ with basepoint $1 \in \mathbb{C}$, we have a pinch map from $\mathbb{S}^1$ to $\mathbb{S}^1 \vee \mathbb{S}^1$ by identifying $(-1, 0)$ and $(1, 0)$, that is, by collapsing the embedded image of $\mathbb{S}^0$ in $\mathbb{S}^1$ to a point. If we compose this with the fold map, we obtain a self-map of $\mathbb{S}^1$.

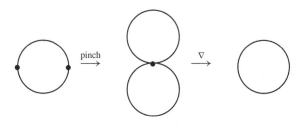

By taking pinch maps $\mathbb{S}^1 \rightarrow \bigvee_k \mathbb{S}^1$, using reversal of loops and fold maps $\bigvee_k \mathbb{S}^1 \rightarrow \mathbb{S}^1$ for $k \geq 1$, we can produce self-maps of $\mathbb{S}^1$ of any desired degree. This generalizes to higher-dimensional spheres.

Example 3.1.11 Let G_1 and G_2 be discrete groups. The coproduct of G_1 and G_2 in the category of groups exists and is called the *free product of G_1 and G_2*. It is denoted by $G_1 * G_2$. Elements in $G_1 * G_2$ are words $w = w_1 \ldots w_n$ of finite length, where the w_is are elements of G_1 or G_2. A word w is equivalent to a word w' if w and w' are connected via a finite chain of reduction moves. The only reduction rules are the omission of neutral elements and the reduction of neighbors. If w_i and w_{i+1} are elements of the same group, then you can evaluate their product. The group structure is given by concatenation. The empty word is the neutral element in $G_1 * G_2$. For instance, the fundamental group of $\mathbb{S}^1 \vee \mathbb{S}^1$ is $\mathbb{Z} * \mathbb{Z}$, which is the free group generated by two elements.

Exercise 3.1.12 Does the category of finite groups have coproducts?

In addition to the fold map, we have a symmetry property of coproducts. If σ is a bijection of the underlying set of $\mathcal{D}$, then σ induces a functor $G_\sigma : \mathcal{D} \rightarrow \mathcal{D}$ of discrete categories. By the universal property of colimits, the coproducts $\bigsqcup_\mathcal{D} F(D)$ and $\bigsqcup_\mathcal{D} F \circ G_\sigma(D)$ are isomorphic. For instance, this gives a natural isomorphism of binary coproducts

$$X \sqcup Y \rightarrow Y \sqcup X$$

for any two objects X and Y of $\mathcal{C}$ as long as their coproduct exists.

Exercise 3.1.13 Let X be a partially ordered set viewed as a category and x and y be elements of X. What is the coproduct of x and y? Does it always exist?

Exercise 3.1.14 Assume that $\mathcal{C}$ has an initial object $\varnothing$ and that binary coproducts exist in $\mathcal{C}$. Show that $C \coprod \varnothing \cong C \cong \varnothing \coprod C$ for all objects C of $\mathcal{C}$.

Definition 3.1.15 If $\mathcal{D}$ is a small discrete category and $F \colon \mathcal{D} \to \mathcal{C}$ is a constant functor with value C, then the coproduct $\sqcup_{\mathcal{D}} F(D)$ is called a *copower* and is denoted by $\mathcal{D} \cdot C$.

Example 3.1.16 If X is a finite set and A is a commutative k-algebra for some commutative ring with unit k, then $X \cdot A$ is nothing but the X-fold tensor power of A over k:

$$X \cdot A = A^{\otimes_k X}.$$

For instance, if $X = \{1, \ldots, n\}$, then

$$\{1, \ldots, n\} \cdot A = \underbrace{A \otimes_k \cdots \otimes_k A}_{n} = A^{\otimes_k n}.$$

The fold map $\nabla \colon \{1, \ldots, n\} \cdot A \to A$ is given by the multiplication in A.

3.1.3 Pushouts

You get *pushouts* as colimits over a diagram category $\mathcal{D}$ of the form

$$D_1 \leftarrow D_0 \to D_2.$$

For instance, if A is a topological space, together with continuous maps $f \colon A \to X$ and $g \colon A \to Y$, then the colimit over the diagram $X \xleftarrow{\;f\;} A \xrightarrow{\;g\;} Y$ is the union of X and Y along the image of A in X and Y. For instance, you get the real projective plane $\mathbb{R}P^2$ as the pushout of $f \colon \mathbb{S}^1 \to \mathbb{D}^2$ and $g \colon \mathbb{S}^1 \to \mathbb{S}^1$, where $\mathbb{S}^1$ denotes the unit 1-sphere, $\mathbb{D}^2$ is the unit ball in $\mathbb{R}^2$, f is the inclusion, and g is a map of degree 2.

Pushouts of groups are given by amalgamated products. For a diagram of groups

$$
\begin{array}{ccc}
G_0 & \xrightarrow{\;f\;} & G_1, \\
{\scriptstyle h}\big\downarrow & & \\
G_2 & &
\end{array}
$$

the pushout is given by $G_1 *_{G_0} G_2$, which is the quotient of the free product $G_1 * G_2$ by the normal subgroup generated by words of the form $f(g_0)h(g_0)^{-1}$ for $g_0 \in G_0$.

The Seifert–van Kampen theorem tells you how to calculate the fundamental group of a nice union of two spaces X and Y (X, Y have to be path-connected and open in $X \cup Y$, and $X \cap Y$ is nonempty and path-connected) as $\pi_1(X, z_0) *_{\pi_1(X \cap Y, z_0)} \pi_1(Y, z_0)$ for $z_0 \in X \cap Y$.

3.1.4 Coequalizers

Another important class of examples comprises *coequalizers*. These are colimits of diagrams of the form

$$F(D_0) \underset{\alpha}{\overset{\beta}{\rightrightarrows}} F(D_1).$$

Here, our category $\mathcal{D}$ is $D_0 \underset{g}{\overset{f}{\rightrightarrows}} D_1$, where f and g are the only nonidentity morphisms of $\mathcal{D}$. The corresponding colimit object is called the coequalizer of $\alpha = F(f)$ and $\beta = F(g)$.

Remark 3.1.17 If

$$F(D_0) \underset{\alpha}{\overset{\beta}{\rightrightarrows}} F(D_1) \overset{\psi}{\longrightarrow} C$$

is a coequalizer diagram, then ψ is an epimorphism. For any two morphisms $h_1, h_2 \colon C \to C'$ with $h_1 \circ \psi = h_2 \circ \psi$, the precomposition with β and α agrees, and therefore, the universal property of the colimit guarantees that $h_1 = h_2$.

Consider the category of abelian groups, Ab, and fix an associative ring with unit R. Let M be a right R-module and N be a left R-module. The tensor product $M \otimes_R N$ is the coequalizer of

$$M \otimes R \otimes N \underset{\nu \otimes \mathrm{id}}{\overset{\mathrm{id} \otimes \nu'}{\rightrightarrows}} M \otimes N.$$

Here, the unadorned tensor products are taken over the ring of integers, the map ν is given by the structure map of the right R-module structure of M, and ν' is given by the left R-module structure of N.

The *cokernel of a homomorphism* f is the coequalizer of the diagram $A \underset{f}{\overset{0}{\rightrightarrows}} B$ in the category Ab.

Usually, functors won't preserve coequalizers, but sometimes, a coequalizer satisfies an extra condition that forces this behavior.

Definition 3.1.18 A coequalizer $\xi \colon C_1 \to C$ in a category $\mathcal{C}$ of

$$C_0 \underset{\alpha}{\overset{\beta}{\rightrightarrows}} C_1 \tag{3.1.2}$$

is an *absolute coequalizer of* (3.1.2) if for any functor $F \colon \mathcal{C} \to \mathcal{E}$, the diagram

$$F(C_0) \underset{F(\alpha)}{\overset{F(\beta)}{\rightrightarrows}} F(C_1) \xrightarrow{F(\xi)} F(C)$$

identifies $F(\xi) \colon F(C_1) \to F(C)$ as the coequalizer of the image of the diagram (3.1.2) under F.

This just defines a strong property; however, it is *a priori* not clear that such things exist, but they do.

Definition 3.1.19 A diagram in a category $\mathcal{C}$

$$C_0 \underset{\alpha}{\overset{\beta}{\rightrightarrows}} C_1 \xrightarrow{\xi} C \tag{3.1.3}$$

with the property that $\xi \circ \alpha = \xi \circ \beta$ is called *split* if there are morphisms $s \colon C \to C_1$ and $t \colon C_1 \to C_0$, such that

$$\xi \circ s = 1_C, \quad \alpha \circ t = 1_{C_1} \text{ and } \beta \circ t = s \circ \xi.$$

$$C_0 \underset{\alpha}{\overset{\beta}{\rightrightarrows}} C_1 \xrightarrow{\xi} C$$

Lemma 3.1.20 *In a split diagram* (3.1.3), *the morphism* $\xi \colon C_1 \to C$ *is the coequalizer of* α *and* β.

Proof   Assume that $f \in \mathcal{C}(C_1, T)$ for some object T of $\mathcal{C}$ and that $f \circ \alpha = f \circ \beta$. Define $\bar{f} \colon C \to T$ as $f \circ s$. Then,

$$\bar{f} \circ \xi = f \circ s \circ \xi = f \circ \beta \circ t = f \circ \alpha \circ t = f,$$

and hence, $\bar{f}$ extends f over ξ.

If we have any other $g \in \mathcal{C}(C, T)$ with $f = g \circ \xi$, then

$$f \circ s = g \circ \xi \circ s = g,$$

and thus, $g = \bar{f}$, which shows that $\bar{f}$ is uniquely determined by its defining property. $\qquad\square$

Definition 3.1.21 A *split coequalizer* is a coequalizer $\xi \colon C_1 \to C$ of a split diagram, as in (3.1.3).

Corollary 3.1.22 *Split coequalizers are absolute coequalizers.*

Proof Split coequalizers are described in terms of morphisms and compositions of morphisms. These data are preserved by any functor. □

Remark 3.1.23 Being a split coequalizer is something special. Note that, in particular, two retractions are part of the data ($\xi \circ s = 1_C$ and $\alpha \circ t = 1_{C_1}$). That alone is a strong requirement. For instance, in the category of groups, you usually do not have a morphism from a quotient G/N back to G unless G is a semi-direct product.

We will meet split coequalizers again when we consider algebras over monads in 6.1.

3.1.5 Limits

Limits are defined dually to colimits.

Definition 3.1.24 A *limit of* $F \colon \mathcal{D} \to \mathcal{C}$ is a pair consisting of

- an object of $\mathcal{C}$, that we denote by $\lim_{\mathcal{D}} F$; and
- a natural transformation $\tau \colon \Delta(\lim_{\mathcal{D}} F) \Rightarrow F$.
- The pair $(\lim_{\mathcal{D}} F, \tau)$ is universal with respect to all natural transformations $\tau' \colon \Delta(C) \Rightarrow F$; that is, for all such τ', there is a unique morphism $\xi \colon C \to \lim_{\mathcal{D}} F$, such that

$$\tau' = \tau \circ \Delta(\xi):$$

Thus, the limit of F, $\lim_{\mathcal{D}} F$, is an object of $\mathcal{C}$ that is "as close to the diagram F as possible" with respect to morphisms *into* the diagram. The limit over the empty category is the terminal object of the category, if this object exists.

3.1.6 Sequential Limits

The simplest kind of sequential limits are those limits indexed over the natural numbers.

Let $(X_n)_{n \in \mathbb{N}_0}$ be a family of sets with $X_{n+1} \subset X_n$. Then, the limit of the system

$$\cdots \subset X_{n+1} \subset X_n \subset \cdots \subset X_1 \subset X_0$$

is the intersection of the sets X_n.

Let p be a fixed prime. The inverse limit of the diagram

$$\mathbb{Z}/p^1\mathbb{Z} \xleftarrow{\ p_2\ } \mathbb{Z}/p^2\mathbb{Z} \xleftarrow{\ p_3\ } \mathbb{Z}/p^3\mathbb{Z} \xleftarrow{\ p_4\ } \cdots$$

is the ring of p-adic integers, $\mathbb{Z}_p$. Here, the maps p_i are the canonical projection maps. An explicit model of the limit is

$$\{(x_1, x_2, x_3, \ldots) \in \prod_{n \geq 1} \mathbb{Z}/p^n\mathbb{Z} \mid p_i(x_i) = x_{i-1} \text{ for all } i \geq 2\}. \tag{3.1.4}$$

This carries a ring structure, where addition and multiplication are defined coordinatewise.

3.1.7 Products

Products are limits over a discrete category. If $\mathcal{D}$ is such a category, then we denote the product of the $F(D)$ by $\prod_{\mathcal{D}} F(D)$.

Definition 3.1.25
- We call the structure maps $\tau_{D'} \colon \prod_{\mathcal{D}} F(D) \to F(D')$ the *projection maps*, and we denote them by $\mathrm{pr}_{D'}$.
- Bijections of the underlying set of the discrete category $\mathcal{D}$ yield a symmetry isomorphism on the corresponding products, in particular, for two objects X and Y in $\mathcal{C}$, there is a natural symmetry isomorphism $X \times Y \cong Y \times X$.
- Dual to the fold map, there is a *diagonal morphism* $\delta \colon C \to \prod_{\mathcal{D}} C$, induced by the identity of C, and for $C = F(D)$, we have $\mathrm{pr}_D \circ \delta = 1_C$.

Exercise 3.1.26 Dual to Exercise 3.1.13, what is a product of two elements in a partially ordered set? Does it always exist?

Exercise 3.1.27 Dual to Exercise 3.1.14, show that $C \times * \cong C \cong * \times C$ for all objects C of $\mathcal{C}$ if binary products exist in $\mathcal{C}$ and if $*$ is a terminal object.

3.1.8 Pullbacks

Pullbacks are limits for diagrams of the form $D_1 \to D_0 \leftarrow D_2$.

3.1.9 Equalizers

Dual to the notion of coequalizers is the notion of *equalizers*, that is, limits of diagrams

$$F(D_0) \underset{f}{\overset{g}{\rightrightarrows}} F(D_1).$$

Remark 3.1.28 As in Remark 3.1.17, we note that for every equalizer diagram

$$C \xrightarrow{\varphi} F(D_0) \underset{f}{\overset{g}{\rightrightarrows}} F(D_1),$$

the morphism $\varphi \colon C \to F(D_1)$ is a monomorphism.

If M and N are left R-modules for a ring R, then the abelian group of R-module homomorphisms is the equalizer in the category Ab of the diagram

$$\mathrm{Hom}(M, N) \underset{f}{\overset{g}{\rightrightarrows}} \mathrm{Hom}(R \otimes M, N).$$

Here, Hom denotes the homomorphisms of abelian groups, f is the map that sends an $h \colon M \to N$ to

$$R \otimes M \xrightarrow{1 \otimes h} R \otimes N \xrightarrow{\nu_N} N,$$

g is the composition

$$R \otimes M \xrightarrow{\nu_M} M \xrightarrow{h} N,$$

and the maps ν_N, ν_M are the structure maps of the R-module structures of N and M.

Exercise 3.1.29 Describe the equalizer in Ab of a diagram $A \underset{f}{\overset{0}{\rightrightarrows}} B$.

Let X, Y be sets. What is the equalizer of the diagram $X \underset{f}{\overset{g}{\rightrightarrows}} Y$?

Example 3.1.30 Let F be a presheaf on a space X with values in Sets, so F assigns to every open subset of X a set $F(U)$, but we only require a compatibility condition with respect to restrictions. In certain situations, we want that F is determined by compatible local data. This can be expressed as an equalizer condition.

The presheaf F is a *sheaf* if for every $U \in \mathfrak{U}(X)$ and for every open covering $(U_i)_{i \in I}$ of U, the following diagram is an equalizer:

$$F(U) \longrightarrow \prod_{i \in I} F(U_i) \rightrightarrows \prod_{i,j \in I} F(U_i \cap U_j).$$

Here, the first map is induced by the restriction maps $\mathrm{res}_U^{U_i}$, and the second pair of arrows is induced by two sets of restriction maps. $U_i \cap U_j$ is a subset of U_i and of U_j.

Sheaves form a category as a full subcategory of the category of presheaves.

The examples we had before, continuous real-valued functions, smooth real-valued functions, and smooth sections, were actually all examples of sheaves. For instance, a continuous real-valued function is determined by its values on the open subsets of a covering as long as the values on intersections are compatible.

A different example is the *skyscraper sheaf* on a point $x \in X$. We fix a set S and choose a one-point set $\{*\}$. One defines $F_x(U)$ to be S if $x \in U$ and let $F_x(U) = \{*\}$ otherwise. So, this is like a sheaf version of the Dirac distribution. What are the restriction maps? Why is this a sheaf?

3.1.10 Fiber Products

An important class of examples of limits in the category of sets, Sets, or the category of topological spaces, Top, is the *fiber product*. This is nothing but the pullback of a diagram

$$
\begin{array}{ccc}
 & & X \\
 & & \downarrow p \\
Z & \xrightarrow{\;f\;} & Y.
\end{array}
$$

A concrete model for this pullback in these categories is

$$
Z \times_Y X := \{(z, x) \subset Z \times X \mid f(z) = p(x)\},
$$

that is, we collect pairs of elements that meet in Y. Often, we denote $Z \times_Y X$ as $f^*(p)$, in order to stress the dependence of the fiber product on the maps involved. An important special case of a fiber product is the case of an inclusion $f : A \hookrightarrow Y$ of a subset or a subspace. Then, the fiber product can be identified with the preimage of A under p.

Exercise 3.1.31 What is the natural transformation τ in the case of a fiber product?

Fiber products give examples of limits that are empty. Consider two embeddings of topological spaces $A \hookrightarrow X$ and $B \hookrightarrow X$ whose images in X are disjoint. Then, the fiber product $A \times_X B$ is empty.

3.1.11 Kernel Pairs

Let $f: A \rightarrow B$ be a morphism in the category of sets. Then, the pullback of

is the fiber product

$$A \times_B A = \{(a_1, a_2) \in A \times A, f(a_1) = f(a_2)\}. \tag{3.1.5}$$

It measures the extent of the noninjectivity of f. This concept makes sense in an arbitrary category.

Definition 3.1.32 Let $\mathcal{C}$ be a category and $f \in \mathcal{C}(A, B)$. The *kernel pair of f* is the pullback of the diagram

$$
\begin{array}{ccc}
P & \xrightarrow{\ p_1\ } & A \\
{\scriptstyle p_2}\big\downarrow & & \big\downarrow{\scriptstyle f} \\
A & \xrightarrow{\ f\ } & B
\end{array}
$$

if this pullback exists.

Remark 3.1.33

- Pullbacks are unique up to unique isomorphism, but we will sometimes encounter different models. Then, we will denote the above pullback by (P, p_1, p_2).
- Note that the morphisms p_1, p_2 described previously are automatically epimorphisms. The identity of A gives rise to a unique morphism $s: A \rightarrow P$ with $p_1 \circ s = 1_A$ and $p_2 \circ s = 1_A$; hence, s is a section for p_1 and p_2, and these morphisms are retractions, hence epimorphisms.

Kernel pairs help to detect monomorphisms.

Proposition 3.1.34 *Given $f \in \mathcal{C}(A, B)$, the following are equivalent:*

(1) The kernel pair of f exists and is $(A, 1_A, 1_A)$.
(2) The kernel pair of f exists and is of the form (A, p, p).
(3) The morphism f is a monomorphism.

Proof   Of course, (1) implies (2). Assuming (2), let $g, h \in \mathcal{C}(C, A)$ with $f \circ g = f \circ h$. Then, g and h induce a unique morphism $\xi: C \rightarrow A$ with $f = p \circ \xi = g$, thus (3) holds.

If f is a monomorphism, then, by definition, $(A, 1_A, 1_A)$ is a kernel pair of f. □

Exercise 3.1.35 Dualize the notion of kernel pairs to define cokernel pairs. Do they detect epimorphisms?

3.2 Existence of Colimits and Limits

It is not always true that colimits and/or limits exist. If they *do* exist, then this deserves an extra name.

Definition 3.2.1
- A category C is called *complete* if for all small categories D and for all functors $F : D \to C$, the limit $\lim_D F$ exists. Similarly, if the colimit exists for all small D and for all functors $F : D \to C$, then C is called *cocomplete*.
- A category C is *bicomplete* if it is complete and cocomplete.
- If limits exist for all finite categories D and all functors $F : D \to C$, then C is *finitely complete*. The dual notion for colimits is called *finitely cocomplete*.

If a category is finitely complete, then a terminal object exists, and dually, finitely cocomplete implies the existence of an initial object.

The categories Ab and Sets are examples of complete and cocomplete categories. The category of fields is not complete as it does not possess products.

The description of the p-adic integers as a limit as in (3.1.4) is typical, as we will see now.

Theorem 3.2.2 *A category C is complete if and only if C has products and equalizers. Dually, C is cocomplete if and only if C has coproducts and coequalizers.*

Proof We show the first claim and leave the dual second one as an exercise.

If C is complete, then it has products and equalizers. Assume that C has products and equalizers, and let $F : D \to C$ a small diagram. Then, all the morphisms in D constitute a set, and we also have a set of objects $\mathrm{Ob}(D)$ of D. Therefore, by assumption, the products $X = \prod_{D \in \mathrm{Ob}(D)} F(D)$ and $Y := \prod_{f \in D(D, D')} F(D)$ exist, and these products have canonical projection morphisms

$$\pi_{D'} : X = \left(\prod_{D \in \mathrm{Ob}(D)} F(D) \right) \to F(D')$$

for objects D of $\mathcal{D}$, and

$$\pi_f : Y = \left(\prod_{f \in \mathcal{D}(D, D')} F(D) \right) \to F(D'),$$

which is given by the projection to the coordinate corresponding to f, followed by the morphism $F(f)$.

Again, we denote the source of a morphism $f : D \to D'$ by $s(f) = D$ and the target by $t(f) = D'$.

We construct two morphisms $\varphi, \psi : X \to Y$. A morphism into a product is determined by a family of morphisms into the coordinates. We define

$$\pi_f \circ \varphi = \pi_{t(f)} : X \to F(t(f)), \quad \pi_f \circ \psi = F(f) \circ \pi_{s(f)} : X \to F(s(f)) \to F(t(f)).$$

If τ' is a natural transformation $\tau' : \Delta(C) \Rightarrow F$, then τ' induces a morphism $T : C \to X$ with $\pi_D \circ T = \tau'_D$. We can control the effect of T, and it holds that

$$\varphi \circ T = \psi \circ T,$$

because

$$\pi_{t(f)} \circ T = F(f) \circ \pi_{s(f)} \circ T.$$

This translates to

$$\tau'_{t(f)} = F(f)\tau'_{s(f)},$$

and this is precisely the naturality condition of τ'. We consider the equalizer

$$E \xrightarrow{\tau} X \underset{\varphi}{\overset{\psi}{\rightrightarrows}} Y.$$

This equalizer satisfies the universal property of the limit, $\lim_{\mathcal{D}} F$. $\qquad \square$

Remark 3.2.3 In Theorem 3.2.2, we assumed that all products (coproducts) exist. If binary products (coproducts) and equalizers (coequalizers) exist, then all finite limits (colimits) exist.

If you do not like equalizers, then you can replace equalizers by pullbacks.

Lemma 3.2.4 *Equalizers exist if binary products and pullbacks exist. Coequalizers exist if binary coproducts and pushouts exist.*

Proof Again, I only show the first claim and leave the second one as an exercise in dualization.

We want to show that the equalizer $E \xrightarrow{\tau} X \overset{g}{\underset{f}{\rightrightarrows}} Y$ exists. Consider the diagram

$$Y$$

$$\downarrow{\scriptstyle\delta_Y}$$

$$X \xrightarrow{\ \delta_X\ } X \times X \xrightarrow{\ f \times g\ } Y \times Y,$$

where the δs are the diagonal morphisms. By assumption, the products $X \times X$ and $Y \times Y$ exist, as well as the pullback of the above diagram:

$$
\begin{array}{ccc}
P & \cdots\cdots\cdots\xrightarrow{\varrho_Y}\cdots\cdots\cdots & Y \\[4pt]
{\scriptstyle\varrho_X}\Big\downarrow & & \Big\downarrow{\scriptstyle\delta_Y} \\[4pt]
X & \xrightarrow[\ \delta_X\ ]{} X \times X \xrightarrow[\ f \times g\ ]{} & Y \times Y.
\end{array}
$$

Then, I claim that the pair (P, ϱ_X) has the universal property of the equalizer. If C is an arbitrary object with a morphism $\alpha\colon C \to X$ with $f \circ \alpha = g \circ \alpha$, then we take the morphism α from C to X and the morphism $f \circ \alpha$ from C to Y. These morphisms are compatible because

$$(f \times g) \circ \delta_X \circ \alpha = (f \circ \alpha, g \circ \alpha) = (f \circ \alpha, f \circ \alpha) = \delta_Y \circ (f \circ \alpha),$$

and thus, there is a unique morphism ξ from C to P with $\varrho_X \circ \xi = \alpha$. $\qquad\square$

3.3 Colimits and Limits in Functor Categories

The (co)completeness of a category C ensures the (co)completeness of all functor categories of the form $\mathsf{Fun}(\mathcal{D}, C)$, where $\mathcal{D}$ is an arbitrary small category. If $\mathcal{E}$ is a small category and if $F\colon \mathcal{E} \to \mathsf{Fun}(\mathcal{D}, C)$ is an $\mathcal{E}$-diagram in the functor category, then F corresponds to a functor $\tilde{F}\colon \mathcal{E} \times \mathcal{D} \to C$. If we fix an object D of $\mathcal{D}$, we obtain a functor $\tilde{F}(-, D)\colon \mathcal{E} \to C$.

Proposition 3.3.1 *Let $\mathcal{E}$ be a small category and $F\colon \mathcal{E} \to \mathsf{Fun}(\mathcal{D}, C)$ be an $\mathcal{E}$-diagram in $\mathsf{Fun}(\mathcal{D}, C)$. Assume that for all objects D of $\mathcal{D}$, the diagram $\tilde{F}(-, D)\colon \mathcal{E} \to C$ possesses a (co)limit. Then, the (co)limit of F exists. In particular, if C is complete or cocomplete, then so is $\mathsf{Fun}(\mathcal{D}, C)$ for all small $\mathcal{D}$.*

We will show that one can actually compute (co)limits of diagrams in functor categories pointwise.

Proof We prove the statement for limits; the case of colimits is dual.

Every morphism $\varphi\colon D \to D'$ induces a natural transformation $\tilde{F}(-, \varphi)\colon \tilde{F}(-, D) \Rightarrow \tilde{F}(-, D')$.

By assumption, the limit $\lim_{\mathcal{E}} \tilde{F}(-, D)$ of $\tilde{F}(-, D)\colon \mathcal{E} \to C$ exists. We denote the natural transformation from $\Delta(\lim_{\mathcal{E}} \tilde{F}(-, D))$ to $\tilde{F}(-, D)$ by τ^D. By the universal property of a limit, a morphism $\varphi\colon D \to D'$

induces a morphism $\tilde{F}(-,\varphi)\colon \lim_{\mathcal{E}}\tilde{F}(-,D) \to \lim_{\mathcal{E}}\tilde{F}(-,D')$. The natural transformation τ^D depends naturally on D, so that the diagram

$$
\begin{array}{ccc}
\lim_{\mathcal{E}}\tilde{F}(-,D) & \xrightarrow{\tau_E^D} & \tilde{F}(E,D) \\
\lim_{\mathcal{E}}\tilde{F}(-,\varphi)\downarrow & & \downarrow \tilde{F}(E,\varphi) \\
\lim_{\mathcal{E}}\tilde{F}(-,D') & \xrightarrow{\tau_E^{D'}} & \tilde{F}(E,D')
\end{array}
$$

commutes.

If $\psi \in \mathcal{D}(D',D'')$, then

$$
\begin{aligned}
\tau_E^{D''} \circ \lim_{\mathcal{E}}\tilde{F}(-,\psi) \circ \lim_{\mathcal{E}}\tilde{F}(-,\varphi) &= \tilde{F}(E,\psi) \circ \tau_E^{D'} \circ \lim_{\mathcal{E}}\tilde{F}(-,\varphi) \\
&= \tilde{F}(E,\psi \circ \varphi) \circ \tau_E^D \\
&= \tau_E^{D''} \circ \lim_{\mathcal{E}}\tilde{F}(-,\psi \circ \varphi).
\end{aligned}
$$

Hence, $\lim_{\mathcal{E}}\tilde{F}(-,\psi \circ \varphi) = \lim_{\mathcal{E}}\tilde{F}(-,\psi) \circ \lim_{\mathcal{E}}\tilde{F}(-,\varphi)$. Similarly, we see that $\lim_{\mathcal{E}}\tilde{F}(-,1_D)$ is the identity morphism on the limit object; thus, $\lim_{\mathcal{E}}\tilde{F}(-,-)$ is a functor from $\mathcal{D}$ to $\mathcal{C}$. $\qquad\square$

Corollary 3.3.2 *Let $F, G \in \mathsf{Fun}(\mathcal{D},\mathcal{C})$ and assume that $\mathcal{C}$ has pullbacks. Then, a natural transformation $\eta\colon F \Rightarrow G$ is a monomorphism in $\mathsf{Fun}(\mathcal{D},\mathcal{C})$ if and only if every component $\eta_D\colon F(D) \to G(D)$ is a monomorphism in $\mathcal{C}$.*

Proof   We can test the property of being a monomorphism by using kernel pairs (see Definition 3.1.32). If pullbacks exist, then so do kernel pairs. As pullbacks are computed pointwise, the result follows from Proposition 3.1.34. $\qquad\square$

3.4 Adjoint Functors and Colimits and Limits

Let $\mathcal{D}$ be a small category and $F\colon \mathcal{D} \to \mathcal{C}$ be a functor. Assume that F has a colimit $(\mathrm{colim}_{\mathcal{D}}, \tau_D\colon F(D) \to \mathrm{colim}_{\mathcal{D}} F)$ in $\mathcal{C}$. If $H\colon \mathcal{C} \to \mathcal{E}$ is any functor, then we always get morphisms

$$
H(\tau_D)\colon H(F(D)) \to H(\mathrm{colim}_{\mathcal{D}} F).
$$

But, in general, these morphisms will not have any universal property. If they do, then this is a particular property of the functor H.

Definition 3.4.1
- A functor $H\colon \mathcal{C} \to \mathcal{E}$ *preserves colimits* if for all functors $F\colon \mathcal{D} \to \mathcal{C}$ from a small category $\mathcal{D}$ for which the colimit $(\mathrm{colim}_{\mathcal{D}} F, \tau_D\colon F(D) \to \mathrm{colim}_{\mathcal{D}} F)$

of F exists, the pair $(H(\text{colim}_{\mathcal{D}} F), H(\tau_D): H(F(D)) \to H(\text{colim}_{\mathcal{D}} F))$ is the colimit of $H \circ F$.

- Dually, a functor $H: \mathcal{C} \to \mathcal{E}$ *preserves limits* if for all functors $F: \mathcal{D} \to \mathcal{C}$ from a small category $\mathcal{D}$ for which the limit $(\lim_{\mathcal{D}} F, \tau_D: \lim_{\mathcal{D}} F \to F(D))$ of F exists, the pair $(H(\lim_{\mathcal{D}} F), H(\tau_D): H(\lim_{\mathcal{D}} F) \to H(F(D)))$ is the limit of $H \circ F$.

Note that we require the structure maps $H(\tau_D)$ to be the limiting cocone or cone. It does not suffice to have abstract isomorphisms in $\mathcal{E}$ between $H(\text{colim}_{\mathcal{D}} F)$ and $\text{colim}_{\mathcal{D}} H \circ F$ or between $H(\lim_{\mathcal{D}} F)$ and $\lim_{\mathcal{D}} H \circ F$.

Limits and colimits are compatible with the fitting half of an adjoint pair of functors.

Theorem 3.4.2 *If $C' \underset{R}{\overset{L}{\rightleftarrows}} C$ is an adjoint pair of functors, then the functor L preserves colimits and R preserves limits.*

Proof Let $\tau': \Delta(D) \Rightarrow R \circ F$. The morphisms $\tau'_{D'}: D \to RF(D')$ correspond to the morphisms $\varphi^{-1}_{D,F(D')}(\tau'_{D'}) = \sigma_{D'}: LD \to F(D')$ under the natural bijections of the adjunction. If $f: D_1 \to D_2$ is a morphism in $\mathcal{D}$, then naturality implies that $\sigma_{D_2} = F(f) \circ \sigma_{D_1}$. Therefore, there is a unique $\xi: LD \to \lim_{\mathcal{D}} F$, such that for all objects D of $\mathcal{D}$, we have the equality $\tau_D \circ \xi = \sigma_D$:

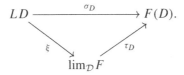

Hence, globally, we get $\tau \circ \Delta(\xi) = \sigma$. The morphism ξ corresponds to a morphism $\tilde{\xi}: D \to R(\lim_{\mathcal{D}} F)$ under the adjunction. By the naturality of the bijections on morphism sets, we obtain that

$$R(\tau_D) \circ \tilde{\xi} = \tau'_D \text{ for all objects } D \text{ of } \mathcal{D},$$

and hence, also,

$$R(\tau) \circ \Delta(\tilde{\xi}) = \tau'.$$

The case of left adjoint functors and colimits is dual to the one above. $\qquad\square$

Examples 3.4.3

- Forgetful functors preserve products if there is a corresponding left adjoint functor, and dually free functors preserve coproducts. Hence, without knowing any specifics about Lie algebras, you know that the n-fold coproduct of a free Lie algebra generated by a 1-dimensional vector space is the free Lie algebra on an n-dimensional vector space. Similarly, the underlying set of the product of groups is the product of the underlying sets of the groups.

- Left adjoint functors preserve initial objects and right adjoints preserve terminal objects.
- If Fr_n denotes the free group on n generators and $G * G'$ denotes the coproduct of the groups G and G', then we obtain

$$*_{i=1}^{n} \mathrm{Fr}_1 \cong \mathrm{Fr}_n.$$

3.5 Exchange Rules for Colimits and Limits

Assume that we have a functor $F: \mathcal{D} \times \mathcal{D}' \to \mathcal{C}$, where $\mathcal{D}$ and $\mathcal{D}'$ are small and $\mathcal{C}$ is complete and cocomplete. For a fixed object D of $\mathcal{D}$, we obtain a functor $F(D, -): \mathcal{D}' \to \mathcal{C}$, and similarly, any object D' of $\mathcal{D}'$ gives rise to a functor $F(-, D'): \mathcal{D} \to \mathcal{C}$. We can form partial (co)limits and obtain functors

$$\mathrm{colim}_{\mathcal{D}} F: \mathcal{D}' \to \mathcal{C}, \quad \mathrm{colim}_{D'}: \mathcal{D} \to \mathcal{C}$$

and

$$\lim_{\mathcal{D}} F: \mathcal{D}' \to \mathcal{C}, \quad \lim_{D'}: \mathcal{D} \to \mathcal{C}.$$

Chasing universal properties gives the following immediate result:

Proposition 3.5.1 *There are canonical isomorphisms*

$$colim_{\mathcal{D}'} colim_{\mathcal{D}} F \cong colim_{\mathcal{D}} colim_{\mathcal{D}'} F \cong colim_{\mathcal{D} \times \mathcal{D}'} F$$

and

$$lim_{\mathcal{D}'} lim_{\mathcal{D}} F \cong lim_{\mathcal{D}} lim_{\mathcal{D}'} F \cong lim_{\mathcal{D} \times \mathcal{D}'} F.$$

The situation is more involved if you try to interchange a limit with a colimit.

Lemma 3.5.2 *Let F be as above. Then, there is a canonical interchange morphism in $\mathcal{C}$*

$$\chi: colim_{\mathcal{D}'} lim_{\mathcal{D}} F \to lim_{\mathcal{D}} colim_{\mathcal{D}'} F.$$

Proof For fixed objects D and D', there is a morphism in $\mathcal{C}$

$$\lim_{\mathcal{D}} F(-, D') \to F(D, D'),$$

and this can be composed with the canonical morphism

$$F(D, D') \to \mathrm{colim}_{\mathcal{D}'} F(D, -).$$

This yields a morphism

$$\psi_{D'} : \lim_{\mathcal{D}} F(-, D') \to \lim_{\mathcal{D}} \mathrm{colim}_{\mathcal{D}'} F$$

for every object D' of $\mathcal{D}'$. For every $f \in \mathcal{D}'(D', \tilde{D}')$, we get

$$\psi_{\tilde{D}'} \circ \lim_{\mathcal{D}} F(-, f) = \psi_{D'},$$

and hence, we get χ as the unique induced morphism

$$\chi : \mathrm{colim}_{\mathcal{D}'} \lim_{\mathcal{D}} F \to \lim_{\mathcal{D}} \mathrm{colim}_{\mathcal{D}'} F. \qquad \square$$

The morphism χ won't be an isomorphism in general.

Examples 3.5.3
- If $\mathcal{D}$ and $\mathcal{D}'$ are empty, then $\mathrm{colim}_{\mathcal{D}'} \lim_{\mathcal{D}} F$ is the initial object, whereas $\lim_{\mathcal{D}} \mathrm{colim}_{\mathcal{D}'} F$ is the terminal object and χ is the only possible morphism.
- If $\mathcal{D}$ is the discrete category with objects $\{1, 2\}$ and $\mathcal{D}'$ is also discrete with objects $\{3, 4\}$ and we take the category of sets as the target category and $F : \mathcal{D} \times \mathcal{D}' \to \mathsf{Sets}$ be a functor, then

$$\chi : (F(1, 3) \times F(2, 3)) \sqcup (F(1, 4) \times F(2, 4)) \to (F(1, 3) \sqcup F(1, 4))$$
$$\times (F(2, 3) \sqcup F(2, 4))$$

will, in general, not be a bijection.

Exercise 3.5.4 Find a concrete counterexample for the example above.

Definition 3.5.5 A small category $\mathcal{D}$ is *filtered* if it is not empty and if

- for every pair of objects D_1, D_2 of $\mathcal{D}$, there is an object D of $\mathcal{D}$ and morphisms $f : D_1 \to D$ and $g : D_2 \to D$; and
- for every pair of morphisms $f_1, f_2 : D_1 \to D_2$, there is a morphism $h : D_2 \to D$ with $h \circ f_1 = h \circ f_2$.

Theorem 3.5.6 *Let $\mathcal{D}$ and $\mathcal{D}'$ be nonempty and small. Assume that $\mathcal{D}$ is finite, $\mathcal{D}'$ is filtered, and $F : \mathcal{D} \times \mathcal{D}' \to \mathsf{Sets}$. Then, χ is an isomorphism.*

Proof　In the category Sets, we have explicit model for limits and colimits. For a fixed object D, we have

$$\mathrm{colim}_{\mathcal{D}'} F(D, -) = \bigsqcup_{D' \in \mathrm{Ob}(\mathcal{D}')} F(D, D') / \sim,$$

where $\sim$ identifies elements that can be compared via morphisms of the kind $F(D, f)$ for f in $\mathcal{D}'$. As $\mathcal{D}'$ is filtered, a finite collection of elements $x_{D'_i} \in F(D, D'_i)$, $1 \le i \le n$ can be compared in a common $F(D, D')$, and we can

decide whether they are equivalent in the colimit by passing to some $F(D, \tilde{D}')$
for some object $\tilde{D}'$ of $\mathcal{D}'$.

As $\mathcal{D}$ is finite, this yields an isomorphism $\lim_{\mathcal{D}} \text{colim}_{\mathcal{D}'} F \rightarrow$
$\text{colim}_{\mathcal{D}'} \lim_{\mathcal{D}} F$. An element in the limit of the colimit consists of a finite
tuple of elements in the colimit that satisfies the coherence condition coming
from $\mathcal{D}$. Every such tuple is equivalent to a coherent tuple, where the elements in the colimit come from a component labelled by a common object
$\tilde{D}'$: $(x_{D_1, \tilde{D}'}, \ldots, x_{D_N, \tilde{D}'})$. This is an element in $\lim_{\mathcal{D}} F(-, \tilde{D}')$ and therefore in
$\text{colim}_{\mathcal{D}'} \lim_{\mathcal{D}} F$, where no further identification takes place in the colimit. $\square$

4

Kan Extensions

Kan extensions take a given functor and extend it to a different category. There are two ways of doing that, via colimits and via limits. These extensions don't have to exist, and even if they exist, they might not have nice properties. But in controlled situations, they are extremely useful and they are actually ubiquitous. If you want to learn more about them, then you could consult [**Bo94-1**, §3.7], [**ML98**, Chapter X], [**Rie14**, Chapter 1], or [**Du70**] in the enriched context.

4.1 Left Kan Extensions

Important constructions that can be carried out with the help of colimits are left Kan extensions.

Definition 4.1.1 Let $G: \mathcal{C} \to \mathcal{D}$ and $F: \mathcal{C} \to \mathcal{E}$ be functors. The *left Kan extension of F along G* is a pair (K, α), where

- $K: \mathcal{D} \to \mathcal{E}$ is a functor; and
- $\alpha: F \Rightarrow K \circ G$ is a natural transformation.

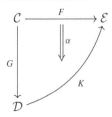

- This pair (K, α) has the universal property that for all pairs (H, β), where $H: \mathcal{D} \to \mathcal{E}$ is a functor and $\beta: F \Rightarrow H \circ G$ is a natural transformation, there is a unique natural transformation $\gamma: K \Rightarrow H$ with the property that $\gamma_G \circ \alpha = \beta$.

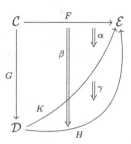

Of course, it is not clear in general whether left Kan extensions exist. If they do, then one can extend the functor F to the category $\mathcal{D}$. But, beware, it is *not* true, in general, that $K \circ G$ coincides with F. However, the natural transformation α compares the two. In certain situations, we can ensure that α is a natural isomorphism (see 4.1.10, for example).

Remark 4.1.2 You might ask yourself why these extensions are called *left* Kan extensions. Assume that the categories $\mathcal{C}$ and $\mathcal{D}$ are small. We abbreviate the functor category $\mathsf{Fun}(\mathcal{D}, \mathcal{E})$ by $\mathcal{E}^{\mathcal{D}}$. We can consider the functor

$$\mathcal{E}^{\mathcal{C}}(F, (-) \circ G) \colon \mathcal{E}^{\mathcal{D}} \to \mathsf{Sets}$$

that sends a functor $H \in \mathcal{E}^{\mathcal{D}}$ to the set of natural transformations from F to $H \circ G$. Any pair (H, β) with $H \colon \mathcal{D} \to \mathcal{E}$ and $\beta \colon F \Rightarrow H \circ G$ corresponds to a natural transformation

$$\beta \colon \mathcal{E}^{\mathcal{D}}(H, -) \Rightarrow \mathcal{E}^{\mathcal{C}}(F, (-) \circ G)$$

via the Yoneda lemma. The universal property of the left Kan extension (K, α) of F along G translates to the fact that α induces a natural isomorphism

$$\alpha \colon \mathcal{E}^{\mathcal{D}}(K, -) \Rightarrow \mathcal{E}^{\mathcal{C}}(F, (-) \circ G).$$

Here, K is on the left in $\mathcal{E}^{\mathcal{D}}(K, -)$. For the dual notion of right Kan extensions, one gets a contravariant representable functor.

Before we start to investigate statements about the existence of left Kan extensions, we construct an auxiliary category.

Definition 4.1.3 Let $G \colon \mathcal{C} \to \mathcal{D}$ be a functor and let D be an object of $\mathcal{D}$. The category $G \downarrow D$ has as objects pairs (C, h), where C is an object of $\mathcal{C}$ and $h \in \mathcal{D}(G(C), D)$. A morphism $f \colon (C, h) \to (C', h')$ is a morphism $f \in \mathcal{C}(C, C')$, such that $h' \circ G(f) = h$:

$$G(C) \xrightarrow{\quad G(f) \quad} G(C')$$

with h going down-right from $G(C)$ to D, and h' going down-left from $G(C')$ to D, meeting at

$$D.$$

This category is a special case of a comma category, and you will learn more about those in Section 5.1.

Theorem 4.1.4 *Let $G: \mathcal{C} \to \mathcal{D}$ and $F: \mathcal{C} \to \mathcal{E}$ be functors. Assume that the category $\mathcal{C}$ is small and that $\mathcal{E}$ is cocomplete. Then, the left Kan extension of F along G exists.*

For the next proof, we closely follow [**Bo94-1**, Theorem 3.7.2].

Proof

- First of all, we will construct the value of the left Kan extension on objects. The category $G \downarrow D$ is small because we assume that $\mathcal{C}$ is small.

  There is a canonical functor $U: G \downarrow D \to \mathcal{C}$, sending a pair (C, h) to C. The cocompleteness of $\mathcal{E}$ ensures that the colimit $\mathrm{colim}_{G \downarrow D}(F \circ U)$ exists with structure maps

  $$\tau^D_{(C,h)}: F(C) = F \circ U(C, h) \to \mathrm{colim}_{G \downarrow D} F \circ U.$$

  We define $K(D)$ as $\mathrm{colim}_{G \downarrow D}(F \circ U)$, and this determines the left Kan extension on objects.

- If $f: D \to D'$ is a morphism in $\mathcal{D}$, then we have to define what $K(f)$ should be. If (C, h) is an object of $G \downarrow D$, then the pair $(C, f \circ h)$ is an object of $G \downarrow D'$, because $f \circ h \in \mathcal{D}(G(C), D')$. In addition, we know that for morphisms $g: (C', h') \to (C, h)$ in $G \downarrow D$, we get that g is a morphism $g: (C', f \circ h') \to (C, f \circ h)$ in $G \downarrow D'$:

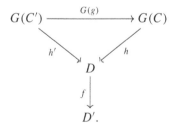

Therefore, we get a uniquely determined morphism

$$K(f): KD = \mathrm{colim}_{G \downarrow D}(F \circ U) \to \mathrm{colim}_{G \downarrow D'}(F \circ U) = KD',$$

such that

$$Kf \circ \tau^D_{(C,h)} = \tau^{D'}_{(C', f \circ h)}.$$

The uniqueness of $K(f)$ ensures that K respects the composition of morphisms and that $K(1_D) = 1_{K(D)}$ for all objects D of $\mathcal{D}$.

- In order to define the natural transformation $\alpha : F \Rightarrow K \circ G$, we set

$$\alpha_C = \tau^{G(C)}_{(C,1_{G(C)})} : F(C) = U(F(C, 1_{G(C)})) \to \mathrm{colim}_{G \downarrow G(C)} F \circ U = K(G(C)).$$

As we have to check the naturality of α_C, we consider a morphism $h \in \mathcal{C}(C, C')$ and the diagram

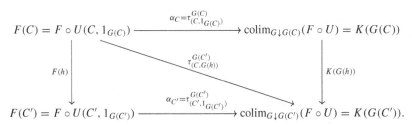

For the morphism $G(h)$, we obtain by construction that $K(G(h)) \circ \tau^{G(C)}_{(C,1_{G(C)})} = \tau^{G(C')}_{(C,G(h))}$. As $1_{G(C')} \circ G(h) = G(h)$, the morphism $h : (C, G(h)) \to (C', 1_{G(C')})$ is a morphism in $G \downarrow G(C')$, and we get

$$\tau^{G(C')}_{(C',1_{G(C')})} \circ F(h) = \tau^{G(C')}_{(C',1_{G(C')})} \circ (F \circ U)(h) = \tau^{G(C')}_{(C,G(h))},$$

so the preceding diagram commutes.

- Let $H : \mathcal{D} \to \mathcal{E}$ be a functor, together with a natural transformation $\beta : F \Rightarrow H \circ G$. We have to establish the existence of a natural transformation $\gamma : K \Rightarrow H$, such that $\gamma_G \circ \alpha = \beta$. Fix an object D of $\mathcal{D}$ and an object (C, h) of $G \downarrow D$. We obtain the composition

$$F \circ U(C, h) = F(C) \xrightarrow{\beta_C} (H \circ G)(C) \xrightarrow{H(h)} H(D),$$

because $h : G(C) \to D$. Again, you should check that the composition is natural for morphisms in $G \downarrow D$, and thus, we obtain a unique factorization

$$\gamma_D : K(D) = \mathrm{colim}_{G \downarrow D}(F \circ U) \to H(D), \quad \gamma_D \circ \tau^{D}_{(C,h)} = H(h) \circ \beta_C. \tag{4.1.1}$$

If $f : D \to D'$ is a morphism in $\mathcal{D}$, then (4.1.1) yields

$$H(f) \circ \gamma_D \circ \tau^{D}_{(C,h)} = H(f) \circ H(h) \circ \beta_C = H(f \circ h) \circ \beta_C,$$

which in turn agrees with

$$\gamma_{D'} \circ \tau^{D'}_{(C,f \circ h)} = \gamma_{D'} \circ K(f) \circ \tau^{D}_{(C,h)},$$

thanks to (4.1.1), and thus, we obtain

$$H(f) \circ \gamma_D = \gamma_{D'} \circ K(f).$$

- It remains to show that $\gamma_G \circ \alpha = \beta$ holds. As $\alpha_C = \tau^{G(C)}_{(C,1_{G(C)})}$ for all C, the claim follows from (4.1.1), because for all C,

$$\gamma_{G(C)} \circ \tau^{G(C)}_{(C,1_{G(C)})} = H(1_{G(C)}) \circ \beta_C = \beta_C. \qquad \square$$

Definition 4.1.5 A functor $F: \mathcal{C} \to \mathcal{E}$ has a *pointwise left Kan extension* along $G: \mathcal{C} \to \mathcal{D}$ if, for all objects D of $\mathcal{D}$, the colimit

$$\operatorname{colim}_{G\downarrow D}(F \circ U)$$

exists. This colimit is then the value of the left Kan extension of F along G, and we denote it by $LKE_G(F)(D)$.

Beware that this is not the standard definition of pointwise left Kan extension. We will reconcile this later in Theorem 4.3.3.

Example 4.1.6 Let $\mathcal{C} = \mathsf{Fin}$ be a small skeleton of the category of finite sets, let $\mathcal{D}$ be the category of sets, Sets, and let I be the canonical inclusion functor of the subcategory Fin into the category Sets. If $\mathcal{E}$ is an arbitrary cocomplete category, then Theorem 4.1.4 ensures that you can extend any functor $F: \mathsf{Fin} \to \mathcal{E}$ as a left Kan extension to the category Sets, and you actually have an explicit formula for doing it.

The following example is a special case of a Day convolution product, that we will meet again later (see 9.8.1).

Example 4.1.7 Let Σ be the category of finite sets of the type $\mathbf{n} = \{1, \ldots, n\}$ for $n \in \mathbb{N}_0$ with the convention that $\mathbf{0} = \varnothing$, and with bijections as morphisms. There is a functor $\sqcup: \Sigma \times \Sigma \to \Sigma$ that is given by the disjoint union of sets with $\mathbf{n} \sqcup \mathbf{m} = \mathbf{n} + \mathbf{m}$. Let $X, Y: \Sigma \to \mathsf{Sets}$ be functors. Then, we can consider the functor

$$X \times Y: \Sigma \times \Sigma \to \mathsf{Sets}, \quad (\mathbf{n}, \mathbf{m}) \mapsto X(\mathbf{n}) \times Y(\mathbf{m}),$$

with the natural behavior on morphisms. The left Kan extension of $X \times Y$ along $\sqcup$ is then given as the functor that sends an object $\mathbf{n}$ to

$$\bigsqcup_{p+q=n} \Sigma_n \times_{\Sigma_p \times \Sigma_q} X(\mathbf{p}) \times Y(\mathbf{q}).$$

Here, $\Sigma_n \times_{\Sigma_p \times \Sigma_q} X(\mathbf{p}) \times Y(\mathbf{q})$ is the quotient of $\Sigma_n \times X(\mathbf{p}) \times Y(\mathbf{q})$ by the equivalence relation that declares an element $(\sigma \circ (\tau_p, \tau_q), (x, y))$ to be equivalent to $(\sigma, (\tau_p(x), \tau_q(y)))$ for all $\sigma \in \Sigma_n$, $\tau_i \in \Sigma_i$, $x \in X(\mathbf{p})$ and $y \in Y(\mathbf{q})$.

Exercise 4.1.8 Check the details of the preceding example. Let $U : \Sigma \to$ Sets be the functor that sends $\mathbf{0}$ to a one-point set and all other objects to the empty set. Can you describe this functor as a representable functor? What is the left Kan extension of $U \times Y$ along $\sqcup$ for an arbitrary $Y : \Sigma \to$ Sets?

Exercise 4.1.9 Let G be a finite group and let H be a subgroup of G. Consider the inclusion of the category $\mathcal{C}_H$ with one object and morphisms H into the category $\mathcal{C}_G$, $i : \mathcal{C}_H \to \mathcal{C}_G$. A functor $F : \mathcal{C}_H \to$ Ab is nothing but a $\mathbb{Z}[H]$-module. $M = F(*)$ carries a linear H-action. What is the left Kan extension of a given F along i?

Although we cannot expect $K \circ G = F$ to hold, quite often, we can at least control the values of K. As a sample result in this direction, we mention the following:

Lemma 4.1.10 *If $\mathcal{C}$ is a small category, $\mathcal{E}$ is a cocomplete category and $G : \mathcal{C} \to \mathcal{D}$ is a fully faithful functor. Then, for all functors $F : \mathcal{C} \to \mathcal{E}$, the natural transformation $\alpha : F \Rightarrow K \circ G$ is a natural isomorphism.*

Proof We first show that the object $(C, 1_{G(C)})$ is terminal in the category $G \downarrow G(C)$. Let (C', h) be an arbitrary object of $G \downarrow G(C)$, that is, $h : G(C') \to G(C)$. As G is a full functor, we know that h is of the form $G(f)$ for an $f : C' \to C$ in $\mathcal{C}$. This f gives rise to a morphism from (C', h) to $(C, 1_{G(C)})$ in $G \downarrow G(C)$. If $f' : (C', h) \to (C, 1_{G(C)})$ is another morphism in $G \downarrow G(C)$, then we obtain

$$G(f) = 1_{G(C)} \circ G(f) = h = 1_{G(C)} \circ G(f') = G(f'),$$

and the faithfulness of G implies that $f = f'$. Colimits over a diagram that possesses a terminal object agree with the value of the functor on this very terminal object (see Exercise 3.1.5). Therefore, we get

$$K(G(C)) = \mathrm{colim}_{G \downarrow G(C)}(F \circ U) \cong F \circ U(C, 1_{G(C)}) = F(C). \qquad \square$$

If both categories $\mathcal{C}$ and $\mathcal{D}$ are small, then left Kan extensions can be interpreted via left adjoint functors. If $G : \mathcal{C} \to \mathcal{D}$ is a functor between small categories, then the precomposition with G yields a functor between the functor category $\mathsf{Fun}(\mathcal{D}, \mathcal{E})$ and $\mathsf{Fun}(\mathcal{C}, \mathcal{E})$.

If F is an object of $\mathsf{Fun}(\mathcal{D}, \mathcal{E})$, then $G^*(F) = F \circ G$ is an object of $\mathsf{Fun}(\mathcal{C}, \mathcal{E})$. If $\mathcal{E}$ is a cocomplete category, then we know that Kan extensions exist. These give rise to a map from $\mathsf{Fun}(\mathcal{C}, \mathcal{E})$ to $\mathsf{Fun}(\mathcal{D}, \mathcal{E})$, and we get the following:

Theorem 4.1.11 *For small categories $\mathcal{C}$, $\mathcal{D}$ and $G : \mathcal{C} \to \mathcal{D}$ and a cocomplete category $\mathcal{E}$ the functor,*

$$G^* : \mathsf{Fun}(\mathcal{D}, \mathcal{E}) \to \mathsf{Fun}(\mathcal{C}, \mathcal{E})$$

has a left adjoint, and this adjoint is given by the left Kan extension.

Proof   If (K, α) is a left Kan extension of F along G, then (K, α) is a reflection of G^*. $\square$

Example 4.1.12 Assume that $f : X \to Y$ is a continuous map between topological spaces and that $\mathcal{F}$ is a presheaf on Y. One could try to pull $\mathcal{F}$ back via f by defining $f^{-1}\mathcal{F}(U) = \mathcal{F}(f(U))$, but, of course, $f(U)$ doesn't have to be open, so instead, one defines the *inverse image presheaf* as the left Kan extension

$$f^{-1}\mathcal{F}(U) = \mathrm{colim}_{f(U) \subset V \text{ open}} \mathcal{F}(V). \tag{4.1.2}$$

Even if $\mathcal{F}$ was a sheaf, $f^{-1}\mathcal{F}$ might not be one, so for sheaves, $f^{-1}\mathcal{F}$ is defined as the sheafification of (4.1.2).

Left Kan extensions have many applications. We will discuss them later again, for instance, when we define the Day convolution product (see Definition 9.8.1) in nice functor categories.

4.2 Right Kan Extensions

The definition of right Kan extension is, of course, dual to the one for left Kan extensions, but what does "dual" mean?

Definition 4.2.1 Let $G : \mathcal{C} \to \mathcal{D}$ and $F : \mathcal{C} \to \mathcal{E}$ be functors. The *right Kan extension of F along G* is a pair (K, α), where

- $K : \mathcal{D} \to \mathcal{E}$ is a functor; and
- $\alpha : K \circ G \Rightarrow F$ is a natural transformation.

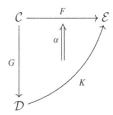

- This pair (K, α) has the universal property that for all pairs (H, β), where $H : \mathcal{D} \to \mathcal{E}$ is a functor and $\beta : H \circ G \Rightarrow F$ is a natural transformation, there is a unique natural transformation $\gamma : H \Rightarrow K$, with the property that $\beta = \alpha \circ \gamma_G$.

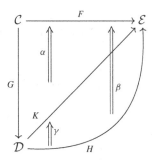

The results for left Kan extensions dualize. Instead of $G \downarrow D$, one uses the category $D \downarrow G$, whose objects are pairs (C, h) with C an object of $\mathcal{C}$ and $h \in \mathcal{D}(D, G(C))$. Morphisms in $(G \downarrow D)((C, h), (C', h'))$ are those $f \in \mathcal{C}(C, C')$ with $G(f) \circ h = h'$.

Theorem 4.2.2

- *If $\mathcal{C}$ is small and $\mathcal{E}$ is complete, then right Kan extensions of F along G, $RKE_G(F)$, exist, and for all objects D of $\mathcal{D}$, we get*

$$RKE_G(F)(D) = \lim_{D \downarrow G} F \circ U.$$

- *We call a right Kan extension* pointwise *if the above limit exists for all objects D of $\mathcal{D}$.*
- *If $\mathcal{C}$ is small, $\mathcal{E}$ is complete, and G is fully faithful, then for all $F : \mathcal{C} \to \mathcal{E}$, the natural transformation $\alpha : K \circ G \Rightarrow F$ is a natural isomorphism.*

Exercise 4.2.3 As in Exercise 4.1.9, let H be a subgroup of a group G. Identify the right Kan extension of an H-module $F : \mathcal{C}_H \to$ Ab along $i : \mathcal{C}_H \to \mathcal{C}_G$.

The preceding exercise can be generalized as follows:

Exercise 4.2.4 Let $\mathcal{C}$ and $\mathcal{D}$ be two small categories and let $\mathcal{E}$ be a bicomplete category. Let $G : \mathcal{C} \to \mathcal{D}$ be a functor. Show that the restriction functor $G^* = (-) \circ G : \text{Fun}(\mathcal{D}, \mathcal{E}) \to \text{Fun}(\mathcal{C}, \mathcal{E})$ has a left adjoint and a right adjoint.

There is a plethora of notation for the preceding adjoints, for instance, $G^!$ and $G_!$.

4.3 Functors Preserving Kan Extensions

Consider a diagram

$$\mathcal{C} \xrightarrow{\ F\ } \mathcal{E} \xrightarrow{\ H\ } \mathcal{F}$$
$$\overset{G}{\searrow}$$
$$\mathcal{D}.$$

If the left Kan extension of F along G exists, then we could be lucky, and the composition with H gives a left Kan extension of $H \circ F$ along G.

Definition 4.3.1 The functor H *preserves the left Kan extension* (K, α) of F along G if $(H \circ K, H\alpha)$ is a left Kan extension of $H \circ F$ along G.

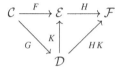

Proposition 4.3.2 *Let $G : \mathcal{C} \to \mathcal{D}$ be a functor between small categories. Left adjoint functors $L : \mathcal{E} \to \mathcal{F}$ preserve left Kan extensions of functors $F : \mathcal{D} \to \mathcal{E}$.*

Proof Let $L : \mathcal{E} \to \mathcal{F}$ be left adjoint to a functor $R : \mathcal{F} \to \mathcal{E}$. Then, we obtain an induced adjunction

$$\mathsf{nat}(L \circ K, J) \cong \mathsf{nat}(K, RJ)$$

for every $J \in \mathsf{Fun}(\mathcal{D}, \mathbb{F})$. We have to check that LK is left adjoint to the precomposition functor G^*. By Theorem 4.1.11, we know that K is left adjoint to G^*, and hence, we get

$$\mathsf{nat}(L \circ K, J) \cong \mathsf{nat}(K, RJ)$$
$$\cong \mathsf{nat}(F, RJG)$$
$$\cong \mathsf{nat}(LF, JG),$$

and this shows the claim. The natural transformation $\alpha : F \Rightarrow KG$ maps to $L\alpha : LF \Rightarrow LKG$. If we consider $J = LK$ and if we start with the identity natural transformation $\mathrm{ID} : LK \Rightarrow LK$, then the chain of bijections described previously transforms ID precisely to $L\alpha$, proving the universality of $L\alpha$. $\square$

Theorem 4.3.3 *Let $\mathcal{C}$ and $\mathcal{D}$ be small. A right Kan extension of $F : \mathcal{C} \to \mathcal{E}$ is pointwise if and only if it is preserved by all representable functors $\mathcal{E}(E, -) : \mathcal{E} \to \mathsf{Sets}$.*

Proof If for all objects D of $\mathcal{D}$, we get that the right Kan extension is given by $\lim_{D\downarrow G} F \circ U$, then

$$\mathcal{E}(E, \lim_{D\downarrow G} F \circ U) \cong \lim_{D\downarrow G} \mathcal{E}(E, F \circ U),$$

and hence, $\mathcal{E}(E, -)$ preserves the right Kan extension.

For the converse, assume that (K, α) is a right Kan extension of F along G, such that $(\mathcal{E}(E, K(-)), \mathcal{E}(E, \alpha))$ is a right Kan extension of $(\mathcal{E}(E, F(-)))$ along G. The Yoneda lemma gives that

$$\mathsf{Fun}(\mathcal{D}, \mathsf{Sets})(\mathcal{D}(D, -), \mathcal{E}(E, K(-))) \cong \mathcal{E}(E, K(D))$$

for all objects D of $\mathcal{D}$. The adjunction property of right Kan extensions dual to Theorem 4.1.11 implies

$$\mathsf{Fun}(\mathcal{D}, \mathsf{Sets})(\mathcal{D}(D, -), \mathcal{E}(E, K(-)))$$
$$\cong \mathsf{Fun}(\mathcal{C}, \mathsf{Sets})(\mathcal{D}(D, G(-)), \mathcal{E}(E, F(-))).$$

We claim that the latter is in bijection with the set of natural transformations

$$\mathsf{Fun}(D \downarrow G, \mathcal{E})(\Delta(E), F \circ U).$$

Consider $\tau \colon \Delta(E) \Rightarrow F \circ U$. Then, τ has components $\tau_{(C,h)} \colon E \to F(C)$ for all objects (C, h) of $D \downarrow G$. Naturality of τ means that the diagram

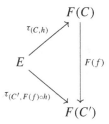

commutes for all $f \in \mathcal{C}(C, C')$.

On the other hand, a natural transformation

$$\xi \colon \mathcal{D}(D, G(-)) \Rightarrow \mathcal{E}(E, F(-))$$

has components $\xi_C \colon \mathcal{D}(D, G(C)) \Rightarrow \mathcal{E}(E, F(C))$, such that the diagram

$$
\begin{array}{ccc}
\mathcal{D}(D, G(C)) & \xrightarrow{\ \xi_C\ } & \mathcal{E}(E, F(C)) \\
{\scriptstyle \mathcal{D}(D, G(f))}\big\downarrow & & \big\downarrow {\scriptstyle \mathcal{E}(E, F(f))} \\
\mathcal{D}(D, G(C')) & \xrightarrow{\ \xi_{C'}\ } & \mathcal{E}(E, F(C'))
\end{array}
$$

commutes for all $f \in \mathcal{C}(C, C')$. The claimed bijection is given by an exponential map, where we send $\tau_{(C,h)}$ to $\xi_C(h)$ and vice versa.

Hence, we get a bijection

$$\mathcal{E}(E, K(D)) \cong \mathsf{Fun}(D \downarrow G, \mathcal{E})(\Delta(E), F \circ U),$$

and therefore, $K(D)$ is isomorphic to the limit $\lim_{D \downarrow G} F \circ U$. □

Remark 4.3.4 The dual statement is also true, but in that case, we have to consider the representable functors $\mathcal{E}(-, E)$ which transform colimits to limits in Sets^o.

4.4 Ends

Before I start with the definitions, let us consider an example. Let $\mathcal{D}$ be a small category and let $F, G \colon \mathcal{D} \to \mathcal{C}$ be two functors from $\mathcal{D}$ to some category $\mathcal{C}$. We all know what natural transformations from F to G are, but can we describe them in terms of a limit construction? Assume that φ is such a natural transformation, then for every object D of $\mathcal{D}$, we have a morphism in $\mathcal{C}$

$$\varphi_D \in \mathcal{C}(F(D), G(D)),$$

and for every morphism $f \in \mathcal{C}(D, D')$, the naturality condition on φ says that

$$G(f) \circ \varphi_D = \varphi_{D'} \circ F(f).$$

Take the set of natural transformations from F to G, $\mathsf{nat}(F, G)$. There are evaluation maps $\varepsilon_D \colon \mathsf{nat}(F, G) \to \mathcal{C}(F(D), G(D))$ for every object D of $\mathcal{D}$. Then, the naturality condition can be rephrased by saying that the diagram

commutes.

Ends are a generalization of this example. We will see that the set of natural transformations from F to G is an end.

Definition 4.4.1 Let $\mathcal{D}$ and $\mathcal{E}$ be categories. Assume that $H_1 \colon \mathcal{D}^o \times \mathcal{D} \to \mathcal{E}$ and $H_2 \colon \mathcal{D}^o \times \mathcal{D} \to \mathcal{E}$ are functors, and let

$$\tau_D \colon H_1(D, D) \to H_2(D, D)$$

be a family (indexed over the objects of $\mathcal{D}$) of morphisms $\tau_D \in \mathcal{E}(H_1(D, D), H_2(D, D))$. Then, $(\tau_D)_D$ is called a *dinatural transformation* if for all morphisms $f \in \mathcal{D}(D, D')$, the diagram

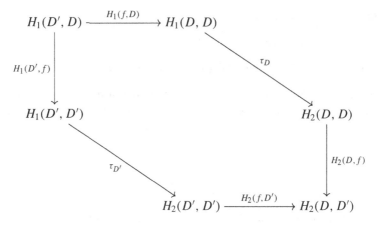

commutes.

Remark 4.4.2 You should think about the word *dinatural* as standing for **natural** on the **dia**gonal and not for natural in two arguments, because τ_D is defined only when both arguments of the H_is agree.

Let us consider some examples of such functors and of dinatural transformations.

Examples 4.4.3
- An important example of a functor $H \colon \mathcal{D}^o \times \mathcal{D} \to \mathcal{E}$ is a natural evaluation map. Fix a K-vector space W, and denote by $L(V, W)$ the vector space of K-linear maps from V to W. Consider the functor

$$L(-, W) \otimes \mathrm{Id} \to \mathrm{vect}^o \times \mathrm{vect} \to \mathrm{vect}, \quad (V_1, V_2) \mapsto L(V_1, W) \otimes V_2.$$

A dinatural transformation from this functor to the constant functor on W, κ_W, consists of a family of linear maps

$$\tau_V \colon L(V, W) \otimes V \to W,$$

which transform naturally in V.
- Let V and W be K-vector spaces, and denote by $\mathrm{Iso}(V, W)$ the vector space of K-linear isomorphisms from V to W. Then,

$$\mathrm{Iso} \colon \mathrm{vect}^o \times \mathrm{vect} \to \mathrm{vect}$$

is a functor, and $\mathrm{Iso}(V, V)$ is the group of automorphisms of V. For instance, if $K = \mathbb{R}$, we can consider the orientation preserving automorphisms of V, $\mathrm{Aut}^+(V)$. The inclusion of $\mathrm{Aut}^+(V)$ into $\mathrm{Aut}(V)$ is then a τ_V, where τ is a dinatural transformation.

- In fact, the preceding example generalizes to any category. For two objects C_1 and C_2 of a category $\mathcal{C}$, we can always consider the set of isomorphisms from C_1 to C_2, $\mathrm{Iso}(C_1, C_2)$, and $\mathrm{Aut}(C_1) = \mathrm{Iso}(C_1, C_1)$, the group of automorphisms of the object C_1. If this group has interesting subgroups that transform naturally in C_1, then the inclusion of such a subgroup into $\mathrm{Aut}(C_1)$ gives rise to a dinatural transformation.
- Last but not least, we fix an object E of $\mathcal{E}$ and consider the constant functor on E, κ_E, as a functor

$$\kappa_E \colon \mathcal{D}^o \times \mathcal{D} \to \mathcal{E}.$$

Ends are universal with respect to constant bifunctors.

Definition 4.4.4 Let $H \colon \mathcal{D}^o \times \mathcal{D} \to \mathcal{E}$ be a functor. An *end of H* is a pair (E, τ), where E is an object of $\mathcal{E}$ and τ is a dinatural transformation from κ_E to H with the property that for all other objects E' of $\mathcal{E}$ with a dinatural transformation ν from $\kappa_{E'}$ to H, there is a unique $\xi \in \mathcal{E}(E', E)$, such that $\nu_D = \tau_D \circ \xi$ for all D.

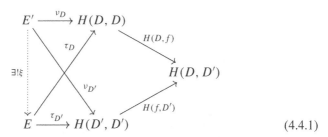

$$(4.4.1)$$

Example 4.4.5 Let $\mathcal{D}$ be a small category, let $\mathcal{E}$ be an arbitrary category, and assume F and G are functors from $\mathcal{D}$ to $\mathcal{E}$. We consider

$$\mathcal{E}(F(-), G(-)) \colon \mathcal{D}^o \times \mathcal{D} \to \mathsf{Sets}$$

as a functor. An end of this functor is a set X, together with a universal dinatural transformation

$$\varepsilon_D \colon X \to \mathcal{E}(F(D), G(D))$$

for all objects D of $\mathcal{D}$, which satisfies the coherence condition, as illustrated in the diagram (4.4.1). It is clear that the set of all natural transformations satisfies this condition. If X' is another set with a dinatural transformation ν from $\kappa_{X'}$ to $\mathcal{E}(F(-), G(-))$, then for every element $x \in X'$, $\nu_D(x)$ is actually a natural transformation because of the naturality of ν, but then, we obtain a function $f \colon X' \to X = \mathsf{nat}(F, G)$ with $f(x)_D = \nu_D(x)$.

As a special case, we obtain that the abelian group of R-module homomorphism between two left R-modules M and N is an end.

As always, coends are dual to ends. Natural transformations are the morphisms in the functor category $\mathsf{Fun}(\mathcal{D}, \mathcal{E})$. Dual to this, coends generalize tensor products.

Definition 4.4.6 A *coend* of a functor $H \colon \mathcal{D}^o \times \mathcal{D} \to \mathcal{E}$ is a pair (E, τ), where E is an object of $\mathcal{E}$ and τ is a dinatural transformation from H to the constant functor κ_E, such that for every object E' of $\mathcal{E}$ with a dinatural transformation ν from H to $\kappa_{E'}$, there is a morphism $f \in \mathcal{E}(E, E')$, such that $f \circ \tau_D = \nu_D$ for all objects D of $\mathcal{D}$.

Example 4.4.7 Let $\mathcal{D}$ be a small category and let $F \colon \mathcal{D}^o \to k\text{-mod}$ and $G \colon \mathcal{D} \to k\text{-mod}$ be functors. Here, k is an arbitrary commutative ring with unit, and $k\text{-mod}$ denotes the category of k-modules and k-linear maps. Then, we can build the *tensor product of F and G* as

$$F \otimes_{\mathcal{D}} G := \bigoplus_D F(D) \otimes_k G(D)/ \sim,$$

where the sum is indexed by all objects D of $\mathcal{D}$ and where we divide out by the k-submodule of $\bigoplus_D F(D) \otimes_k G(D)$ generated by

$$F(f)(x) \otimes y - x \otimes G(f)(y), \quad x \in F(D'), y \in G(D), f \in \mathcal{D}(D, D'). \tag{4.4.2}$$

We claim that $F \otimes_{\mathcal{D}} G$, together with the dinatural transformation τ that sends $F(D) \otimes_k G(D)$ to the class of the summand in $F \otimes_{\mathcal{D}} G$, is the coend of the functor $F \otimes_k G \colon \mathcal{D}^o \times \mathcal{D} \to k\text{-mod}$ that sends (D_1, D_2) to $F(D_1) \otimes_k G(D_2)$ and $(f, g) \in \mathcal{D}(D_1, D_2) \times \mathcal{D}(D_3, D_4)$ to $F(f) \otimes_k G(g)$.

By construction, τ is dinatural. If ν is another dinatural transformation from F to some other k-module M, then M receives a map ν_D from all the $F(D) \otimes_k G(D)$. The dinaturality of ν then guarantees that these maps are compatible, and thus, the submodule corresponding to (4.4.2) is in the kernel of the map $\bigoplus_D \nu_D$ and we get the desired map from M to $F \otimes_{\mathcal{D}} G$.

As a special case, we get the tensor product of two R-modules M and N by considering $\mathcal{D} = \mathcal{C}_R$, $F(*) = M$ and $G(*) = N$.

You will find other instances of the tensor product of functors later, when we learn about symmetric monoidal categories (Definition 15.1.1) and about geometric realizations of simplicial sets (see Remark 10.6.2). These tensor products are crucial for functor homology, and we will meet them again in full generality in Section 15.1.

4.5 Coends as Colimits and Ends as Limits

We want to clarify the question of when ends and coends exist. To this end, we will express ends as limits. The case of coends as colimits is dual.

The following definition is an important variant of the morphism category:

Definition 4.5.1 Let $\mathcal{D}$ be a category. The *twisted arrow category of $\mathcal{D}$, $\mathcal{D}^\tau$* has the morphisms of $\mathcal{D}$ as objects. A morphism in $\mathcal{D}^\tau$ from $f : D_1 \to D_2$ to $g : D_3 \to D_4$ is a pair of morphisms (h_1, h_2) of $\mathcal{D}$, such that $h_1 \in \mathcal{D}(D_3, D_1)$ and $h_2 \in \mathcal{D}(D_2, D_4)$ with the property that $g = h_2 \circ f \circ h_1$:

$$
\begin{array}{ccc}
D_1 & \xrightarrow{\ f\ } & D_2 \\
{\scriptstyle h_1}\big\uparrow & & \big\downarrow{\scriptstyle h_2} \\
D_3 & \xrightarrow{\ g\ } & D_4.
\end{array}
$$

Definition 4.5.2 Let $\mathcal{D}$ be a category. We define the functor $\chi : \mathcal{D}^\tau \to \mathcal{D}^o \times \mathcal{D}$ as $\chi(f) = (s(f), t(f))$ and $\chi(h_1, h_2) = (h_1^o, h_2)$. Here, $s(f)$ denotes the source of f and $t(f)$ denotes its target.

With the help of these auxiliary data, we can express an end of a functor H as a limit.

Proposition 4.5.3 *Let $\mathcal{D}$ and $\mathcal{E}$ be categories and $H : \mathcal{D}^o \times \mathcal{D} \to \mathcal{E}$ be a functor. An end of H is isomorphic to the limit of $H \circ \chi$ over $\mathcal{D}$. In particular, if $\mathcal{E}$ is complete and $\mathcal{D}$ is small, then ends always exist.*

Proof   Given a dinatural transformation ν from κ_E to H with components $\nu_D : E \to H(D, D)$, we define a natural transformation $\tau : \Delta(E) \Rightarrow H \circ \chi$ for every object $f \in \mathcal{D}(D, D')$ of $\mathcal{D}^\tau$ as $\tau_f = H(D, f) \circ \nu_D = H(f, D') \circ \nu_{D'}$:

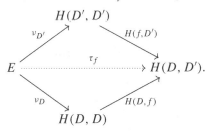

We have to check that τ is natural, that is, $H \circ \chi(\alpha, \beta) \circ \tau_f = \tau_g$ for $\alpha : \tilde{D} \to D$, $\beta : \tilde{D}' \to D'$, such that

$$
\begin{array}{ccc}
D & \xleftarrow{\ \alpha\ } & \tilde{D} \\
{\scriptstyle f}\big\downarrow & & \big\downarrow{\scriptstyle g} \\
D' & \xrightarrow{\ \beta\ } & \tilde{D}'
\end{array}
$$

commutes. Consider the diagram

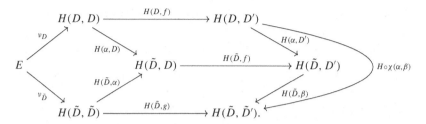

The left rhombus commutes and actually defines τ_α. The top parallelogram commutes because H is a functor and the bottom one commutes because $g = \beta \circ f \circ \alpha$. The right wing describes the definition of $H \circ \chi(\alpha, \beta)$. Therefore, the whole diagram commutes, giving

$$\tau_g = H(\tilde{D}, g) \circ \nu_{\tilde{D}} = H \circ \chi(\alpha, \beta) \circ H(D, f) \circ \nu_D = H \circ \chi(\alpha, \beta) \circ \tau_f.$$

Conversely, given $\tau \colon \Delta(E) \Rightarrow H \circ \chi$, we can evaluate τ on identity morphisms to get $\nu_D := \tau_{1_D}$. As $(1_D, f)$ is a morphism from 1_D to f and as $(f, 1_{D'})$ is a morphism from $1_{D'}$ to f in $\mathcal{D}^\tau$, the binaturality of ν follows from the commutativity of

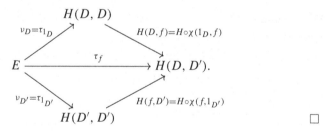

4.6 Calculus Notation

It is a tradition to denote ends and coends by integrals. If you consider the examples, then the heuristics behind that is that you integrate over the diagram category $\mathcal{D}$ and get an object E in the target category, where the contribution of $\mathcal{D}$ is subsumed into E. Another reason for this notation is that it helps to guess some correct statements about ends and coends – there is, for instance, a Fubini theorem for ends and coends.

Notation 4.6.1 Let $\mathcal{D}$ and $\mathcal{E}$ be categories and let $H \colon \mathcal{D}^o \times \mathcal{D} \to \mathcal{E}$ be a functor. We denote

- by $\int_{\mathcal{D}} H$, the end of the functor H; and
- by $\int^{\mathcal{D}} H$, the coend of the functor H.

Remark 4.6.2 This notation is the traditional one. Jacob Lurie uses exactly the opposite notation in [**Lu09**].

With Proposition 3.5.1 and Proposition 4.5.3, we obtain the following result:

Proposition 4.6.3 (Fubini theorem for ends) *Let* $H : (\mathcal{D} \times \mathcal{D}')^o \times (\mathcal{D} \times \mathcal{D}') \to \mathcal{E}$ *be a functor. If the ends* $\int_{\mathcal{D}} H(D, D'_1, D, D'_2)$ *exist for all objects* D'_1, D'_2 *of* $\mathcal{D}'$ *and if the ends* $\int_{\mathcal{D}'} H(D_1, D', D_2, D')$ *exist for all objects* D_1 *and* D_2 *of* $\mathcal{D}$, *then*

$$\int_{\mathcal{D}} \int_{\mathcal{D}'} H(D, D', D, D') \cong \int_{\mathcal{D}'} \int_{\mathcal{D}} H(D, D', D, D') \cong \int_{\mathcal{D} \times \mathcal{D}'} H(D, D', D, D'),$$

and if one of them exists, then the others do as well. ☐

There is a dual version for coends.

4.7 "All Concepts are Kan Extensions"

This slogan is taken from [**ML98**, X.7]. We will explain how to express (co)limits, the Yoneda lemma, and adjoint functors in terms of Kan extensions. As these are the main building blocks of category theory, this justifies the slogan.

Proposition 4.7.1 *Colimits (and limits) are special cases of left (and right) Kan extensions.*

Proof We prove the case of colimits. Consider the terminal category [0]. A left Kan extension of a functor $F : \mathcal{C} \to \mathcal{E}$ from a small category $\mathcal{C}$ to the unique functor $G : \mathcal{C} \to [0]$ is a functor $K : [0] \to \mathcal{E}$, together with a natural transformation $\alpha : F \Rightarrow K \circ G$. The functor K is nothing but a choice of an object in $\mathcal{E}$, say E. Precomposition with G just gives the constant functor

$$K \circ G = \Delta(E) : \mathcal{C} \to \mathcal{E},$$

and $\alpha : F \Rightarrow \Delta(E)$ is a cone. The universality of (K, α) then implies that any other cone receives a map from $K \circ G = \Delta(E)$. Thus, E is the colimit of $F : \mathcal{C} \to \mathcal{E}$. ☐

Proposition 4.7.2 *The Yoneda lemma follows from the existence of right Kan extensions for functors* $F : \mathcal{C} \to$ Sets *for small* $\mathcal{C}$.

Proof   The category of sets is complete; hence, Kan extensions can be determined pointwise. We consider the identity functor $\text{Id}_\mathcal{C} \colon \mathcal{C} \to \mathcal{C}$. The right Kan extension of F along $\text{Id}_\mathcal{C}$ is F with $\alpha = \text{ID}$. In particular,

$$F(C) = RKE_{\text{Id}_\mathcal{C}}(F)(C). \tag{4.7.1}$$

On the other hand, we obtain

$$RKE_{\text{Id}}(F)(C) = \lim_{C \downarrow \text{Id}} F \circ U.$$

The category $C \downarrow \text{Id}_\mathcal{C}$ has objects (C', h), where C' is an object of $\mathcal{C}$ and $h \in \mathcal{C}(C, C')$; thus, $C \downarrow \text{Id}_\mathcal{C}$ is isomorphic to the category of morphisms h in $\mathcal{C}$ with source C and $F \circ U(h) = F(C')$. Thus, the functor $F \circ U \colon C \downarrow \text{Id}_\mathcal{C} \to \mathcal{E}$ is an assignment $\eta_{C'} \colon \mathcal{C}(C, C') \to F(C')$, such that $F(f) \circ \eta_{C'} = \eta_{C''}$ for all $f \in \mathcal{C}(C', C'')$. These $\eta_{C'}$s are the components of a natural transformation $\eta \in \text{nat}(\mathcal{C}(C, -), F)$, which is the limit $\lim_{C \downarrow \text{Id}} F \circ U$. Hence, using (4.7.1), we get

$$F(C) \cong \text{nat}(\mathcal{C}(C, -), F). \qquad \square$$

Adjoint functors can be expressed as Kan extensions as follows:

Proposition 4.7.3 *Let $L \colon \mathcal{C} \to \mathcal{E}$ and $R \colon \mathcal{E} \to \mathcal{C}$ be functors.*

(1) Assume that (L, R) is an adjoint pair of functors with unit $\eta \colon \text{Id}_\mathcal{C} \Rightarrow RL$ and counit $\varepsilon \colon LR \Rightarrow \text{Id}_\mathcal{E}$. Then, (R, η) is a left Kan extension of $\text{Id}_\mathcal{C}$ along L and is preserved by L, and (L, ε) is a right Kan extension of $\text{Id}_\mathcal{E}$ along R and is preserved by R.

(2) Conversely, if the left Kan extension (K, α) of $\text{Id}_\mathcal{C}$ along L exists, then K is a right adjoint for L, and $\alpha = \eta$ is the unit of the adjunction. If the right Kan extension (K', α') of $\text{Id}_\mathcal{E}$ along R exists, then K' is left adjoint to R with counit $\varepsilon = \alpha'$.

Proof   We first show (1). Assume first that L has a right adjoint R and that $\eta \colon \text{Id}_\mathcal{C} \Rightarrow RL$ and $\varepsilon \colon LR \Rightarrow \text{Id}_\mathcal{E}$ are the unit and counit of the adjunction, respectively. Assume that $H \colon \mathcal{C} \to \mathcal{C}$ is a functor with a natural transformation $\beta \colon \text{Id}_\mathcal{C} \Rightarrow HL$.

- We need a natural transformation $\xi \colon R \Rightarrow H$ and define it as the composite

$$\xi \colon R = \text{Id}_\mathcal{C} \circ R \xrightarrow{\beta_R} HLR \xrightarrow{H(\varepsilon)} H.$$

- We claim that ξ is unique with $\xi_L \circ \eta = \beta$, so let $\zeta \colon R \Rightarrow H$ be another natural transformation with $\zeta_L \circ \eta = \beta$.

Then, the diagram

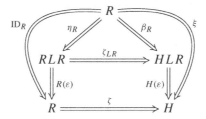

commutes, identifying ξ with ζ.

Thus, (R, η) is a left Kan extension of $\mathrm{Id}_\mathcal{C}$ along L.

• In order to show that $(LR, L\eta)$ is a left Kan extension of L along L, we consider a functor $J : \mathcal{E} \to \mathcal{E}$, together with $\gamma : L \Rightarrow JL$.

For $\tau : LR \Rightarrow J$, we take

$$\tau : LR \overset{\gamma R}{\Longrightarrow} JLR \overset{J(\varepsilon)}{\Longrightarrow} J.$$

As the diagram

$$
\begin{array}{ccc}
L & \overset{\gamma}{\Longrightarrow} & JL \\
L(\eta) \big\Downarrow & JL\eta \big\Downarrow & \diagdown \\
LRL & \underset{\gamma RL}{\Longrightarrow} JLRL & \underset{J(\varepsilon_L)}{\Longrightarrow} JL
\end{array}
$$

commutes, we obtain $\tau_L \circ L(\eta) = \gamma$.

The uniqueness of τ is shown as previously.

For (2), assume that the left Kan extension (K, α) of $\mathrm{Id}_\mathcal{C}$ along L exists and is preserved by L. Then, $\alpha : \mathrm{Id}_\mathcal{C} \Rightarrow RL$ and $L\alpha : L \Rightarrow LRL$. We set $\eta := \alpha$.

• Consider the functor $\mathrm{Id}_\mathcal{E}$, together with the natural transformation $\beta = \mathrm{ID}_L$. As $(LK, L\eta)$ is a left Kan extension, there is a unique $\varepsilon : LR \Rightarrow \mathrm{Id}_\mathcal{E}$, such that $\varepsilon_L \circ L(\eta) = \mathrm{ID}_L$.

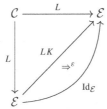

It remains to show that $K(\varepsilon) \circ \eta_K$ is also the identity transformation.

The left Kan extension property of (K, η) ensures that the functor $K : \mathcal{E} \to \mathcal{C}$, together with $\beta : \mathrm{Id}_\mathcal{C} \Rightarrow KL$, $\beta := (K\varepsilon_L) \circ \eta_{KL} \circ \eta$, gives rise to a unique natural transformation $\tau : K \Rightarrow K$ with $\tau_L \circ \eta = \beta$.

The commutativity of the diagram

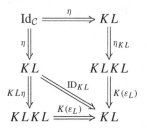

tells us that $\beta = \eta$ and hence that $\tau = \mathrm{ID}_K$. The uniqueness of τ then implies that $\mathrm{ID}_K = K(\varepsilon) \circ \eta_K$. $\square$

5

Comma Categories and the Grothendieck Construction

In this chapter, we introduce the notion of comma categories and related concepts. We present a criterion based on comma categories that tells us when colimits can be determined via simplified diagram categories. We briefly mention sifted colimits and their characterization. The Yoneda lemma expresses values of a functor with the help of representable functors. There is a dual statement, the co-Yoneda lemma, that is related to the concept of density.

Grothendieck constructions play an important role in applications in homotopy theory and in other areas of mathematics. We define them and discuss some examples.

5.1 Comma Categories: Definition and Special Cases

Definition 5.1.1 Given two functors $\mathcal{C} \xrightarrow{F} \mathcal{D} \xleftarrow{G} \mathcal{E}$, the *comma category* (F, G) has as objects triples (C, f, E), where C is an object of $\mathcal{C}$, E is an object of $\mathcal{E}$, and $f \in \mathcal{D}(F(C), G(E))$.

A morphism in (F, G) from (C_1, f_1, E_1) to (C_2, f_2, E_2) consists of a pair of morphisms $g \in \mathcal{C}(C_1, C_2)$ and $h \in \mathcal{E}(E_1, E_2)$, such that $G(h) \circ f_1 = f_2 \circ F(g)$:

$$\begin{array}{ccc} F(C_1) & \xrightarrow{f_1} & G(E_1) \\ {\scriptstyle F(g)}\downarrow & & \downarrow{\scriptstyle G(h)} \\ F(C_2) & \xrightarrow{f_2} & G(E_2). \end{array} \qquad (5.1.1)$$

The composition of morphisms in (F, G) is induced by the composition of morphisms in $\mathcal{C}$ and $\mathcal{E}$.

You only allow objects of C and $\mathcal{E}$ as objects in the comma category that talk to each other in $\mathcal{D}$ via a morphism. But note that C and $\mathcal{E}$ are not quite on equal footing. The morphisms always have to run from some $F(C)$ to some $G(E)$.

There are canonical functors $p_C \colon (F, G) \to C$ and $p_{\mathcal{E}} \colon (F, G) \to \mathcal{E}$, given by projecting to the first and last component:

$$p_C(C, f, E) = C, \, p_C(g, h) = g; \quad p_{\mathcal{E}}(C, f, E) = E, \, p_{\mathcal{E}}(g, h) = h.$$

Lemma 5.1.2 *There is a natural transformation* $\tau \colon F \circ p_C \Rightarrow G \circ p_{\mathcal{E}}$.

$$
\begin{array}{ccc}
(F, G) & \xrightarrow{\; p_C \;} & C \\
{\scriptstyle p_{\mathcal{E}}} \downarrow & {\scriptstyle \tau}\!\!\nearrow\!\!\!\!\!\diagup & \downarrow {\scriptstyle F} \\
\mathcal{E} & \xrightarrow[\; G \;]{} & \mathcal{D}
\end{array}
$$

Proof Define $\tau_{(C, f, E)}$ as f. The compatibility condition in (5.1.1) ensures that this indeed defines a natural transformation. $\square$

Exercise 5.1.3 Consider the partially ordered set $[1] = 0 < 1$. Show that

$$
\begin{array}{ccc}
(F, G) & \longrightarrow & \mathsf{Fun}([1], \mathcal{E}) \\
\downarrow & & \downarrow {\scriptstyle p} \\
C \times \mathcal{D} & \xrightarrow[\; F \times G \;]{} & \mathcal{E} \times \mathcal{E}
\end{array}
$$

is a pullback diagram. Here, $\mathsf{Fun}([1], \mathcal{E})$ is the category of functors from $[1]$ to $\mathcal{E}$ and p collects the values of such a functor on 0 and 1 and forgets the morphism.

This is a mapping object interpolating between F and G. If C and $\mathcal{D}$ are the category $[0]$ and if $F = G$, then you get a categorical model of a based loop space.

Example 5.1.4 If $C = \mathcal{D} = \mathcal{E}$ and if both functors are the identity functors, then the objects of $(\mathrm{Id}_C, \mathrm{Id}_C)$ are morphisms in C, and the morphisms between two morphisms $f_1 \colon C_1 \to C_2$ and $f' \colon C_1' \to C_2'$ are pairs of morphisms (g, h) in C that render the diagram

$$
\begin{array}{ccc}
C_1 & \xrightarrow{\; g \;} & C_1' \\
{\scriptstyle f_1} \downarrow & & \downarrow {\scriptstyle f_2} \\
C_2 & \xrightarrow[\; h \;]{} & C_2'
\end{array}
$$

commutative. This gives the *category of morphisms* of C, $\mathsf{Fun}([1], C)$, as a special case of a comma category.

We saw an example of a comma category already in Definition 4.1.3, when we constructed Kan extensions.

Definition 5.1.5 Let $F: \mathcal{C} \to \mathcal{D}$ be a functor and let D be an object of $\mathcal{D}$. The *comma category* $F \downarrow D$ is $(F, \kappa_D: [0] \to \mathcal{D})$, where κ_D sends 0 to D.

Hence, objects of $F \downarrow D$ are pairs (C, f), where C is an object of $\mathcal{C}$ and $f \in \mathcal{D}(F(C), D)$. A morphism from (C, f) to (C', f') in $F \downarrow D$ is a morphism $g: C \to C'$ in $\mathcal{C}$, such that $f' \circ F(g) = f$.

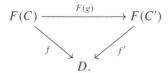

Dually, we can define the comma category $D \downarrow F$ in such a way that the morphisms start in D and end in images of F, that is, $D \downarrow F = (\kappa_D: [0] \to \mathcal{D}, F: \mathcal{C} \to \mathcal{D})$.

Other important special cases are over and under categories (define one functor to be the identity and the other one to be a functor from the category with one object and one morphism).

Definition 5.1.6 Let $\mathcal{C}$ be a category and let C be an object on $\mathcal{C}$.

- The *category of objects over C in $\mathcal{C}$* has as objects morphisms $f \in \mathcal{C}(C', C)$, and a morphism from $f \in \mathcal{C}(C', C)$ to $f' \in \mathcal{C}(C'', C)$ is a morphism $g \in \mathcal{C}(C', C'')$, such that $f' \circ g = f$:

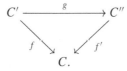

We denote this category by $\mathcal{C} \downarrow C$.
- The *category of objects under C* has as objects morphisms $f \in \mathcal{C}(C, C')$, and a morphism from $f \in \mathcal{C}(C, C')$ to $f' \in \mathcal{C}(C, C'')$ is a morphism $g \in \mathcal{C}(C', C'')$, such that $g \circ f = f'$:

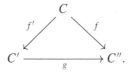

We denote this category by $C \downarrow \mathcal{C}$.

Note that we can describe the morphism sets in $\mathcal{C} \downarrow C$ and in $C \downarrow \mathcal{C}$ as pullbacks. Morphisms in $\mathcal{C} \downarrow C$ from $f : C' \to C$ to $g : C'' \to C$ are elements in the pullback of

$$
\begin{array}{ccc}
 & & \{f\} \\
 & & \downarrow \\
\mathcal{C}(C', C'') & \xrightarrow{\;\mathcal{C}(C',g)\;} & \mathcal{C}(C', C),
\end{array}
$$

and morphisms in $C \downarrow \mathcal{C}$ from $f : C \to C'$ to $g : C \to C''$ are elements in the pullback of

$$
\begin{array}{ccc}
 & & \{g\} \\
 & & \downarrow \\
\mathcal{C}(C', C'') & \xrightarrow{\;\mathcal{C}(f,C'')\;} & \mathcal{C}(C, C'').
\end{array}
$$

Similarly, for a fixed morphism $\xi \in \mathcal{C}(C_1, C_2)$, the *category of objects under C_1 and over C_2* has as objects pairs of morphisms (f, g) with $f \in \mathcal{C}(C_1, C)$ and $g \in \mathcal{C}(C, C_2)$, such that $g \circ f = \xi$, and a morphism from (f, g) to (f', g') is a morphism $h \in \mathcal{C}(C, C')$, such that $h \circ f = f'$ and $g' \circ h = g$:

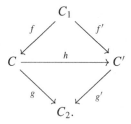

We use the notation $C_1 \downarrow \mathcal{C} \downarrow C_2$ for this category, suppressing ξ from the notation. Such categories feature prominently in the concept of André–Quillen (co)homology [**Q70**]. If k is a commutative ring and B is a commutative k-algebra, then a commutative k-algebra A is an object of the category of commutative k-algebras over B if there is a morphism of k-algebras $f : A \to B$, that is, if A is augmented over B. As k is initial, A is also an object of the category of commutative k-algebras under k and over B (so ξ is the unit map of B). A commutative k-algebra C is an object of the category of commutative k-algebras under A if and only if C is a commutative A-algebra.

Remark 5.1.7 It is common to call categories of the form $\mathcal{C} \downarrow C$ (or $C \downarrow \mathcal{C}$) *slice categories* (*coslice categories*).

Exercise 5.1.8 Show that $C \downarrow C$ has the identity of C (which we identify with C) as a terminal object and $C \downarrow C$ has the identity of C as initial object. If $C_1 = C_2 = C$ and $\xi = 1_C$, then $C \downarrow C \downarrow C$ has the pair $(1_C, 1_C)$ as a zero object.

If C has a terminal object $*$, then $C \downarrow *$ is isomorphic to C and $* \downarrow C$ consists of objects in C with a chosen map from $*$. For instance, the category of pointed topological spaces, Top_*, is the category of topological spaces with a chosen basepoint, and this choice is nothing but a map from a one-point space to a space. Morphisms are continuous maps preserving the basepoint.

Exercise 5.1.9 In the category of small categories, $[0]$ is a terminal object; thus, $\mathsf{cat} \downarrow [0] \cong \mathsf{cat}$. Let C and $\mathcal{D}$ be two categories, then show that their join $C \to \mathcal{D}$ has a functor to $[0] * [0] \cong [1]$. Prove that the join gives rise to a functor

$$\mathsf{cat} \times \mathsf{cat} \to \mathsf{cat} \downarrow [1].$$

Example 5.1.10 Consider the category of sets and fix a nonempty set S. An object in the category $\mathsf{Sets} \downarrow S$ is a set X, together with a function $f_X \colon X \to S$, and morphisms in $\mathsf{Sets} \downarrow S$ are functions that preserve the anchor map to S. This category is nothing but the *category of S-graded sets*: The preimage $f^{-1}(s)$ of an element $s \in S$ is the s-component of X, X_s. We can write $X = \bigsqcup_{s \in S} X_s$, and morphisms in the category preserve these decompositions; that is, a $g \colon X \to Y$ for an object $f_Y \colon Y \to S$ of $\mathsf{Sets} \downarrow S$ satisfies $g(X_s) \subset Y_s$. Note that the anchor maps do not have to be surjective, so some of the summands in $X = \bigsqcup_{s \in S} X_s$ might be empty.

5.1.1 The Categories $F \backslash C$

Given a functor $F \colon C \to \mathsf{Sets}$, we can build a category $F \backslash C$ out of F, C and the values of F.

Definition 5.1.11 For a given F, we let $F \backslash C$ be the category whose objects are pairs (C, x) with C an object of C and $x \in F(C)$. A morphism from (C, x) to (C', x') is an $f \in C(C, C')$ with $F(f)(x) = x'$.

There is a forgetful functor $\rho \colon F \backslash C \to C$ that omits the element x.

We will see later that for certain functors F, the functor ρ gives rise to a covering map on the level of classifying spaces (see Theorem 11.5.5).

There are variants of this construction. Note that in the previous definition, $F(C)$ is a set, and this was a crucial ingredient because we require an equality between $F(f)(x)$ and x'. We can change the target category into any category that has a notion of elements.

Definition 5.1.12 A category $\mathcal{D}$ together with a faithful functor $U : \mathcal{D} \to$ Sets is called *concrete*.

Note that there could be several faithful functors from a given category to the category of sets, but in our examples, it should be clear which functor we mean. If D is an object of $\mathcal{D}$, then $U(D)$ is a set, and hence, we can talk about the elements of $U(D)$. By abuse of language, we might also call them the elements of D.

There are many examples of concrete categories: every category that has objects that are sets with some extra structure and whose morphisms are functions respecting that extra structure is concrete. So, for instance, the categories of groups, of topological spaces, of R-modules for a fixed ring R, and of k-algebras for a fixed commutative ring with unit k are examples of concrete categories. In all of these cases, we can use the obvious forgetful functor to the category of sets as the functor U.

If $\mathcal{C}$ is a category with a functor $F : \mathcal{C} \to \mathcal{D}$ and $(\mathcal{D}, U)$ is a concrete category, then $F \backslash \mathcal{C}$ is short for $(U \circ F) \backslash \mathcal{C}$.

5.1.2 Category of Cones

Let $F : \mathcal{D} \to \mathcal{C}$ be a functor. If $\mathcal{D}$ were small and we wanted to build the limit of F over $\mathcal{D}$, then we would consider objects C of $\mathcal{C}$, together with compatible maps into the system $(F(D))_{D \in \mathrm{Ob}(\mathcal{D})}$. The corresponding category is an important slice category and features prominently in the work of Joyal [**Jo-c∞**] and Lurie [**Lu09, Lu∞**]. Note that for the join of two categories $\mathcal{E} * \mathcal{D}$, there are inclusion functors

$$\mathcal{E} \xrightarrow{\ i_{\mathcal{E}}\ } \mathcal{E} * \mathcal{D} \xleftarrow{\ i_{\mathcal{D}}\ } \mathcal{D}.$$

We can consider functors $G : \mathcal{E} * \mathcal{D} \to \mathcal{C}$ that restrict to a given functor on $\mathcal{E}$ or $\mathcal{D}$. For simplicity, we assume that all categories that are involved are small.

Proposition 5.1.13 *Let* $F : \mathcal{D} \to \mathcal{C}$ *be a functor. Denote by* $\mathsf{Fun}_F(\mathcal{E} * \mathcal{D}, \mathcal{C})$ *the set of functors* $G : \mathcal{E} * \mathcal{D} \to \mathcal{C}$ *with* $G \circ i_{\mathcal{D}} = F$. *Then, there is a category* $\mathcal{C}_{/F}$, *such that*

$$\mathsf{Fun}(\mathcal{E}, \mathcal{C}_{/F}) \cong \mathsf{Fun}_F(\mathcal{E} * \mathcal{D}, \mathcal{C}).$$

Dually, given a functor $H : \mathcal{E} \to \mathcal{C}$, *there is a category* $\mathcal{C}_{H/}$, *such that*

$$\mathsf{Fun}(\mathcal{E}, \mathcal{C}_{H/}) \cong \mathsf{Fun}_H(\mathcal{E} * \mathcal{D}, \mathcal{C}),$$

where $\mathsf{Fun}_H(\mathcal{E} * \mathcal{D}, \mathcal{C})$ *denotes the set of functors* $G: \mathcal{E} * \mathcal{D} \rightarrow \mathcal{C}$ *that restrict to H under* $i_{\mathcal{E}}$.

Proof We prove the first case. Assume that G is a functor $G: \mathcal{E} * \mathcal{D} \rightarrow \mathcal{C}$ that extends $F: \mathcal{D} \rightarrow \mathcal{C}$. Then, there are values $G(E)$ in $\mathcal{C}$ for all objects E of $\mathcal{E}$, and morphisms $f \in \mathcal{E}(E_1, E_2)$ induce morphisms $G(f) \in \mathcal{C}(G(E_1), G(E_2))$ as usual. The extra datum comes from the fact that for every object E of $\mathcal{E}$ and for every object D of $\mathcal{D}$, there is a singleton morphism set $\{*\} = (\mathcal{E}*\mathcal{D})(E, D)$, and therefore, we obtain induced morphism $G(E) \rightarrow G(D)$. But as G extends F on $\mathcal{D}$, the latter term is $G(D) = F(D)$. For a fixed E, we obtain such a morphism $G(E) \rightarrow F(D)$ for every object D of $\mathcal{D}$, and for any morphism $h \in \mathcal{D}(D_1, D_2)$, the composite $h \circ *$ is $*$; thus, we obtain a morphism from every $G(E)$ to the diagram $(F(D))_{D \in \mathrm{Ob}(\mathcal{D})}$.

Therefore, the category $\mathcal{C}_{/F}$ has all cones $(G(E) \rightarrow F(D))_{D \in \mathrm{Ob}(\mathcal{D})}$ as objects and morphisms of cones as morphisms. $\qquad\square$

Remark 5.1.14 Proposition 5.1.13 can and should be read as the fact that for a fixed small category $\mathcal{D}$, the functor $(-) * \mathcal{D}: \mathsf{cat} \rightarrow \mathcal{D} \downarrow \mathsf{cat}$ has a right adjoint. Dually, the functor $\mathcal{E} * (-): \mathsf{cat} \rightarrow \mathcal{E} \downarrow \mathsf{cat}$ also has a right adjoint.

André Joyal calls $\mathcal{C}_{/F}$ the *lower slice of $\mathcal{C}$ by F* and $\mathcal{C}_{H/}$ the *upper slice of $\mathcal{C}$ by H*.

You have already seen a special case of these categories in Definition 1.5.7, where we considered the case where one of the categories is $[0]$. Functors $G: [0] * \mathcal{D} \rightarrow \mathcal{C}$ extending a given $F: \mathcal{D} \rightarrow \mathcal{C}$ choose an object $G(0)$ of $\mathcal{C}$ and build the cone to the values $F(D)$. Dually, functors $G: \mathcal{D} * [0] \rightarrow \mathcal{C}$ extending a given $H: \mathcal{D} \rightarrow \mathcal{C}$ also choose an object $G(0)$ of $\mathcal{C}$ and build the cocone of all morphisms $F(D) \rightarrow G(0)$.

5.2 Changing Diagrams for Colimits

Often, it can be quite tricky to explicitly determine a colimit or limit of a functor. Sometimes, it is easier to identify the colimit or limit of a modified diagram category. If $\phi: \mathcal{D}' \rightarrow \mathcal{D}$ is a functor between small categories, then the precomposition with ϕ is a functor

$$\phi^*: \mathsf{Fun}(\mathcal{D}, \mathcal{C}) \rightarrow \mathsf{Fun}(\mathcal{D}', \mathcal{C})$$

for every category $\mathcal{C}$. We will use the categories $D \downarrow \phi$ in order to calculate colimits over $\mathcal{D}$ via colimits over $\mathcal{D}'$ for nice-enough ϕs. The idea is that $\mathcal{D}'$ captures enough of information about $\mathcal{D}$ via ϕ.

If the colimits $\mathrm{colim}_{\mathcal{D}} F$ and $\mathrm{colim}_{\mathcal{D}'} \phi^*(F)$ exist, then ϕ induces a morphism

$$\Phi \colon \mathrm{colim}_{\mathcal{D}'} \phi^*(F) \to \mathrm{colim}_{\mathcal{D}} F$$

as follows. We need to define morphisms $F(\phi(D')) \to \mathrm{colim}_{\mathcal{D}} F$ for all objects D' of $\mathcal{D}'$ that define a cone. Let $\tau' \colon \phi^*(F) \Rightarrow \Delta(\mathrm{colim}_{\mathcal{D}'} \phi^*(F))$ and $\tau \colon F \Rightarrow \Delta(\mathrm{colim}_{\mathcal{D}} F)$ be the cones for $F \circ \phi$ and F. Then, Φ is defined as the unique morphism with $\Phi \circ \tau'_{D'} = \tau_{\phi(D')}$ for all objects D' of $\mathcal{D}'$:

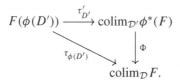

Definition 5.2.1 A functor $\phi \colon \mathcal{D}' \to \mathcal{D}$ is called *cofinal* (or *terminal*) if for all objects D of $\mathcal{D}$, the category $D \downarrow \phi$ is not empty and connected.

Example 5.2.2 The name *terminal functor* comes from the example where t is a terminal object in $\mathcal{D}$, and $\phi \colon \{t\} \to \mathcal{D}$ is the inclusion functor of the category $\{t\}$ with object t and only 1_t as morphism into $\mathcal{D}$. This inclusion functor is terminal, and we already know from Exercise 3.1.5 that $\mathrm{colim}_{\mathcal{D}} F \cong F(t)$ if $\mathcal{D}$ possesses a terminal object.

Example 5.2.3 If $(X, \le)$ is a linearly ordered set viewed as a category and if Y is a subset of X, then the inclusion $Y \to X$ is a cofinal functor if and only if Y is a cofinal subset of X; that is, for all x in X, there is a y in Y with $x \le y$. That's the reason for the name *cofinal functor*. For instance, every infinite subset of the natural numbers is cofinal.

Exercise 5.2.4 Let H be a subgroup of a finite group G. A functor $F \colon \mathcal{C}_G \to k\text{-mod}$ is a G-representation $M = F(*)$. Determine $\mathrm{colim}_{\mathcal{C}_G} F$. When is the inclusion functor $\phi \colon \mathcal{C}_H \to \mathcal{C}_G$ cofinal?

Theorem 5.2.5 *If the functor ϕ is cofinal, then the morphism*

$$\Phi \colon colim_{\mathcal{D}'} \phi^*(F) \to colim_{\mathcal{D}} F$$

is an isomorphism.

Proof   We define morphisms $\psi_D \colon F(D) \to \mathrm{colim}_{\mathcal{D}'} \phi^*(F)$ by choosing an object D' of $\mathcal{D}'$ and a morphism $f_D \colon D \to \phi(D')$ for every D. This is possible because $D \downarrow \phi$ is not empty. If $D = \phi(D')$ for an object D' of $\mathcal{D}'$, then we choose $f_D = 1_{\phi(D')}$. We set $\psi_D := \tau'_{D'} \circ F(f_D)$:

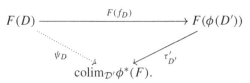

If $D = \phi(D')$ for some object D' of $\mathcal{D}'$, then we obtain

$$\psi_{\phi(D')} = \tau'_{D'}. \tag{5.2.1}$$

As the category $D \downarrow \phi$ is connected for every object D of $\mathcal{D}$, the choice of f_D does not matter. Assume that $\tilde{f}_D \colon D \to \phi(\tilde{D}')$ is another morphism, then as $D \downarrow \phi$ is connected, we know that there is a finite zigzag of morphisms between $\phi(D')$ and $\phi(\tilde{D}')$, indicated by the dashed line in the following diagram:

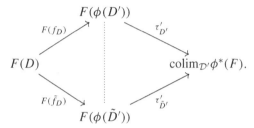

Therefore, $\tau'_{D'} \circ F(f_D) = \tau'_{\tilde{D}'} \circ F(\tilde{f}_D)$.

The family of morphisms ψ_D are components of a natural transformation

$$\psi \colon F \Rightarrow \Delta(\mathrm{colim}_{\mathcal{D}'}\phi^*(F)),$$

and hence, we get an induced morphism

$$\Psi \colon \mathrm{colim}_{\mathcal{D}} F \to \mathrm{colim}_{\mathcal{D}'}\phi^*(F)).$$

We claim that Ψ is inverse to Φ. Note that by the definition of Ψ, we have that $\Psi \circ \tau_D = \psi_D$ and

$$\Phi \circ \psi_D = \tau_{\phi(D')} \circ F(f_D) = \tau_D,$$

where the latter equality follows from the naturality of τ. Hence, the following diagram is commutative:

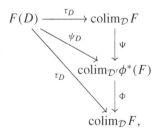

and we get $\Phi \circ \Psi \circ \tau_D = \tau_D$; hence, $\Phi \circ \Psi = 1_{\mathrm{colim}_{\mathcal{D}} F}$.

As Φ satisfies $\Phi \circ \tau'_{D'} = \tau_{\phi(D')}$, we obtain with the help of (5.2.1)

$$\Psi \circ \Phi \circ \tau'_{D'} = \Psi \circ \tau_{\phi(D')} = \psi_{\phi(D')} = \tau'_{D'}$$

thus $\Psi \circ \Phi = 1_{\mathrm{colim}_{D'} \phi^*(F)}$.   □

Remark 5.2.6 You could and *should* ask whether the condition of being cofinal for a functor is the best one can get, and the answer is yes. If $\phi \colon D' \to D$ is a functor between small categories, such that for all functors $F \colon D \to C$ with cocomplete C, the canonical morphism $\Phi \colon \mathrm{colim}_{D'} \phi^*(F) \to \mathrm{colim}_D F$ is an isomorphism, then ϕ is cofinal. It suffices to take $C = \mathrm{Sets}$. We've seen in Example 3.1.6 that $\mathrm{colim}_D D(D, -)$ is a one-point set, thus by assumption, $\mathrm{colim}_{D'} D(D, \phi(-))$ is also a one-point set. But in the category of sets, the explicit formula for a colimit is

$$\mathrm{colim}_{D'} D(D, \phi(-)) = \bigsqcup_{D' \text{ an object of } D'} D(D, \phi(D'))/ \sim,$$

where $f \in D(D, \phi(D'_1))$ is equivalent to $g \in D(D, \phi(D'_2))$ if and only if D'_1 and D'_2 are connected via a zigzag of morphisms in D' that relate f to g. If $D \downarrow \phi$ were empty, then the colimit would be empty. If $D \downarrow \phi$ were not connected, then the colimit would consist of more than one point.

5.3 Sifted Colimits

Interchanging limits and colimits can be difficult or impossible. Sifted colimits interact well with finite products.

Definition 5.3.1

(1) A small category D is called *sifted* if colimits of diagrams $F \colon D \to \mathrm{Sets}$ commute with finite products; that is, if E is a finite set viewed as a discrete category, then for every $F \colon D \times E \to \mathrm{Sets}$,

$$\mathrm{colim}_D \left(\prod_{x \in E} F(-, x) \right) \cong \prod_{x \in E} \mathrm{colim}_D F(-, x).$$

(2) Colimits of diagrams over sifted categories are called *sifted colimits*.

Gabriel and Ulmer [**GU71**] proved a recognition principle for sifted categories.

Proposition 5.3.2 *A nonempty small category D is sifted if and only if the diagonal functor $\Delta \colon D \to D \times D$ is cofinal.*

Proof Assume that the diagonal functor is cofinal. It suffices to show that colimits commute with binary products. Thus, we aim at a bijection

$$\chi : \mathrm{colim}_{\mathcal{D}}(F(-, 1) \times F(-, 2)) \cong (\mathrm{colim}_{\mathcal{D}} F(-, 1)) \times (\mathrm{colim}_{\mathcal{D}} F(-, 2)).$$

A direct inspection shows that

$$(\mathrm{colim}_{\mathcal{D}} F(-, 1)) \times (\mathrm{colim}_{\mathcal{D}} F(-, 2)) \cong \mathrm{colim}_{\mathcal{D} \times \mathcal{D}}(F(-, 1)$$
$$\times (\mathrm{colim}_{\mathcal{D}} F(-, 2)). \tag{5.3.1}$$

If the diagonal is cofinal, then by Theorem 5.2.5, we can reduce the latter to $\mathrm{colim}_{\mathcal{D}}(F(-, 1) \times (\mathrm{colim}_{\mathcal{D}} F(-, 2)) \circ \Delta$.

Conversely, if $\mathcal{D}$ is sifted, then Remark 5.2.6 combined with the bijection of (5.3.1) implies that Δ has to be terminal. □

Examples 5.3.3

(1) As we showed in Theorem 3.5.6 that filtered colimits commute with finite limits for all functors to the category Sets, nonempty filtered categories are sifted.
(2) Split coequalizer diagrams are sifted categories. See [**ARV10**, Example 2.2] for a proof. The splitting is crucial.
(3) Every small category with finite coproducts is sifted. Such a category is not empty because it has an initial object, and for every object (D_1, D_2) of $\mathcal{D} \times \mathcal{D}$, the object $((D_1, D_2), (D_1, D_2) \to (D_1 \sqcup D_2, D_1 \sqcup D_2))$ is an object of $(D_1, D_2) \downarrow \Delta$. For morphisms $(f, g) : (D_1, D_2) \to (D, D)$ and $(u, v) : (D_1, D_2) \to (D', D')$, we can connect these two objects with

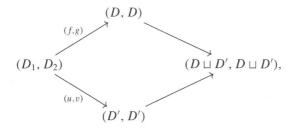

hence, $(D_1, D_2) \downarrow \Delta$ is connected for all (D_1, D_2).

Exercise 5.3.4 Dualize the concept of sifted categories and sifted colimits to cosifted categories and cosifted limits.

We will later see in Proposition 10.12.14 that the opposite of the simplicial category, Δ^o, is sifted.

5.4 Density Results

In the category of sets, Sets, every set X can be expressed as the coproduct of its elements $X \cong \bigsqcup_{x \in X} \{x\}$, and an element $x \in X$ corresponds to a morphism $x : \{*\} \to X$ from the one-point set to X. In this sense, we can describe every set in terms of the one-point set, and the categorical concept axiomatizing this situation is density.

For a functor $F : \mathcal{D} \to \mathcal{C}$, recall the definition of the comma category $F \downarrow \mathcal{C}$ from Definition 5.1.5. We use the projection functor $U : F \downarrow \mathcal{C} \to \mathcal{D}$ given by $U(D, f) = D$.

Definition 5.4.1

(1) Let $\mathcal{D}$ be a small category, $\mathcal{C}$ an arbitrary category, and $F : \mathcal{D} \to \mathcal{C}$ a functor. The functor F is *dense* if for all objects C of $\mathcal{C}$, the natural transformation

$$\psi_F^C : F \circ U \Rightarrow \Delta_C, \quad (\psi_F^C)_{(D,f)} = f$$

is universal in the sense that it induces an isomorphism

$$\mathrm{colim}_{F \downarrow C} F \circ U \cong C.$$

(2) A small subcategory $\mathcal{D}$ of $\mathcal{C}$ is *dense* if the inclusion functor $\mathcal{D} \hookrightarrow \mathcal{C}$ is dense.

So, if a functor $F : \mathcal{D} \to \mathcal{C}$ with $\mathcal{D}$ small is dense, then the pair $(\mathrm{Id}_{\mathcal{C}}, \mathrm{ID}_F)$ is a pointwise left Kan extension of F along F.

Examples 5.4.2

- As every set can be written as the disjoint union of its elements, any subcategory of Sets with one point is dense in Sets.
- If R is an associative ring with unit, then the full subcategory consisting of the object R is *not* dense in the category of R-modules. For instance, for $R = \mathbb{Z}$, the object $\mathbb{Z} \oplus \mathbb{Z}$ cannot be written as a colimit, as in Definition 5.4.1. However, taking the full subcategory with object $R \oplus R$ gives a dense subcategory of the category of R-modules.

One can check density with the help of the following criterion. We denote by $Y_{\mathcal{C}}^F$ the functor from $\mathcal{C}$ to $\mathrm{Sets}^{\mathcal{D}^o}$ that sends an object C to the functor $\mathcal{C}(F(-), C)$, and a morphism $f \in \mathcal{C}(C, C')$ is sent to the postcomposition with f. We can view $Y_{\mathcal{C}}^F$ as a representable functor that is twisted by F.

Theorem 5.4.3 *Let $\mathcal{D}$ be a small category and let $F : \mathcal{D} \to \mathcal{C}$ be a functor. Then, F is dense if and only if the functor*

$$\mathcal{C} \xrightarrow{\quad Y^F_\mathcal{C} \quad} \mathsf{Sets}^{\mathcal{D}^o}$$

is full and faithful.

Proof We define two auxiliary functions for two arbitrary objects C and C' of $\mathcal{C}$. First, we consider $\xi \colon \mathcal{C}(C, C') \to \mathrm{nat}(F \circ U, \Delta_{C'})$ that sends a morphism $g \colon C \to C'$ in $\mathcal{C}$ to the natural transformation of functors from $F \downarrow C$ to Sets, given by postcomposition with g, that is,

$$\xi(g)_{(D,f)} = g \circ f \colon F(D) \to C \to C'.$$

We also consider

$$\Theta^C_{C'} \colon \mathrm{nat}(F \circ U, \Delta_{C'}) \to \mathrm{nat}(\mathcal{C}((F(-), C), \mathcal{C}(F(-), C'))),$$

which maps a natural transformation $\varphi \colon F \circ U \Rightarrow \Delta_{C'}$ to the natural transformation $\Theta^C_{C'}(\varphi)$ with components $\Theta^C_{C'}(\varphi)_D$ that send a $g \colon F(D) \to C$ to $\varphi_{(D,g)}$.

Note that $\Theta^C_{C'}$ is a bijection. A natural transformation $\varphi \colon F \circ U \Rightarrow \Delta_{C'}$ has components

$$\varphi_{(D,f)} \colon F(D) \to C',$$

and for any $h \in F \downarrow C((D, f), (D', f'))$, we have that

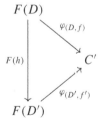

commutes. This is nothing but a natural transformation of functors from $\mathcal{D}$ to Sets from $\mathcal{C}((F(-), C)$ to $\mathcal{C}(F(-), C')$. An explicit inverse of $\Theta^C_{C'}$ is given by sending a natural transformation

$$\alpha \colon \mathcal{C}((F(-), C) \Rightarrow \mathcal{C}((F(-), C')$$

with components $\alpha_D \colon \mathcal{C}((F(D), C) \to \mathcal{C}((F(D), C')$ to the natural transformation τ with components $\tau_{(D,f)} := \alpha_D(f)$.

We fit ξ and $\Theta^C_{C'}$ into a diagram

$$
\begin{array}{ccc}
\mathcal{C}(C, C') & \xrightarrow{\quad (-)_* \quad} & \mathrm{nat}(\mathcal{C}((F(-), C), \mathcal{C}(F(-), C'))). \\
& \searrow{\scriptstyle \xi} \qquad \nearrow{\scriptstyle \Theta^C_{C'}} & \\
& \mathrm{nat}(F \circ U, \Delta_{C'}) &
\end{array}
$$

Here, the top map is taking a $g \in \mathcal{C}(C, C')$ to postcomposition with g. The diagram commutes because

$$(\Theta^C_{C'} \circ \xi(g))(\alpha)_D = \xi(g)_{(D, \alpha_D)} = g \circ \alpha_D = g_*(\alpha)_D$$

for every $\alpha \in \mathrm{nat}(\mathcal{C}(F(-), C), \mathcal{C}(F(-), C'))$.

As $\Theta^C_{C'}$ is a bijection, ξ is a bijection if and only if $(-)_*$ is a bijection, that is, if and only if Y^F_C is fully faithful. But ξ being bijective is equivalent to ψ^C_F being universal. $\qquad\square$

With the help of the density criterion, it is easy to see that the Yoneda embeddings are dense functors. Here, $Y_{\mathcal{D}} : \mathcal{D} \to \mathrm{Sets}^{\mathcal{D}^o}$ sends an object D of $\mathcal{D}$ to $\mathcal{D}(-, D)$ and $Y^{\mathcal{D}} : \mathcal{D}^o \to \mathrm{Sets}^{\mathcal{D}}$ sends D to $\mathcal{D}(D, -)$.

Corollary 5.4.4 *Let $\mathcal{D}$ be a small category. Then, both the Yoneda functors $Y_{\mathcal{D}}$ and $Y^{\mathcal{D}}$ are dense.*

Proof   We prove the claim for $Y_{\mathcal{D}}$. The argument for $Y^{\mathcal{D}}$ is dual.

We have to show that the functor

$$Y^{Y_{\mathcal{D}}}_{\mathrm{Sets}^{\mathcal{D}^o}} : \mathrm{Sets}^{\mathcal{D}^o} \to \mathrm{Sets}^{\mathcal{D}^o}, \quad F \mapsto \mathrm{Sets}^{\mathcal{D}^o}(Y_{\mathcal{D}}, F) : \mathcal{D}^o \to \mathrm{Sets}$$

is fully faithful. The Yoneda lemma 2.2.2 yields that the bijection

$$\mathrm{Sets}^{\mathcal{D}^o}(Y_{\mathcal{D}}(D'), F) = \mathrm{nat}(\mathcal{D}(-, D'), F) \cong F(D')$$

is natural in D', and hence, the functor F is naturally isomorphic to the functor $\mathrm{Sets}^{\mathcal{D}^o}(Y_{\mathcal{D}}, F)$ via $Y^{Y_{\mathcal{D}}}_{\mathrm{Sets}^{\mathcal{D}^o}}$. As $Y^{Y_{\mathcal{D}}}_{\mathrm{Sets}^{\mathcal{D}^o}}$ is an isomorphism of categories, it is fully faithful. $\qquad\square$

Remark 5.4.5 Thus, we get that every set-valued (co- or contravariant) functor from a small category is a canonical colimit of representable functors.

Density helps us to recover objects from certain coends.

Proposition 5.4.6 *Let $\mathcal{D}$ be a small category and let $F : \mathcal{D} \to \mathcal{C}$ be a functor. Assume that all copowers $\mathcal{C}(F(D'), C) \cdot F(D)$ exist. Then, F is dense if and only if every object C is a coend*

$$\int^{\mathcal{D}} \mathcal{C}(F(D), C) \cdot F(D),$$

such that the canonical morphism $\lambda \colon C(F(D), C) \cdot F(D) \to C$ *restricted to the component of an* $f \in C(F(D), C)$ *is* $\lambda_f = f \colon F(D) \to C$.

Proof The fact that C is such a coend with this particular λ is equivalent to the fact that the components of the structure map λ restricted to

$$\bigsqcup_{C(F(D),C)} F(D) \to C$$

are given by the natural evaluation map that sends $F(D)$ via $f \in C(F(D), C)$ to C and that this family is universal and natural in D. This is nothing but the universality of the natural transformation ψ_F^C from Definition 5.4.1. $\square$

Remark 5.4.7 Proposition 5.4.6 is a reason why density is sometimes called *co-Yoneda property*. The Yoneda lemma identifies an object with the set of natural transformations from a representable functor. Natural transformations are ends. Here, we express an object as the coend involving a twisted version of a representable functor.

For representable functors, one gets the co-Yoneda lemma.

Theorem 5.4.8 (co-Yoneda lemma) *Let* $\mathcal{D}$ *be a small category and let* $F \colon \mathcal{D}^o \to \mathsf{Sets}$ *be a functor. Then, there is a natural bijection*

$$\int^{\mathcal{D}} F(D) \cdot \mathcal{D}(D_1, D) \cong F(D_1).$$

Proof As the category Sets is cocomplete, the existence of the copowers and the coend is clear. We show that there is an isomorphism of functors

$$\int^{\mathcal{D}} F(D) \cdot \mathcal{D}(-, D) \cong F.$$

Natural transformations from the above coend to a functor $G \colon \mathcal{D}^o \to \mathsf{Sets}$ are families of functions

$$\psi_D^{D_1} \colon F(D) \cdot \mathcal{D}(D_1, D) \to G(D_1),$$

such that for all $g \colon D_1 \to D_2$ and all $h \colon \tilde{D} \to D$, the diagrams

and

commute. In particular, every ψ_D is a natural transformation

$$\psi_D : F(D) \cdot \mathcal{D}(-, D) \Rightarrow G.$$

By adjunction, we get

$$F(D) \to \int_{\mathcal{D}} \mathsf{Sets}(\mathcal{D}(D_1, D), G(D_1)) \cong G(D).$$

Therefore, the set of natural transformations from $\int^{\mathcal{D}} F(D) \cdot \mathcal{D}(-, D)$ to G is isomorphic to the set of natural transformations from F to G, hence

$$\int^{\mathcal{D}} F(D) \cdot \mathcal{D}(-, D) \cong F. \qquad \square$$

We will see an enriched version of the co-Yoneda lemma later in Proposition 9.3.10.

There is a dual notion to density, and we mention it briefly.

Definition 5.4.9 Let $\mathcal{D}$ be a small category and let $F : \mathcal{D} \to \mathcal{C}$ be a functor. Then, F is *codense* if for all objects C, the natural transformation

$$\psi_C^F : \Delta_C \Rightarrow F \circ U$$

is universal in the sense that it induces an isomorphism

$$C \cong \lim_{C \downarrow F} F \circ U.$$

Note that here, $C \downarrow F$ is the comma category with objects (D, f) with $f : C \to F(D)$ and $U(D, f) = D$.

Exercise 5.4.10 Dualize Theorem 5.4.3. Is the dualization of 5.4.4 true?

5.5 The Grothendieck Construction

Recall from Definition 1.4.15 that cat denotes the category of all small categories.

Definition 5.5.1 Let $\mathcal{C}$ be a category and let $F : \mathcal{C} \to$ cat be a functor. The *Grothendieck construction* $\mathcal{C} \int F$ is the category whose objects are pairs (C, X), where C is an object of $\mathcal{C}$ and X is an object of the category $F(C)$. A morphism in $\mathcal{C} \int F$ from (C_1, X_1) to (C_2, X_2) is a pair (f, g) of morphisms $f \in \mathcal{C}(C_1, C_2)$ and $g \in F(C_2)(F(f)(X_1), X_2)$.

If $(f_1, g_1) \in (\mathcal{C} \int F) ((C_1, X_1), (C_2, X_2))$ and $(f_2, g_2) \in (\mathcal{C} \int F) ((C_2, X_2), (C_3, X_3))$, then the composition of these two morphisms is given by $(f_2 \circ f_1, g_2 \circ (F(f_2)(g_1)))$:

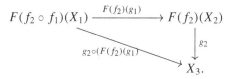

Remark 5.5.2

- As the composition of morphisms in $C \int F$ in the first component is just the composition of morphism in C, we get a projection functor $U : C \int F \to C$.
- If $\tau : F \Rightarrow G$ is a natural transformation of functors $F, G : C \to$ cat, then τ induces a functor $C \int \tau : C \int F \to C \int G$. On objects, $C \int \tau$ sends (C, X) to $(C, \tau_C(X))$. For $f \in C(C, C')$ and $g \in F(C')(F(f)(X), X')$, we keep $f \in C(C, C')$. Then, we need a morphism from $G(f)(\tau_C(X))$ to $\tau_{C'}(X')$. But $G(f)(\tau_C(X)) = \tau_{C'} F(f)(X)$, and therefore, we take $\tau_{C'}(g)$ as the second component, so $\tau(f, g) = (f, \tau_{C'}(g))$:

$$G(f)(\tau_C(X)) = \tau_{C'} F(f)(X) \xrightarrow{\tau_{C'}(g)} \tau_{C'}(X').$$

There are several other notations in use for the Grothendieck construction, for instance, $\int_C F$ or $F \rtimes C$. We avoid them because the first one clashes with the notation for ends and the second one would clash with other notation that we will use later. The symbol $F \rtimes C$ has its origin in the example in the next exercise.

Exercise 5.5.3 Let N and H be groups, and assume that we have a homomorphism $\varphi : H \to \mathrm{Aut}(N)$. Consider the associated categories C_H and C_N with one object $*$.

(1) Show that there is a functor $F : C_H \to$ cat with $F(*) = C_N$.
(2) Prove that the Grothendieck construction $C_H \int F$ is the category $C_{N \rtimes_\varphi H}$, where $N \rtimes_\varphi H$ denotes the (external) semidirect product of the groups N and H.

Let C be a small category. The Grothendieck construction $C \int F$ associated with a functor $F : C \to$ cat has the following universal property [**T79**, 1.3.1], describing functors out of the categories that are in the image of F.

Proposition 5.5.4 *Let C and D be small categories. The set of all functors $G : C \int F \to D$ is in bijection with the following data:*

- *For every object C of C, there is a functor $G_C : F(C) \to D$.*
- *For every morphism $f \in C(C_1, C_2)$, there is a natural transformation*

$$G_f \colon G_{C_1} \Rightarrow G_{C_2},$$

such that $G_{1_C} = ID_{G_C}$ *and such that* G_- *respects composition; that is, for* $f \in C(C_1, C_2)$, $g \in C(C_2, C_3)$, *the natural transformations* $G_{g \circ f}$ *and* $G_g \circ G_f$ *are equal.*

Proof Given a functor $G \colon C \int F \to D$, we get the functor G_C by setting $G_C(X) = G(C, X)$ for X an object of $F(C)$ and by defining $G_C(g)$ as $G(1_C, g)$ for $g \in F(C)(X, Y)$. The natural transformation G_f is given by $(G_f)_X = G(f, 1_X)$.

Conversely, given the data, we can define a functor $G \colon C \int F \to D$ by setting $G(C, X) := G_C(X)$ on objects, and a morphism (f, g) in $(C \int F)((C_1, X_1), (C_2, X_2))$ is sent to

$$G(f, g) \colon G(C_1, X_1) = G_{C_1}(X_1) \xrightarrow{\ G_f(X_1)\ } G_{C_2}(X_1) \xrightarrow{\ G_{C_2}(g)\ } G_{C_2}(X_2)$$

$$= G(C_2, X_2). \qquad \square$$

Remark 5.5.5 For the definition of a Grothendieck construction, it is not necessary to have a functor on the nose, but one can loosen the assumptions and consider *op-lax 2-functors*, as mentioned in Remark 9.6.5. See Thomason's thesis [**T77**] for the full story.

6

Monads and Comonads

6.1 Monads

Let $\mathcal{C}$ be a category. We can consider the endofunctors of $\mathcal{C}$, $T : \mathcal{C} \to \mathcal{C}$. We can compose such functors, and the identity functor on $\mathcal{C}$, $\mathrm{Id}_{\mathcal{C}}$, is a unit for this composition. We are interested in endofunctors T that possess a multiplicative structure.

In the following, we will often omit composition signs of functors in order to ease notation.

Definition 6.1.1 A *monad* in a category $\mathcal{C}$ consists of an endofunctor $T : \mathcal{C} \to \mathcal{C}$, together with two natural transformations, $\eta : \mathrm{Id} \Rightarrow T$ and $\mu : T \circ T \Rightarrow T$, such that the following diagrams commute:

$$
\begin{array}{ccc}
T^3 \xRightarrow{T\mu} T^2 \\
\mu_T \Big\Downarrow \qquad \Big\Downarrow \mu \\
T^2 \xRightarrow{\mu} T
\end{array}
\qquad \text{and} \qquad
\mathrm{Id} \circ T \xRightarrow{\eta T} T^2 \xLeftarrow{T\eta} T \circ \mathrm{Id}
$$

$$
\begin{array}{c}
\Big\Downarrow \mu \\
T.
\end{array}
$$

For a monad T, the natural transformation η is called the *unit* of the monad and μ is called the *multiplication*. In addition to the name *monad*, the notion of a *triple* is quite common.

Definition 6.1.2 Let $\mathcal{C}$ be a category and let (T, μ, η) and (T', μ', η') be two monads on $\mathcal{C}$. A *morphism of monads from T to T'* is a natural transformation $\tau : T \Rightarrow T'$, which is compatible with the structure maps: τ commutes with the unit, that is, $\tau \circ \eta = \eta'$, and the diagram

$$
\begin{array}{ccc}
T \circ T & \xRightarrow{T'(\tau) \circ \tau T} & T' \circ T' \\
\mu \Big\Downarrow & & \Big\Downarrow \mu' \\
T & \xRightarrow{\tau} & T'
\end{array}
$$

commutes.

The identity functor on $\mathcal{C}$ is always a monad. You know nontrivial examples of monads because every pair of adjoint functors gives rise to a monad. We will see later that the converse is also true: every monad gives rise to an adjoint pair of functors.

Theorem 6.1.3 *Let* $\mathcal{C} \underset{R}{\overset{L}{\rightleftarrows}} \mathcal{D}$ *be an adjoint pair of functors. Then, the endofunctor* $T = R \circ L : \mathcal{C} \to \mathcal{C}$ *is a monad. The transformation* $\mu : RLRL \Rightarrow RL$ *is given by* $R\varepsilon_L$, *where* ε *is the counit of the adjunction and the transformation* $\eta : Id \Rightarrow T$ *is the unit of the adjunction* $\eta : Id \Rightarrow RL$.

Proof   We have to show that the diagrams

$$
\begin{array}{ccc}
RLRLRL & \xrightarrow{\;RLR\varepsilon_L\;} & RLRL \\[2pt]
\left\Vert{\scriptstyle R\varepsilon_{LRL}}\right. \downarrow & & \downarrow {\scriptstyle R\varepsilon_L} \\[2pt]
RLRL & \xrightarrow[\;R\varepsilon_L\;]{} & RL
\end{array}
\tag{6.1.1}
$$

and

$$
\begin{array}{ccccc}
Id \circ RL & \xRightarrow{\;\eta_{RL}\;} & RLRL & \xLeftarrow{\;RL\eta\;} & RL \circ Id \\
& \searrow & \downarrow {\scriptstyle R\varepsilon_L} & \swarrow & \\
& & RL & &
\end{array}
\tag{6.1.2}
$$

commute. In diagram (6.1.1), the outer copy of R and the inner copy of L are not involved; that is, we can reduce the diagram to the relevant part and that is

$$
\begin{array}{ccc}
LRLR & \xRightarrow{\;LR\varepsilon\;} & LR \\[2pt]
\left\Vert{\scriptstyle \varepsilon_{LR}}\right. \downarrow & & \downarrow {\scriptstyle \varepsilon} \\[2pt]
LR & \xrightarrow[\;\varepsilon\;]{} & Id.
\end{array}
$$

This diagram commutes because it consists of the two-fold application of the counit.

The commutativity of (6.1.2) can be shown by considering both triangular-shaped subdiagrams separately. For these diagrams, the claim follows, because for any adjunction, the identities $ID = R\varepsilon \circ \eta_R : R \Rightarrow R$ and $ID = \varepsilon_L \circ L\eta : L \Rightarrow L$ hold.   □

As we know plenty of examples of adjunctions, we get examples of monads.

Examples 6.1.4

- We have the forgetful functor U from the category of groups to the category of sets, and this has the free group functor, Fr, as a left adjoint. Hence, $U \circ Fr$ is a monad on the category Sets.

- Let k be a commutative ring with unit. The tensor algebra functor sends a k-module M to the tensor algebra

$$T(M) = \bigoplus_{i \geq 0} M^{\otimes_k i},$$

with $M^{\otimes_k 0} = k$. We endow $T(M)$ with the multiplication given by the concatenation of tensors. This functor is left adjoint to the forgetful functor U' from associative unital k-algebras to k-modules, and $U' \circ T$ is a monad.
- Similar to the above, let $\mathrm{Sym}(M)$ be the free unital associative and commutative algebra on M:

$$\mathrm{Sym}(M) = \bigoplus_{i \geq 0} M^{\otimes_k i} / \Sigma_i.$$

Here, we divide by the action of the symmetric group Σ_i on the i-fold tensor power of M. For $\sigma \in \Sigma_i$ and $m_1 \otimes \cdots \otimes m_i$, the action is given by

$$\sigma.(m_1 \otimes \cdots \otimes m_i) = m_{\sigma^{-1}(1)} \otimes \cdots \otimes m_{\sigma^{-1}(i)}.$$

The composite with the forgetful functor from commutative unital algebras to k-modules, U, is a monad

$$U \circ \mathrm{Sym} \colon k\text{-mod} \to k\text{-mod}.$$

There is a morphism of monads from $U' \circ T$ to $U \circ \mathrm{Sym}$ that is induced by the projection maps $M^{\otimes_k i} \to M^{\otimes_k i} / \Sigma_i$.
- The loop-suspension adjunction is crucial in algebraic topology. The suspension functor

$$\Sigma \colon \mathsf{cg}_* \to \mathsf{cg}_*, \quad X \mapsto \mathbb{S}^1 \wedge X$$

is left adjoint to the based loop functor

$$\Omega \colon \mathsf{Top}_* \to \mathsf{Top}_*, \quad X \mapsto k\underline{\mathsf{Top}}_*(\mathbb{S}^1, X).$$

Here, cg is the category of compactly generated weak Hausdorff spaces, which we will discuss in detail in 8.5. Similarly, for $n > 1$, we can consider the adjunction of the n-fold suspension

$$\Sigma^n \colon \mathsf{cg}_* \to \mathsf{cg}_*, \quad X \mapsto \mathbb{S}^n \wedge X$$

and the n-fold based loop functor

$$\Omega^n \colon \mathsf{Top}_* \to \mathsf{Top}_*, \quad X \mapsto k\underline{\mathsf{Top}}_*(\mathbb{S}^n, X).$$

Exercise 6.1.5 Let P be a partially ordered set viewed as a category. What is a monad on P?

Exercise 6.1.6 Consider the assignment that sends a set to its power set. Is this a monad?

6.2 Algebras over Monads

Adjoint pairs of functors give rise to monads and we will now show the converse: For every monad T there are (at least) two pairs of adjoint functors that can be associated with T. We consider one of it here and another one later (Corollary 6.3.6).

Definition 6.2.1 Let $T : \mathcal{C} \to \mathcal{C}$ be a monad with multiplication μ and unit η. We call an object C of $\mathcal{C}$ a *T-algebra* if there is a morphism $\xi : TC \to C$, such that the following diagrams commute:

$$\begin{array}{ccc} T^2C \xrightarrow{T(\xi)} TC & \text{and} & C \xrightarrow{\eta_C} TC \\ \mu_C \downarrow \qquad \downarrow \xi & & {}^{1_C}\searrow \quad \downarrow \xi \\ TC \xrightarrow{\quad \xi \quad} C & & C. \end{array}$$

The morphism ξ is often called the *structure map* of the T-algebra C, and we often denote a T-algebra by (C, ξ). The left diagram is an associativity condition, whereas the right one poses a condition on the unit. A *morphism of T-algebras* is a morphism in $\mathcal{C}$ respecting the structure maps. We use $T\text{-alg}_\mathcal{C}$ to denote the category of T-algebras in $\mathcal{C}$.

Remark 6.2.2 A monad T is an endofunctor that is a monoid. The defining diagrams of a T-algebra then mean that the monoid T acts on C, so one could equally well call a T-algebra a left T-module.

Examples 6.2.3

- Let $\mathrm{Fr} \colon \mathsf{Sets} \to \mathsf{Gr}$ be the functor that assigns to a set S the free group generated by the set S. Then, *every* group G is an algebra of the monad $T := U \circ \mathrm{Fr}$. Here, U again denotes the forgetful functor from the category of groups to the category of sets. The morphism $\xi \colon U\mathrm{Fr}(G) \to G$ takes a formal word in the elements of G and considers it as a word in G. This example is prototypical for monads that arise from a free-forgetful adjunction.

- Let $(-)_+ \colon \mathsf{Sets} \to \mathsf{Sets}$ be the monad that sends a set X to $X_+ := X \sqcup \{+\}$. This defines a monad on Sets. An algebra for this monad is a set X, together with a function $f \colon X_+ \to X$, but the unit condition of the algebra structure implies that f does not move the points in X, so the only datum is the value $f(+) \in X$. Thus, $(-)_+$-algebras are based sets.

- Based loop spaces $\Omega^n X$ for $n \geq 1$ are algebras over the monad $\Omega^n \Sigma^n$.
- If T is a monad, then for every object C of $\mathcal{C}$, the object TC is a T-algebra. The structure map $T(T(C)) \to T(C)$ is given by μ_C, and the unit is induced by the unit of the monad. Such algebras are called *free T-algebras*. In the example described previously, for every set S, $T(S)$ is the underlying set of the free group generated by S.

Exercise 6.2.4 Let $\mathcal{C} \underset{R}{\overset{L}{\rightleftarrows}} \mathcal{D}$ be an adjoint pair of functors and let $T = RL$ be the corresponding monad. Is RD always a T-algebra for all objects D of $\mathcal{D}$?

Theorem 6.2.5 *Let (T, μ, η) be a monad in $\mathcal{C}$. Then, there is an adjoint pair of functors*

$$\mathcal{C} \underset{R}{\overset{L}{\rightleftarrows}} T\text{-alg}_{\mathcal{C}}.$$

Here, R maps a T-algebra (C, ξ) to the object C of $\mathcal{C}$, and L sends an object C of $\mathcal{C}$ to the free T-algebra (TC, μ_C). The corresponding monad of this adjunction is precisely the monad T that we started with.

Remark 6.2.6 The adjunction described in Theorem 6.2.5 is called the *Eilenberg–Moore adjunction of the monad T*. We will see another one later.

Proof If we apply the composite RL to an object C of $\mathcal{C}$, then we obtain

$$RLC = R(TC, \xi \colon T(TC) = (T^2)(C) \xrightarrow{\mu_C} TC) = TC,$$

and the unit is given by $\eta_C \colon C \to RLC = TC$. Conversely, we get that $LR(C, \xi) = L(C) = (TC, \mu_C)$, and the structure map $\xi \colon TC \to C$ is, by definition, a morphism of T-algebras

$$\xi \colon (TC, \mu_C) \to (C, \xi).$$

Thus, the morphism $\varepsilon_{(C,\xi)} \colon LR(C, \xi) \to (C, \xi)$ is given by this morphism of T-algebras. If $f \colon (C, \xi) \to (C', \xi')$ is an arbitrary morphism of T-algebras, then f has to satisfy

$$\xi' \circ T(f) = f \circ \xi,$$

and this is exactly the naturality condition for $\varepsilon_{(C,\xi)}$. The composite $TC \xrightarrow{T\eta_C} TTC \xrightarrow{\mu_C} TC$ and the morphism $C \xrightarrow{\eta_C} TC \xrightarrow{\xi} C$ are the identity on TC and C because of the unit condition for the monad T and the unit condition for the T-algebra (C, ξ). $\qquad\square$

We denote the forgetful functor from the category of T-algebras in $\mathcal{C}$ to $\mathcal{C}$ again by U. Note that U does not change the morphisms but only forgets a coherence property, so we obtain the following:

Proposition 6.2.7 *The forgetful functor* $U = U : T\text{-alg}_{\mathcal{C}} \to \mathcal{C}$ *is faithful.*   □

Examples 6.2.8

- If R is an associative ring with unit, then the functor $R \otimes (-) \colon \mathsf{Ab} \to \mathsf{Ab}$ is a monad in the category of abelian groups. An $R \otimes (-)$-algebra is a left R-module. The forgetful functor U takes an R-module M to its underlying abelian group.
- Similarly, for every group G, the functor

$$G \times (-) \colon \mathsf{Sets} \to \mathsf{Sets}$$

is a monad on the category Sets, and the $G \times (-)$-algebras are G-sets, that is, sets with an action of the group G. In this example, the forgetful functor just forgets the G-action.

If T is a monad on $\mathcal{C}$ and C is a T-algebra, then the associativity condition on the structure map $\xi \colon T(C) \to C$ implies that ξ fits into a diagram

$$T(T(C)) \underset{\mu_C}{\overset{T(\xi)}{\rightrightarrows}} T(C) \overset{\xi}{\longrightarrow} C. \tag{6.2.1}$$

Here, $T(T(C))$ is a T-algebra via $\mu_{T(C)}$, and $T(C)$ has μ_C as structure map. Actually, more is true.

Proposition 6.2.9 *Let T be a monad on $\mathcal{C}$ and let C be a T-algebra with structure map $\xi \colon T(C) \to C$. Then, the diagram (6.2.1) is a coequalizer diagram in the category of T-algebras. Applying the forgetful functor from T-algebras to $\mathcal{C}$ turns this diagram into a split coequalizer diagram in $\mathcal{C}$.*

Proof   All morphisms involved in the diagram are morphisms of T-algebras. If f is a morphism of T-algebras from $(T(C), \mu_C)$ to (D, τ) with the property that $f \circ \mu_C = f \circ T(\xi)$, then we precompose f with the unit of C, $\eta_C \colon C \to T(C)$.

$$T(T(C)) \underset{\mu_C}{\overset{T(\xi)}{\rightrightarrows}} T(C) \overset{\xi}{\longrightarrow} C$$

We have to show that this is a T-algebra morphism, that is, $\tau \circ T(f \circ \eta_C) = f \circ \eta_C \circ \xi$.

As f is a morphism of T-algebras, we have

$$(\tau \circ T(f)) \circ T(\eta_C) = (f \circ \mu_C) \circ T(\eta_C) = f,$$

and as f satisfies the condition that $f \circ \mu_C = f \circ T(\xi)$, we also get

$$f = (f \circ \mu_C) \circ \eta_{T(C)} = (f \circ T(\xi)) \circ \eta_{T(C)} = f \circ \eta_C \circ \xi,$$

and this proves the claim.

In the category $\mathcal{C}$, we set $s = \eta_C$ and $t = \eta_{T(C)}$. The unit conditions yield $\mu_C \circ t = 1_{T(C)}$ and $\xi \circ s = 1_C$. As units are natural, we also obtain $T(\xi) \circ t = s \circ \xi$, and thus, the coequalizer diagram turns into a split coequalizer diagram in $\mathcal{C}$.

By Proposition 6.2.7, the forgetful functor is faithful; this shows that the extension of f as $f \circ \eta_C$ was unique. $\qquad\qquad\square$

Remark 6.2.10 For the splitting in the proof, it was crucial that we worked in the underlying category, because the unit maps η_C and $\eta_{T(C)}$ are, in general, not maps of T-algebras. For instance, if A is an associative unital k-algebra, then $\eta_A \colon A \to \bigoplus_{i \geq 0} A^{\otimes_k i}$ sends A to the $(i = 1)$-part of the tensor algebra. This map is neither unital nor multiplicative.

We now want to study the situation where we have morphisms between different kinds of algebras over monads. Later, we will apply this to lifting left adjoint functors to the level of algebras over monads.

Let (T, μ, η) and (T', μ', η') be two monads. We abbreviate the forgetful functors with U and U' and the free functors with F and F'. Assume that we have a functor $R \colon \mathcal{C} \to \mathcal{D}$ and that it lifts to the level of algebras, that is, there is a functor

$$\tilde{R} \colon T\text{-alg}_{\mathcal{C}} \to T'\text{-alg}_{\mathcal{D}},$$

with

$$U' \circ \tilde{R} = R \circ U. \tag{6.2.2}$$

In particular, every T-algebra (C, ξ) is sent to a T'-algebra $\tilde{R}(C, \xi)$. Consider the following diagram:

$$
\begin{array}{ccc}
T\text{-alg}_{\mathcal{C}} & \xrightarrow{\ \tilde{R}\ } & T'\text{-alg}_{\mathcal{D}} \\
{\scriptstyle F}\big\uparrow\big\downarrow{\scriptstyle U} & & {\scriptstyle F'}\big\uparrow\big\downarrow{\scriptstyle U'} \\
\mathcal{C} & \xrightarrow[\ R\ ]{} & \mathcal{D}.
\end{array}
$$

We do not assume any other commutativity relations in this diagram.

Lemma 6.2.11 *In the previous situation, there is a natural transformation*

$$\alpha : T' \circ R \Rightarrow R \circ T.$$

This α commutes with the unit of T and T' as follows:

$$
\begin{array}{ccc}
R & \xrightarrow{\eta'_R} & T' \circ R \\
{\scriptstyle R(\eta)}\Big\| & \swarrow{\scriptstyle \alpha} & \\
R \circ T. &
\end{array}
\qquad (6.2.3)
$$

There is also a compatibility with the multiplications in the monads T and T' in the sense that

$$
\begin{array}{ccccc}
T' \circ T' \circ R & \xrightarrow{\quad\mu'_R\quad} & T' \circ R & \qquad & (6.2.4)\\
{\scriptstyle T'(\alpha)}\Big\| & & \Big\|{\scriptstyle \alpha} & & \\
T' \circ R \circ T & & R \circ T & & \\
& {\scriptstyle \alpha T}\searrow \quad {\scriptstyle R(\mu)}\swarrow & & & \\
& R \circ T \circ T & & &
\end{array}
$$

commutes.

Proof   We define α as the following composition:

$$T'R = U'F'R \xrightarrow{U'F'R(\eta)} U'F'RUF = U'F'U'\tilde{R}F \xrightarrow{U'\varepsilon'_{\tilde{R}F}} U'\tilde{R}F = RUF = RT.$$

The unit condition (6.2.3) is easy to see because of the commutativity of the diagram

$$
\begin{array}{ccc}
R & \xrightarrow{\quad\eta'_R\quad} & U'F'R = T'R \\
{\scriptstyle R(\eta)}\Big\downarrow & & \Big\downarrow{\scriptstyle U'F'R\eta} \\
RUF & \xrightarrow{\eta'_{RUF}} & U'F'RUF = U'F'U'\tilde{R}F \\
& \searrow & \Big\downarrow{\scriptstyle U'\varepsilon'_{\tilde{R}F}} \\
& & RUF = U'\tilde{R}F.
\end{array}
$$

For the compatibility with the monad multiplication, as in (6.2.4), we consider the diagram

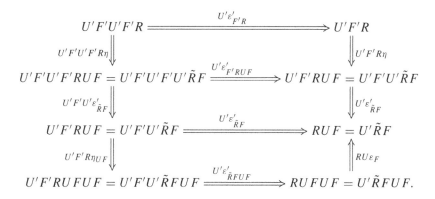

The outer compositions give exactly the morphisms that we are looking for. The horizontal maps make the top two rectangles commute because of naturality; the only thing to check is that the bottom square commutes. Because of the naturality of the transformation ε', we know that

$$U'\varepsilon'_{\tilde{R}F} \circ U'F'RU\varepsilon_F = RU\varepsilon_F \circ U'\varepsilon'_{\tilde{R}FUF}.$$

Precomposition with the transformation $U'F'R\eta_{UF}$ gives the claim, because $U'F'RU\varepsilon_F \circ U'F'R\eta_{UF} = \mathrm{Id}$. Hence, we get

$$RU\varepsilon_F \circ \alpha_{UF} = RU\varepsilon_F \circ U'\varepsilon'_{\tilde{R}FUF} \circ U'F'R\eta_{UF} = U'\varepsilon'_{\tilde{R}F}. \qquad (6.2.5)$$

$\square$

Remark 6.2.12 In particular, if $\mathcal{C} = \mathcal{D}$ and R is the identity functor, then for every functor $\tilde{\mathrm{Id}}_{\mathcal{C}} \colon T\text{-alg}_{\mathcal{C}} \to \mathsf{T}'\text{-alg}$ with $U' \circ \tilde{\mathrm{Id}}_{\mathcal{C}} = U$, we obtain a natural transformation $\alpha \colon T' \Rightarrow T$. The conditions (6.2.3) and (6.2.4) ensure that α is a morphism of monads. In order to abandon the awkward notation $\tilde{\mathrm{Id}}_{\mathcal{C}}$, we use a neutral $G \colon T\text{-alg}_{\mathcal{C}} \to \mathsf{T}'\text{-alg}$ instead.

Example 6.2.13 Every unital associative and commutative k-algebra A is in particular a unital associative k-algebra, so we get a forgetful functor G that takes the underlying unital associative algebra of a unital associative and commutative algebra. Let us denote by As the monad $U' \circ T$ and by Com the monad $U \circ \mathrm{Sym}$. Then, there is a commutative diagram

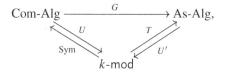

and the corresponding morphism of monads α: As $\to$ Com is the one discussed earlier, induced by the projection maps that divide out by the symmetric group action. In the language of monads, the map from commutative algebras to associative algebras is obtained by precomposing the structure map with the projection map α: As $\to$ Com.

Example 6.2.14 Let A be a unital associative k-algebra. Then, A is a Lie algebra over k, using the commutator bracket

$$[-, -]: A \otimes_k A \to A, \quad [a, b] := ab - ba \text{ for } a, b \in A.$$

Therefore, there must be a morphism from the Lie monad to the associative monad As.

These examples are prototypical.

Lemma 6.2.15 *Assume that (T, μ, η) and (T', μ', η') are monads on the category $\mathcal{C}$ and that we have a functor G from the category of T-algebras to the category of T'-algebras, such that the diagram*

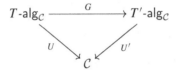

commutes. Then, there exists a morphism of monads α^G from T' to T. Conversely, every morphism of monads α: $T' \Rightarrow T$ induces a functor G_α: T-alg$_\mathcal{C} \to T'$-alg$_\mathcal{C}$, with $U' \circ G_\alpha = U$. The assignments $G \mapsto \alpha^G$ and $\alpha \mapsto G_\alpha$ are inverse to each other.

Proof We established the first fact in Lemma 6.2.11. The inverse to the assignment $G \mapsto \alpha^G$ sends a morphism α of monads to the functor G_α that maps a T-algebra (C, ξ) to $(C, \xi \circ \alpha_C)$. By construction, we have that $U' \circ G_\alpha = U$, and if $f: (C, \xi) \to (D, \tau)$ is a morphism of T-algebras, that is, $\tau \circ T(f) = f \circ \xi$, then f is also a morphism of T'-algebras from $(C, \xi \circ \alpha_C)$ to $(D, \tau \circ \alpha)$, because the diagram

$$
\begin{array}{ccc}
T'(C) & \xrightarrow{T'(f)} & T'(D) \\
\alpha_C \downarrow & & \downarrow \alpha_D \\
T(C) & \xrightarrow{T(f)} & T(D) \\
\xi \downarrow & & \downarrow \tau \\
C & \xrightarrow{f} & D
\end{array}
$$

commutes.

Before we show that these assignments are inverse to each other, we describe G in more detail. As $U' \circ G = U$, we obtain that $G(C, \xi)$ has to be of the form (C, ρ) with $\rho \colon T(C) \to C$ being the T-algebra structure of $G(C, \xi)$. From the proof of Theorem 6.2.5, we know that we can recover this structure map from the counit of the adjunction, and hence,

$$\rho = U'(\varepsilon'_{(C,\rho)}) = U'(\varepsilon'_{G(C,\xi)}).$$

We claim that ρ is always of the form $U\varepsilon_{(C,\xi)} \circ \alpha^G_C = \xi \circ \alpha^G_C$. In the following diagram, we heavily use the fact that $U' \circ G = U$:

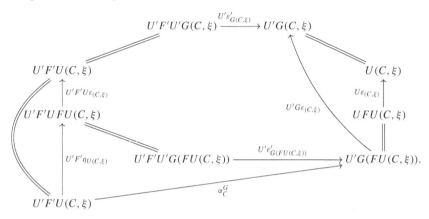

The pentagon commutes because of the naturality of the counit maps. The counit–unit property of the adjunction renders the left half-moon commutative, the small triangle-shaped square in the upper-right commutes because of the identification $U'G = U$, and the lower part of the diagram displays the definition of α_C. Therefore,

$$\rho = U'(\varepsilon'_{G(C,\xi)}) - U\varepsilon_{(C,\xi)} \circ \alpha_C - \xi \circ \alpha_C. \tag{6.2.6}$$

We first show that $\alpha^{G_\alpha} = \alpha$. For this, we note that with the preceding calculation, we get that

$$G_\alpha(TC, \mu_C) = (TC, \mu_C \circ \alpha_{TC}),$$

and, following the notation of [**BaWe05**, Proof of Theorem 6.3], we define $\sigma_C := \mu_C \circ \alpha_{TC}$. This is the T'-algebra structure map of TC, and hence, $\sigma_C \colon T'TC \to TC$. As α is a natural transformation, the diagram

$$
\begin{array}{ccc}
T'C & \xrightarrow{\ \alpha_C\ } & TC \\
{\scriptstyle T'(\eta_C)}\downarrow & \nearrow{\scriptstyle \sigma_C} & \downarrow{\scriptstyle T(\eta_C)} \quad \big)\mu_C \\
T'TC & \xrightarrow{\ \alpha_{TC}\ } & TTC
\end{array}
$$

commutes. Thus, we can express α_C in terms of σ_C, as

$$\alpha_C = \mu_C \circ T(\eta_C) \circ \alpha_C = \mu_C \circ \alpha_{TC} \circ T'(\eta_C) = \sigma_C \circ T'(\eta_C). \qquad (6.2.7)$$

We can compare α and $\alpha^{G\alpha}$ via the diagram

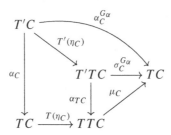

and hence, the two maps agree for all C.

Conversely, given G, we know that $G(C, \xi) = (C, \xi \circ \alpha_C)$ but also $G_{\alpha G}(C, \xi) = (C, \xi \circ \alpha_C^{G\alpha}) = (C, \xi \circ \alpha_C)$, so both agree. □

6.3 Kleisli Category

In addition to the description of a monad in terms of the free-forgetful pair from 6.2.5, there is an alternative description of a monad in terms of its free algebras. We will actually see that all the ways to express a given monad in terms of adjoint functors live between the Kleisli adjunction and the Eilenberg–Moore adjunction.

Definition 6.3.1 Let (T, μ, η) be a monad on a category $\mathcal{C}$. The *Kleisli category* of T, K_T, has the objects of $\mathcal{C}$, as objects and a morphism $f \in K_T(C_1, C_2)$ is a morphism f in $\mathcal{C}$ from C_1 to $T(C_2)$. The identity morphism of an object C is the unit of the adjunction $\eta_C: C \to T(C)$, and the composite of $f \in K_T(C_1, C_2)$ and $g \in K_T(C_2, C_3)$ is

$$C_1 \xrightarrow{f} T(C_2) \xrightarrow{T(g)} T(T(C_3)) \xrightarrow{\mu_{C_3}} T(C_3).$$

Exercise 6.3.2 Check the associativity of the composition and the unit axiom.

Example 6.3.3 The monad $(-)_+ : \mathsf{Sets} \to \mathsf{Sets}$ that adds a disjoint point has as a Kleisli category the category whose objects are sets, and an $f \in K_{(-)_+}(X, Y)$ is a function $f: X \to Y_+$. Such a function is sometimes called a *partial function*. The interpretation is that f is a function on all the points that are *not* sent to $+$. If you have a partial function, that is, a function $\bar{f}: A \to Y$ for some $A \subset X$, then you can extend it to a morphism $f \in K_{(-)_+}(X, Y)$ by declaring $f(x) = +$ for all $x \in X \setminus A$.

Exercise 6.3.4 Work out what the Kleisli category is for the monad corresponding to the free-forgetful pair, where the free functor takes a set X to the free monoid generated by X.

Proposition 6.3.5 *The category K_T is equivalent to the full subcategory $F_T(\mathcal{C})$ of T-alg$_\mathcal{C}$, whose objects are the free T-algebras.*

Proof Define a functor $G \colon K_T \to F_T(\mathcal{C})$ by setting $G(C) = (T(C), \mu_C)$, and for $f \colon C_1 \to T(C_2)$, a morphism in K_T from C_1 to C_2, we set $G(f) = \mu_{C_2} \circ T(f)$:

$$
\begin{array}{ccc}
T(C_1) & \xrightarrow{\quad G(f) \quad} & T(C_2). \\
{\scriptstyle T(f)} \downarrow & \nearrow {\scriptstyle \mu_{C_2}} & \\
TT(C_2) & &
\end{array}
$$

Then, $G(f)$ is a morphism from $(T(C_1), \mu_{C_1})$ to $(T(C_2), \mu_{C_2})$ in $F_T(\mathcal{C})$, because in the following diagram, the two squares commute (because of the naturality and the associativity of μ); hence, the following diagram commutes:

$$
\begin{array}{ccc}
TT(C_1) & \xrightarrow{\quad \mu_{C_1} \quad} & T(C_1) \\
{\scriptstyle TT(f)} \downarrow & & \downarrow {\scriptstyle T(f)} \\
TTT(C_2) & \xrightarrow{\quad \mu_{TC_2} \quad} & TT(C_2) \\
{\scriptstyle T\mu_{C_2}} \downarrow & & \downarrow {\scriptstyle \mu_{C_2}} \\
TT(C_2) & \xrightarrow{\quad \mu_{C_2} \quad} & T(C_2),
\end{array}
$$

and this shows that $\mu_{C_2} \circ TG(f) = G(f) \circ \mu_{C_1}$.

The assignment G is actually a functor. G applied to the identity morphism in the Kleisli category is

$$G(\eta_C) = \mu_C \circ T(\eta_C) = 1_{T(C)},$$

and the diagram

$$
\begin{array}{ccccc}
T(C_1) & \xrightarrow{\;T(f)\;} & TT(C_2) & \xrightarrow{\;\mu_{C_2}\;} & T(C_2) \\
 & & {\scriptstyle TT(g)} \downarrow & & \downarrow {\scriptstyle T(g)} \\
{\scriptstyle G(g \circ f) = T(\mu_{C_3} \circ T(g) \circ f)} & & TTT(C_3) & \xrightarrow{\;\mu_{T(C_3)}\;} & TT(C_3) \\
 & & {\scriptstyle T\mu_{C_3}} \downarrow & & \downarrow {\scriptstyle \mu_{C_3}} \\
 & & TT(C_3) & \xrightarrow{\;\mu_{C_3}\;} & T(C_3)
\end{array}
$$

commutes for all $f \in K_T(C_1, C_2)$ and $g \in K_T(C_2, C_3)$, showing that G respects composition in the Kleisli category.

By construction, G is essentially surjective, and hence, we have to show that G is fully faithful.

Let $\alpha \colon (T(C_1), \mu_{C_1}) \to (T(C_2), \mu_{C_2})$ be any morphism of T-algebras. As the diagram

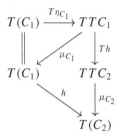

commutes, we can write h as $\mu_{C_2} \circ T(h \circ \eta_{C_1})$ and $h \circ \eta_{C_1} \in K_T(C_1, C_2)$.

Assume that $f, g \in K_T(C_1, C_2)$ with

$$G(f) = \mu_{C_2} \circ T(f) = \mu_{C_2} \circ T(g) = G(g).$$

But then, we also get that the precomposition with η_{C_1} yields the same morphism, but due to the naturality of the unit of T, this is equal to f and g. □

This equivalence of category gives rise to the following adjunction:

Corollary 6.3.6 *The functor $U_T \colon K_T \to C$ that sends an object C of K_T to $U_F(C) = T(C)$ and an $f \in K_T(C_1, C_2) = C(C_1, T(C_2))$ to $\mu_{C_2} \circ T(f)$ has a left adjoint $F_T \colon C \to K_T$, given by $F_T(C) = C$ and $F_T(f) = \eta_{C_2} \circ f$.*

Remark 6.3.7 Note that this adjunction recovers the monad T. We have $U_T \circ F_T(C) = TC$ for all objects C of C, and the composite $U_T \circ F_T$ sends a morphism $f \in C(C_1, C_2)$ to

$$TC_1 \xrightarrow{T(f)} TC_2 \xrightarrow{T\eta_{C_2}} TTC_2 \xrightarrow{\mu_{C_2}} TC_2.$$

This is precisely $T(f)$.

The adjunction (F_T, U_T) is sometimes called the *Kleisli adjunction of the monad*.

In order to distinguish the Eilenberg–Moore adjunction from the Kleisli adjunction, we now use (F^T, U^T) for the Eilenberg–Moore adjoint functor pair. We already know that there is a functor $G\colon K_T \to F_T(\mathcal{C})$, and as the free T-algebras are a full subcategory of T-alg$_\mathcal{C}$, we view G as a functor

$$G\colon K_T \to T\text{-alg}_\mathcal{C}.$$

This is actually a map from an initial object to a terminal object.

Definition 6.3.8 Let (T, μ, η) be a fixed monad on a category $\mathcal{C}$ and let (L, R) be an adjoint functor pair

$$\mathcal{C} \underset{R}{\overset{L}{\rightleftarrows}} \mathcal{D},$$

with unit $\tilde\eta\colon \mathrm{Id}_\mathcal{C} \Rightarrow RL$ and counit $\tilde\varepsilon\colon LR \Rightarrow \mathrm{Id}_\mathcal{D}$, such that $(RL, R\tilde\varepsilon_L, \tilde\eta) = (T, \mu, \eta)$.

These adjunctions form a category, the *category of adjunctions building T*, in which a morphism from $\mathcal{C} \underset{R}{\overset{L}{\rightleftarrows}} \mathcal{D}$ to $\mathcal{C}' \underset{R'}{\overset{L'}{\rightleftarrows}} \mathcal{D}'$ is a functor $H\colon \mathcal{D} \to \mathcal{D}'$, such that

$$HL = L' \text{ and } R'H = R.$$

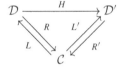

Note that we do not demand that $L'R$ is H.

Exercise 6.3.9 Prove that in the situation of Definition 6.3.8, H carries the adjoint of a morphism $f\colon C \to R(D) = R'H(D)$ to the adjoint of this morphism in $\mathcal{D}'$.

Theorem 6.3.10 *Let (T, μ, η) be a monad. The Kleisli adjunction for T is initial in the category of adjunctions giving T, and the Eilenberg–Moore adjunction is terminal.*

Let us fix the notation before we start a sketch of proof. We have the Kleisli adjunction (F_T, U_T), and we call the unit η_T and the counit ε_T. For the Eilenberg–Moore adjunction, we use superscripts, so (F^T, U^T) with unit η^T and counit ε^T. We only sketch the proof of Theorem 6.3.10. For a full proof, see, for instance, [**Rie16**, Proposition 5.2.12].

Sketch of Proof Let (L, R) be an adjunction building T with unit $\tilde\eta\colon \mathrm{Id}_\mathcal{C} \Rightarrow RL$ and counit $\tilde\varepsilon\colon LR \Rightarrow \mathrm{Id}_\mathcal{D}$. We have to define unique functors A and Z, as in the following diagram:

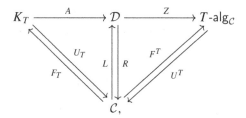

with

$$RA = U_T, U^T Z = R, A F_T = L, \text{ and } F^T Z = L.$$

Let C be an object of K_T. As $A(C)$ has to satisfy $A(C) = A F_T(C) = L(C)$, we can only set $A(C) := L(C)$. For $f \in K_T(C_1, C_2)$, that is, $f \in \mathcal{C}(C_1, T(C_2))$, we have to set $A(f)$ as the composite

$$
\begin{array}{ccc}
L(C_1) & \xrightarrow{A(f)} & L(C_2) \\
\downarrow{\scriptstyle L(f)} & & \uparrow{\scriptstyle \tilde{\varepsilon}_{L(C_2)}} \\
LT(C_1) & = \!\!= & LRL(C_1),
\end{array}
$$

because by Exercise 6.3.9, A has to send adjoints of morphisms to adjoints. This determines A uniquely, and $RA = U_T$ is satisfied.

For Z, we have the relation $U^T Z = R$, so we need a T-algebra structure map for $Z(D)$ for any object D of $\mathcal{D}'$, that is, a morphism $\xi \colon T Z(D) \to Z(D)$ compatible with the multiplication and the unit of T. But the structure map of a T-algebra can be read off the counit of the Eilenberg–Moore adjunction as the morphism $\xi \colon (T Z(D), \mu_{Z(D)}) \to (Z(D), \xi)$ (see the Proof of Theorem 6.2.5). Thus, we have to set $Z(D) := (R(D), R(\tilde{\varepsilon}_D))$ with

$$T R(D) = RLR(D) \xrightarrow{R(\tilde{\varepsilon}_D)} R(D).$$

Also, for morphisms $h \in \mathcal{D}(D_1, D_2)$, we have no choice but to set

$$Z(h) = R(h) \colon (R(D_1), R(\tilde{\varepsilon}_{D_1})) \to (R(D_2), R(\tilde{\varepsilon}_{D_2})).$$

Note that

$$Z(L(C)) = (RL(C), R(\tilde{\varepsilon}_{L(C)})) = (T(C), \mu_C) = F^T(C). \qquad \square$$

Exercise 6.3.11 Show that A and Z given earlier are actually functors.

6.4 Lifting Left Adjoints

We present a diagrammatic proof of the existence of left adjoints here. There are other versions available in the literature; see, for instance, [**Bo94-2**, §4.5] or [**BaWe05**, §3.7].

Assume that we have an adjunction

$$\mathcal{C} \underset{R}{\overset{L}{\rightleftarrows}} \mathcal{D},$$

and let $\nu\colon \mathrm{Id} \Rightarrow RL$ denote the unit and $\pi\colon LR \to \mathrm{Id}$ the counit of the adjunction. Assume further that R lifts to the level of algebras over monads

$$
\begin{array}{ccc}
T\text{-alg}_{\mathcal{C}} & \xrightarrow{\ \tilde{R}\ } & T'\text{-alg}_{\mathcal{D}} \\[4pt]
F \big\updownarrow U & & F' \big\updownarrow U' \\[4pt]
\mathcal{C} & \xrightarrow[\ R\ ]{} & \mathcal{D},
\end{array}
\qquad (6.4.1)
$$

that is, $U' \circ \tilde{R} = RU$. Again, we do not pose any other commutativity constraints in the preceding diagram.

Our aim is to lift the left adjoint L to R to the level of algebras over monads. We want a functor

$$\tilde{L}\colon T'\text{-alg}_{\mathcal{D}} \to \mathsf{T}\text{-alg}_{\mathcal{C}},$$

such that the pair $(\tilde{L}, \tilde{R})$ is an adjoint functor pair.

If D is a free T'-algebra, $D = T'(D')$, then the adjunction properties dictate that we define $\tilde{L}(D)$ as TLD' because

$$T\text{-alg}_{\mathcal{C}}(TLD', C) \cong \mathcal{C}(LD', UC) \cong \mathcal{D}(D', RUC) = \mathcal{D}(D', U'\tilde{R}C)$$

$$\cong T'\text{-alg}_{\mathcal{D}}(T'D', \tilde{R}C).$$

If $\tilde{L}$ exists, then it has to preserve coequalizer. So, according to Proposition 6.2.9, we have no choice but to define $\tilde{L}(D)$ on a general T'-algebra D as the coequalizer of the diagram

$$FLU'F'D \underset{\psi_2}{\overset{\psi_1}{\rightrightarrows}} FLD$$

for every T'-algebra D. Here, the morphisms ψ_1 and ψ_2 are defined as follows. The morphism $\psi_1 = (\psi_1)_D$ is the composition

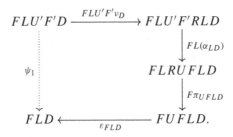

whereas $\psi_2 = (\psi_2)_D$ is just $FL(\theta_D)$, where $\theta_D \colon T'D \to D$ is the T'-algebra structure map of D.

Theorem 6.4.1 *Assume that (L, R) is an adjoint functor pair and that R has a lift to the level of algebras over monads, as in (6.4.1). If the category of T-algebras in $\mathcal{C}$ is cocomplete, then a left adjoint $\tilde{L}$ of the functor $\tilde{R}$ exists.*

Proof The cocompleteness assumption guarantees that we can define $\tilde{L}$ as explained earlier. Our strategy is to show that the following diagram is a commutative diagram of equalizer diagrams in the category of sets.

$$
\begin{array}{ccc}
T\text{-alg}_\mathcal{C}(\tilde{L}D, C) \longrightarrow T\text{-alg}_\mathcal{C}(FLD, C) & \underset{\psi_2^*}{\overset{\psi_1^*}{\rightrightarrows}} & T\text{-alg}_\mathcal{C}(FLU'F'D, C) \\
\varphi_2 \uparrow \qquad\qquad\qquad & & \downarrow \varphi_3 \\
\mathcal{C}(LD, UC) & & \mathcal{C}(LU'F'D, UC) \\
\varphi_1 \downarrow \qquad\qquad\qquad & & \downarrow \varphi_4 \\
\mathcal{D}(D, RUC) & & \mathcal{D}(U'F'D, RUC) \\
\| \qquad\qquad\qquad & & \| \\
\mathcal{D}(D, U'\tilde{R}C) & & \mathcal{D}(U'F'D, U'\tilde{R}C) \\
\alpha_1 \downarrow \qquad\qquad\qquad & & \uparrow \alpha_2 \\
T'\text{-alg}_\mathcal{D}(D, \tilde{R}C) \longrightarrow T'\text{-alg}_\mathcal{D}(F'D, \tilde{R}C) & \underset{\xi_2^*}{\overset{\xi_1^*}{\rightrightarrows}} & T'\text{-alg}_\mathcal{D}(F'U'F'D, \tilde{R}C).
\end{array}
$$

$$(6.4.2)$$

The morphisms ξ_1 and ξ_2 are given by $\varepsilon'_{F'D}$ and $F'(\theta_D)$, respectively. The morphisms φ_i, $1 \le i \le 4$ and α_i, $i = 1, 2$ are the adjunction bijections. We just label them here for later reference.

Naturality ensures that ψ_2^* and ξ_2^* render the diagram commutative. It remains to show that ψ_1^* and ξ_1^* are compatible with the bijections coming from the adjunctions.

We pick an arbitrary $h \in C(LD, UC)$ and show that

$$\varphi_4 \circ \varphi_3(\psi_1^*(\varphi_2(h))) = \alpha_2 \circ \xi_1^*(\alpha_1(\varphi_1(h))).$$

We spell out what these compositions amount to in the following diagram. Here, the left vertical morphism is $\alpha_2 \circ \xi_1^*(\alpha_1(\varphi_1(h)))$, and the remaining U-shaped morphism depicts the composition $\varphi_4 \circ \varphi_3(\psi_1^*(\varphi_2(h)))$.

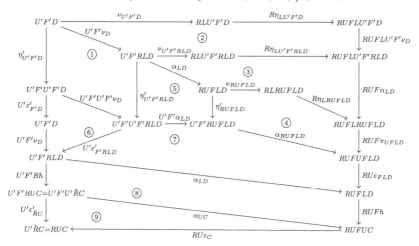

Diagram ① commutes because η' is natural. In ②, there is a sequence of units of adjunctions, but the unit ν_D can be moved along the other two. The naturality of α ensures that ③ and ⑧ commute. The unit condition for α, as in (6.2.3), implies that $\alpha_{RUFLD} \circ \eta'_{RUFLD}$ is equal to $R(\eta_{UFLD})$, and this agrees with the composition in the top row of ④. Naturality ensures that ⑤ and ⑥ commute. As α satisfies the compatibility condition with respect to the multipliction in the monads T and T', as in (6.2.4), we get that ⑦ commutes. We already showed in (6.2.11) that ⑨ commutes.

Hence, in total, the diagram (6.4.2) is a commutative diagram of equalizers, and thus, we get a bijection

$$T\text{-alg}_C(\tilde{L}D, C) \cong T'\text{-alg}_D(D, \tilde{R}C).$$

This bijection is natural in D and C, and thus, we obtain the adjointness of $\tilde{L}$ and $\tilde{R}$. $\square$

6.5 Colimits and Limits of Algebras over a Monad

Of course, you want to build limits and colimits in your favorite category of algebras over a monad, whether these are differential graded Lie algebras or algebras associated with the monad coming from the loop–suspension adjunction. The good news is that the concrete form of your monad does not matter. Limits of algebras over monads are easy to describe and are nothing to worry about.

Theorem 6.5.1 *Let T be a monad on a complete category C. Then, the category T-alg$_C$ is complete.*

Proof Let $G: \mathcal{D} \to T$-alg$_C$ be a small diagram of T-algebras in C and let $(L, (\tau_D: L \to G(D)))$ be the limit of $U \circ G: \mathcal{D} \to C$. We claim that this limit carries a canonical T-algebra structure and is the limit of G in the category of T-algebras.

Denote $G(D)$ as (A_D, ξ_D) with $\xi_D: T(A_D) \to A_D$, and recall that $U(A_D, \xi_D) = A_D$. We have compatible morphisms

$$T(L) \xrightarrow{T(\tau_D)} T(U(G(D)) = T(A_D) \xrightarrow{\xi_D} A_D$$

for all objects D of $\mathcal{D}$, and hence, we get a unique morphism $\xi: T(L) \to L$ with

$$\tau_D \circ \xi = \xi_D \circ T(\tau_D). \tag{6.5.1}$$

We have to show that $\xi \circ \eta_L = 1_L$ and $\xi \circ \mu_L = \xi \circ T(\xi)$.
For the first property, we note that

$$\tau_D \circ \xi \circ \eta_L = \xi_D \circ T(\tau_D) \circ \eta_L,$$

and the naturality of η yields that this is equal to $\xi_D \circ T(\eta_{F(D)}) \circ \tau_D$. As every $F(D)$ is a T-algebra, this reduces to τ_D, and hence, $\xi \circ \eta_L = 1_L$.
For the associativity, we consider the commutative cube:

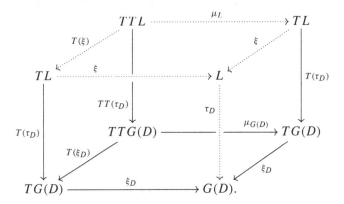

For $\tau_D \circ \xi \circ T(\xi) = \tau_D \circ \xi \circ \mu_L$, you start with the dotted zigzag for $\tau_D \circ \xi \circ T(\xi)$. Sliding down the front square gives $\xi_D \circ T(\tau_D) \circ T(\xi)$, and sliding down the square on the left-hand side gives $\xi_D \circ T(\xi_D) \circ TT(\tau_D)$. Now, you can swop the sides in the bottom square to obtain $\xi_D \circ \mu_{G(D)} \circ TT(\tau_D)$. From here, you just move first up the back square to get $\xi_D \circ T(\tau_D) \circ \mu_L$, and finally, you walk up the right-hand square to end up at $\tau_D \circ \xi \circ \mu_L$.

Here, we have just used (6.5.1) and the fact that $G(D)$ is a T-algebra.

It remains to show that (L, ξ) is the limit of G in the category of T-algebras, so we assume that we have a T-algebra (B, ζ) and a compatible family of morphisms of T-algebras

$$\nu_D \colon (B, \zeta) \to (G(D), \xi).$$

We know that there is a unique morphism in $\mathcal{C}$

$$\phi \colon B = U(B, \zeta) \to U(L, \xi) = L,$$

with $\tau_D \circ \phi = \nu_D$, and we claim that ϕ is a T-algebra morphism.

For all objects D of $\mathcal{D}$, we have $\nu_D \circ \zeta = \xi_D \circ T(\nu_D)$, because every ν_D is a morphism of T-algebras. This yields

$$\tau_D \circ \phi \circ \zeta = \nu_D \circ \zeta = \xi_D \circ T(\nu_D) = \tau_D \circ \xi \circ T(\phi),$$

and hence, $\phi \circ \zeta = \xi \circ T(\phi)$.

As ϕ is unique with $\tau_D \circ \phi = \nu_D$ and as the forgetful functor $U \colon T\text{-alg}_\mathcal{C} \to \mathcal{C}$ is faithful by Proposition 6.2.7, ϕ is unique as a morphism of T-algebras with that property. $\qquad\square$

It is easy to see that the analog of Theorem 6.5.1 for colimits is utterly wrong. For instance, let k be a commutative ring with unit. The coproduct of two k-modules M and N is their direct sum, $M \oplus N$, whereas if M and N happen to be commutative k-algebras, then their coproduct in the category of commutative k-algebras is the tensor product $M \otimes_k N$, so you *cannot* just take the colimit in the underlying category in order to produce a colimit on the level of algebras. Under rather restrictive conditions, there is a transfer result.

Lemma 6.5.2 *Let T be a monad on $\mathcal{C}$, let $\mathcal{D}$ be a small category, and let $G \colon \mathcal{D} \to T\text{-alg}_\mathcal{C}$ be a functor, such that $UG \colon \mathcal{D} \to \mathcal{C}$ possesses a colimit. If T and $T \circ T$ preserve this colimit, then G has a colimit in $T\text{-alg}_\mathcal{C}$ which is preserved by U.*

Proof Let $(\text{colim}_D UG, \tau_D \colon UG(D) \to \text{colim}_\mathcal{D} UG)$ be the colimit of the underlying diagram $UG \colon \mathcal{D} \to \mathcal{C}$. Every $G(D)$ is a T-algebra, so $G(D)$ is of the form $(UG(D), \xi_D)$ with structure maps $\xi_D \colon T(U(G(D))) \to U(G(D))$. The ξ_Ds are components of a natural transformation

$$\xi \colon T \circ U \circ G \Rightarrow U \circ G,$$

because for every $f \in \mathcal{D}(D_1, D_2)$, the morphism $G(f)$ is an element of $T\text{-alg}_C(G(D_1), G(D_2))$; therefore, for all such f, the diagram

$$
\begin{array}{ccc}
TUG(D_1) & \xrightarrow{\ \xi_{D_1}\ } & UG(D_1) \\
{\scriptstyle TUG(f)}\downarrow & & \downarrow{\scriptstyle UG(f)} \\
TUG(D_2) & \xrightarrow{\ \xi_{D_2}\ } & UG(D_2)
\end{array}
$$

commutes.

By assumption, $(T(\mathrm{colim}_\mathcal{D} UG), T(\tau_D)\colon TUG(D) \to T(\mathrm{colim}_\mathcal{D} UG))$ is the colimit of TUG. The morphisms $\tau_D \circ \xi_D \colon TUG(D) \to \mathrm{colim}_\mathcal{D} UG$ induce a unique morphism

$$\Xi \colon T(\mathrm{colim}_\mathcal{D} UG) \cong \mathrm{colim}_\mathcal{D} TUG \to \mathrm{colim}_\mathcal{D} UG,$$

with the property that

$$\tau_D \circ \xi_D = \Xi \circ T(\tau_D). \tag{6.5.2}$$

We claim that $((\mathrm{colim}_\mathcal{D} UG, \Xi), \tau_D \colon (UG(D), \xi_D) \to (\mathrm{colim}_\mathcal{D} UG, \Xi))$ is the colimit of G in $T\text{-alg}_C$. By construction, U preserves the colimit $((\mathrm{colim}_\mathcal{D} UG, \Xi), \tau_D \colon (UG(D), \xi_D) \to (\mathrm{colim}_\mathcal{D} UG, \Xi))$.

We first show that Ξ defines a T-algebra structure on $\mathrm{colim}_\mathcal{D} UG$.

As the diagram

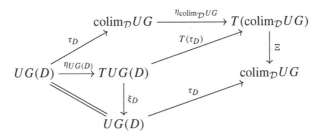

commutes, we get

$$\Xi \circ \eta_{\mathrm{colim}_\mathcal{D} UG} \circ \tau_D = \Xi \circ T(\tau_D) \circ \eta_{UG(D)} = \tau_D,$$

and hence, $\Xi \circ \eta_{\mathrm{colim}_\mathcal{D} UG} = 1_{\mathrm{colim}_\mathcal{D} UG}$, so Ξ is compatible with the unit of T.

For the associativity of Ξ, we use that by assumption, TT preserves $\mathrm{colim}_\mathcal{D} UG$, so in particular, it suffices to check any relation by precomposing with $TT(\tau_D)$ for all objects D of $\mathcal{D}$. Consider the solid part of the diagram

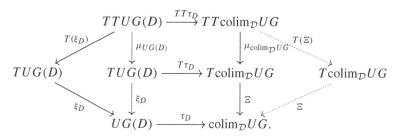

We want to show that the right dashed wing of the diagram commutes. The solid part commutes, and this yields

$$\Xi \circ \mu_{\text{colim}_{\mathcal{D}} UG} \circ TT(\tau_D) = \Xi \circ T(\tau_D)\mu_{UG(D)} = \tau_D \circ \xi_D \circ T(\xi_D).$$

By (6.5.2), we get that this is equal to $\Xi \circ T(\tau_D) \circ T(\xi_D) = \Xi \circ T(\tau_D \circ \xi_D)$, and using (6.5.2) again, this equals

$$\Xi \circ T(\Xi \circ T(\tau_D)) = \Xi \circ T(\Xi) \circ TT(\tau_D).$$

It remains to check that $((\text{colim}_{\mathcal{D}} UG, \Xi), \tau_D \colon (UG(D), \xi_D) \to (\text{colim}_{\mathcal{D}} UG, \Xi))$ has the universal property of the colimit of G, so assume that $(\sigma_D \colon (UG(D), \xi_D) = G(D) \to (C, \zeta))$ is another cocone for G. Then, the σ_Ds are also a cocone for $UG(D)$ when viewed as morphisms $\sigma_D \colon UG(D) \to C$, so we get a unique morphism $\psi \in \mathcal{C}(\text{colim}_{\mathcal{D}} UG, C)$ with

$$\psi \circ \tau_D = \sigma_D. \tag{6.5.3}$$

We claim that ψ is a morphism of T-algebras. As ζ is a morphism of T-algebras, we have $\zeta \circ T(\sigma_D) = \sigma_D \circ \xi_D$ for all objects D of $\mathcal{D}$.

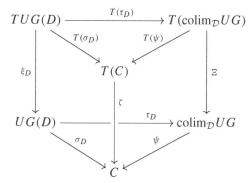

We want to show that the right front face commutes. By assumption, all other faces of the triangular prism commute; thus,

$$\zeta \circ T(\psi) \circ T(\tau_D) = \zeta \circ T(\sigma_D) = \sigma_D \circ \xi_D$$
$$= \psi \circ \tau_D \circ \xi_D = \psi \circ \Xi \circ T(\tau_D),$$

and hence, $\zeta \circ T(\psi) = \psi \circ \Xi$. $\qquad \square$

You know from Theorem 3.2.2 that a category is cocomplete if and only if it has coequalizers and coproducts. In the case of a category of algebras over a monad, we only have to establish the existence of coequalizers.

Theorem 6.5.3 *Let C be a cocomplete category and let T be a monad on C. Then, the category of T-algebras in C, T-alg$_C$, is cocomplete if and only if it has coequalizers.*

Proof   We have to show that the existence of coequalizers implies the existence of coproducts. The proof has a conceptual part, where we actually construct coproducts. Checking the properties of the coproduct is then rather dull but necessary, so you are cordially invited to skip that part.

The idea of the proof is to use the important property of algebras over a monad of being expressible as the coequalizer of free objects. Let $\mathcal{D}$ be a small discrete category and let (C_D, ξ_D) be a $\mathcal{D}$-diagram in T-alg$_C$. The free functor is a left adjoint functor, so it has to preserve coproducts. Hence, we know that if coproducts of T-algebras in C exist, then

$$(T(\coprod_{\mathcal{D}} C_D), \mu_{\coprod_{\mathcal{D}} C_D}) \cong \coprod_{\mathcal{D}} (T(C_D), \mu_{C_D}),$$

where the coproduct on the left-hand side denotes the coproduct in C (and we know that this exists), whereas the coproduct on the right-hand side is the one that we want.

So, we have no choice but to define the coproduct

$$(C, \xi) = \coprod_{\mathcal{D}} (C_D, \xi_D)$$

as a coequalizer of a diagram

$$(T(\coprod_{\mathcal{D}}(T(C_D), \mu_{C_D})), \mu_{\coprod_{\mathcal{D}}(T(C_D))}) \underset{\beta}{\overset{\alpha}{\rightrightarrows}} (T(\coprod_{\mathcal{D}} C_D), \mu_{\coprod_{\mathcal{D}} C_D}),$$

with morphisms of T-algebras α and β that we will specify later.

As coproducts exist in C, we have the coproducts $\coprod_{\mathcal{D}} C_D$ and $\coprod_{\mathcal{D}} T(C_D)$ in C with universal cocones

$$i_{C_D} \colon C_D \to \coprod_{\mathcal{D}} C_D \text{ and } i_{T(C_D)} \colon T(C_D) \to \coprod_{\mathcal{D}} T(C_D).$$

The morphisms $T(i_{C_D}) \colon T(C_D) \to T(\coprod_{\mathcal{D}} C_D)$ are a cocone in C, and hence, there is a unique morphism

$$\theta \colon \coprod_{\mathcal{D}} T(C_D) \to T(\coprod_{\mathcal{D}} C_D)$$

in C with

$$\theta \circ i_{T(C_D)} = T(i_{C_D}). \tag{6.5.4}$$

We consider the two morphisms

$$\alpha = (T(\textstyle\bigsqcup_D T(C_D)), \mu_{\bigsqcup_D T(C_D)}) \xrightarrow{T(\theta)} (T(T(\textstyle\bigsqcup_D C_D)),$$

$$\mu_{T(\bigsqcup_D C_D)}) \xrightarrow{\mu_{\bigsqcup_D C_D}} (T(\textstyle\bigsqcup_D C_D), \mu_{\bigsqcup_D C_D})$$

and

$$\beta = T(\textstyle\bigsqcup_D \xi_D) \colon (T(\textstyle\bigsqcup_D T(C_D)), \mu_{\bigsqcup_D T(C_D)}) \to (T(\textstyle\bigsqcup_D C_D), \mu_{\bigsqcup_D C_D}).$$

We have to show that α and β are morphisms of T-algebras.
For α, consider the diagram

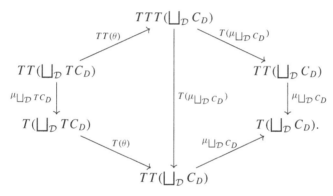

The right half of the diagram trivially commutes, and the left half commutes because of the naturality of μ, and hence, so does the whole diagram. The naturality of μ also implies

$$\mu_{\bigsqcup_D C_D} \circ T(\beta) = \mu_{\bigsqcup_D C_D} \circ TT(\textstyle\bigsqcup_D \xi_D)$$

$$= T(\textstyle\bigsqcup_D \xi_D) \circ \mu_{\bigsqcup_D T(C_D)}$$

$$= \beta \circ \mu_{\bigsqcup_D T(C_D)},$$

and hence, β is a morphism of T-algebras as well.
So, the coequalizer of α and β exists in $T\text{-alg}_C$:

$$(T(\textstyle\bigsqcup_D T(C_D)), \mu_{\bigsqcup_D T(C_D)}) \underset{\beta}{\overset{\alpha}{\rightrightarrows}} (T(\textstyle\bigsqcup_D C_D), \mu_{\bigsqcup_D C_D}) \xrightarrow{\pi} (C, \xi).$$

We claim that (C, ξ) is the coproduct of the (C_D, ξ_D) in $T\text{-alg}_C$. So, we have to

(1) construct the morphisms $\sigma_D \colon (C_D, \xi_D) \to (C, \xi)$ for the universal cocone; and

(2) show that this coequalizer satisfies the universal property of the coproduct.

For (1), we consider the coequalizer diagram for (C_D, ξ_D) in $T\text{-alg}_C$, together with the composition $\pi \circ T(i_{C_D})$:

$$(TT(C_D), \mu_{T(C_D)}) \underset{T(\xi_D)}{\overset{\mu_{C_D}}{\rightrightarrows}} (T(C_D), \mu_{C_D}) \xrightarrow{\xi_D} (C_D, \xi_D).$$

$$\downarrow{\pi \circ T(i_{C_D})}$$

$$(C, \xi)$$

By the coequalizer property of π, we know that

$$\pi \circ \mu_{\bigsqcup_D C_D} \circ T(\theta) = \pi \circ T\left(\bigsqcup_D \xi_D\right).$$

The diagram

commutes by (6.5.4) and by the naturality of μ, and therefore,

$$\pi \circ T(i_{C_D}) \circ \mu_{C_D} = \pi \circ \mu_{\bigsqcup_D C_D} \circ T(\theta) \circ T(i_{T(C_D)}) = \pi \circ T\left(\bigsqcup_D \xi_D\right) \circ T(i_{T(C_D)}),$$

but $T(\bigsqcup_D \xi_D) \circ T(i_{T(C_D)})$ is equal to $T(i_{C_D}) \circ T(\xi_D)$, and hence,

$$\pi \circ T(i_{C_D}) \circ \mu_{C_D} = \pi \circ T(i_{C_D}) \circ T(\xi_D).$$

The universal property of the coequalizer yields a unique morphism $\sigma_D \colon (C_D, \xi_D) \to (C, \xi)$ with

$$\sigma_D \circ \xi_D = \pi \circ T(i_{C_D}). \tag{6.5.5}$$

For (2), we show that $((C, \xi), \sigma_D \colon (C_D, \xi_D) \to (C, \xi))$ satisfies the universal property of the coproduct.

Let $\gamma_D \colon (C_D, \xi_D) \to (A, \zeta)$ be any cocone. We want to construct a unique $\gamma \in T\text{-alg}_C((C, \xi), (A, \zeta))$ with $\gamma \circ \sigma_D = \gamma_D$. We first use the universal property of the coproducts in $\mathcal{C}$.

The morphisms $U(\gamma_D) = \gamma_D \colon C_D \to A$ form a cocone in C, and hence, there is a unique $\psi \in C(\bigsqcup_{\mathcal{D}} C_D, A)$ with

$$\psi \circ i_{C_D} = \gamma_D. \tag{6.5.6}$$

Similarly, we get a unique $\varphi \in C(\bigsqcup_{\mathcal{D}} T(C_D), T(A))$ with

$$\varphi \circ i_{T(C_D)} = T(\gamma_D). \tag{6.5.7}$$

These two morphisms are related. The diagram

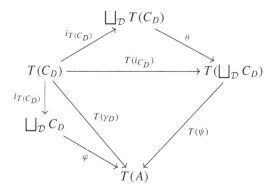

commutes, and therefore,

$$T(\psi) \circ \theta = \varphi. \tag{6.5.8}$$

As the γ_Ds are morphisms of T-algebras and by (6.5.6) and (6.5.7), we obtain that

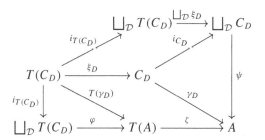

is commutative, and this implies

$$\zeta \circ \varphi = \psi \circ \bigsqcup_{\mathcal{D}} \xi_D. \tag{6.5.9}$$

We collect all these relations to show that

$$\zeta \circ T(\psi) \circ \mu_{\bigsqcup_{\mathcal{D}} C_D} \circ T(\theta) = \zeta \circ T(\psi) \circ T(\bigsqcup_{\mathcal{D}} \xi_D). \tag{6.5.10}$$

The left-hand side is contained in the solid commutative part of the diagram

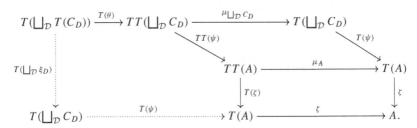

Thus,

$$\zeta \circ T(\psi) \circ \mu_{\bigsqcup_{\mathcal{D}} C_D} \circ T(\theta) = \zeta \circ T(\zeta) \circ TT(\psi) \circ T(\theta) = \zeta \circ T(\zeta \circ T(\psi) \circ \theta).$$

With the relations (6.5.8) and (6.5.9), we have that

$$\zeta \circ T(\zeta \circ T(\psi) \circ \theta) = \zeta \circ T(\zeta \circ \varphi)$$
$$= \zeta \circ T(\psi \circ \bigsqcup_{\mathcal{D}} \xi_D),$$

and this proves the relation (6.5.10), and hence, $\zeta \circ T(\psi)$ coequalizes $\mu_{\bigsqcup_{\mathcal{D}} C_D} \circ T(\theta)$ and $T(\bigsqcup_{\mathcal{D}} \xi_D)$. Therefore, by the universal property of the coequalizer, there is a unique morphism $\gamma : (C, \xi) \to (A, \zeta)$ with

$$\gamma \circ \pi = \zeta \circ T(\psi). \qquad (6.5.11)$$

It remains to show that γ is also unique, with the property that for all objects D of $\mathcal{D}$, $\gamma \circ \sigma_D = \gamma_D$. So, we assume that there is another morphism $f \in T\text{-alg}_{\mathcal{C}}((C, \xi), (A, \zeta))$, with the property that $f \circ \sigma_D = \gamma_D$ for all D. By (6.5.4), we know that

$$f \circ \pi \circ \theta \circ i_{T(C_D)} = f \circ \pi \circ T(i_{C_D}).$$

But by (6.5.5), the latter is

$$f \circ \sigma_D \circ \xi_D = \gamma_D \circ \xi_D = \psi \circ i_{C_D} \circ \xi_D = \psi \circ \bigsqcup_{\mathcal{D}} \xi_D \circ i_{T(C_D)},$$

and therefore, we obtain that

$$f \circ \pi \circ \theta = \psi \circ \bigsqcup_{\mathcal{D}} \xi_D.$$

We embed $T(-)$ applied to the preceding relation as the right part of the roof of the following diagram:

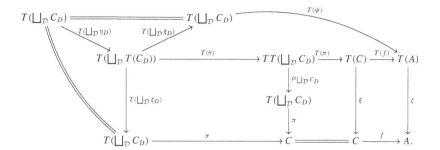

The left-hand triangles commute because $\xi_D \circ \eta_D = 1_{C_D}$ for all objects D of $\mathcal{D}$. The left square commutes because π is a morphism in a coequalizer diagram. The middle and right squares commute because π and f are morphisms of T-algebras.

So, in total, we get $f \circ \pi = \zeta \circ T(\psi)$, but γ was unique with this property, so $f = \gamma$. $\qquad\square$

6.6 Monadicity

We close with a criterion that helps you to decide whether a given category is equivalent to a category of T-algebras for some monad T. There are several versions of such a result in the literature (**[BaWe05]**, **[Bo94-2**, 4.4]), and we present only one of them. The original version is due to Beck **[BaWe05**, Theorem 3.14].

When does a functor "look like" a forgetful functor $U: T\text{-alg}_{\mathcal{C}} \to \mathcal{C}$?

Definition 6.6.1 A functor $R: \mathcal{D} \to \mathcal{C}$ is *monadic* if there exists a monad T on $\mathcal{C}$ and an equivalence of categories $\Phi: \mathcal{D} \to T\text{-alg}_{\mathcal{C}}$, such that $U \circ \Phi$ is isomorphic to R.

Barr and Beck called such functors *tripleable*. Beware that Mac Lane **[ML98**, VI.3] reserves the term monadic only for functors for which Φ is an *isomorphism*.

Theorem 6.6.2 (Barr and Beck monadicity theorem) *Let* $R\colon \mathcal{D} \to \mathcal{C}$ *be a functor. Then, the following conditions are equivalent.*

(1) The functor R is monadic.

(2) • *The functor R has a left adjoint F.*

  • *A morphism* $f \in \mathcal{D}(D_1, D_2)$ *is an isomorphism if and only* $R(f)$ *is an isomorphism.*

  • *For any pair of morphisms,*

$$D_1 \underset{g}{\overset{f}{\rightrightarrows}} D_2 \qquad\qquad (6.6.1)$$

in $\mathcal{D}$ *for which* $R(D_1) \underset{R(g)}{\overset{R(f)}{\rightrightarrows}} R(D_2)$ *has a split coequalizer in* $\mathcal{C}$,

the coequalizer of f and g exists in $\mathcal{D}$ *and is preserved by R.*

Proof   If R is monadic, then we show that the forgetful functor U satisfies the conditions from (2). The left adjoint is F. If $f \in T\text{-alg}_\mathcal{C}((C_1, \xi_1), (C_2, \xi_2))$ and if $U(f)$ is an isomorphism in $\mathcal{C}\colon U(f) = f\colon C_1 \to C_2$, with inverse $f^{-1}\colon C_2 \to C_1$, then f^{-1} is automatically a morphism of monads. As $f \circ \xi_1 = \xi_2 \circ T(f)$, we get

$$f \circ \xi_1 \circ T(f^{-1}) = \xi_2 \circ T(f) \circ T(f^{-1}) = \xi_2 = f \circ f^{-1} \circ \xi_2,$$

and as f is an isomorphism, this implies $\xi_1 \circ T(f^{-1}) = f^{-1} \circ \xi_2$. So, it remains to show the coequalizer property. Let $(C_1, \xi_1) \underset{g}{\overset{f}{\rightrightarrows}} (C_2, \xi_2)$ be a coequalizer in $T\text{-alg}_\mathcal{C}$, which becomes split after the application of U, that is, there are morphisms s, σ, and π, such that

$$C_1 \underset{g}{\overset{f}{\underset{\sigma}{\rightrightarrows}}} C_2 \overset{s}{\underset{\pi}{\to}} C$$

is a split coequalizer in $\mathcal{C}$. Such split coequalizers are preserved by every functor, so in particular, T and TT preserve it. Therefore, we can apply Lemma 6.5.2 and obtain that the coequalizer of f and g exists in $T\text{-alg}_\mathcal{C}$ and is preserved by U.

Conversely, we assume that R satisfies the conditions from (2); in particular, we have a left adjoint L of R and a monad $T = RL$. We define $\Psi\colon \mathcal{D} \to T\text{-alg}_\mathcal{C}$ as

$$\Psi(D) = (R(D), R\varepsilon_D\colon RLR(D) \to R(D)).$$

We claim that Ψ is an equivalence of categories, so we check that it is full, faithful, and essentially surjective.

- Consider the functor $\Phi\colon K_T \to \mathcal{D}$ with $\Phi(C) = L(C)$ and $\Phi(f\colon C_1 \to T(C_2)) = \varepsilon_{LC_2} \circ L(f)$. The composite $\Psi \circ \Phi$ is equal to the functor G from Proposition 6.3.5 that identifies the Kleisli category K_T with the full subcategory of T-alg_C of free objects. Therefore, Ψ is full.

- As $U \circ \Psi = R$, it is enough to show that R is faithful, and by Proposition 2.4.11, we have to show that ε_D is an epimorphism for all objects D of $\mathcal{D}$. To this end, consider the diagram

$$LRLR(D) \underset{\varepsilon_{LRD}}{\overset{LR\varepsilon_D}{\rightrightarrows}} LR(D) \overset{\varepsilon_D}{\longrightarrow} D.$$

If we apply the functor R to this diagram, then we obtain a split coequalizer diagram

because ε is the counit of the adjunction (L, R). By assumption, we get a coequalizer diagram

$$LRLR(D) \underset{\varepsilon_{LRD}}{\overset{LR\varepsilon_D}{\rightrightarrows}} LR(D) \overset{\pi}{\longrightarrow} \tilde{D}.$$

By the universal property, there is a unique morphism $h \in \mathcal{D}(\tilde{D}, D)$ with $h \circ \pi = \varepsilon_D$. By assumption, $R(D)$ is isomorphic to $R(\tilde{D})$ and $R(\pi)$ is equal to $R(\varepsilon_D)$ up to isomorphism. But then, again up to an isomorphism, $R(h) \circ R(\varepsilon_D)$ is $R(\varepsilon_D)$. Thus, $R(h)$ is an isomorphism, and by assumption, this implies that h is an isomorphism, so ε_D is an epimorphism.

- Let (C, ξ) be an object of T-alg_C. Consider the diagram

$$LRL(C) \underset{\varepsilon_{L(C)}}{\overset{L\xi}{\rightrightarrows}} L(C).$$

Applying $R(-)$ again gives a split coequalizer diagram by Proposition 6.2.9.

$$RLRL(C) \underset{R\varepsilon_{L(C)}}{\overset{RL\xi}{\rightrightarrows}} RL(C) \overset{\xi}{\longrightarrow} C, \qquad (6.6.2)$$

and therefore, by assumption, we obtain a coequalizer diagram

$$LRL(C) \underset{\varepsilon_{L(C)}}{\overset{L\xi}{\rightrightarrows}} L(C) \overset{\pi}{\longrightarrow} \tilde{C},$$

such that $R(\tilde{C})$ is isomorphic to C, and up to isomorphism, ξ is $R(\pi)$, as discussed earlier. The splitting of (6.6.2) gives a section for $R(\pi)$, and hence, for $LR(\pi)$. Therefore $LR(\pi)$ is an epimorphism.

As $\pi \circ L(\xi) = \pi \circ \varepsilon_{L(C)}$ and as up to isomorphism, $L(\xi) = LR(\pi)$ and $\varepsilon_{L(C)} = \varepsilon_{LR(\tilde{C})}$, we obtain that up to an isomorphism, $\pi \circ L(\pi) = \varepsilon_{\tilde{C}} \circ LR(\pi)$, because the counit is natural, that is, $\pi \circ \varepsilon_{LR(\tilde{C})} = \varepsilon_{\tilde{C}} \circ LR(\pi)$.

As $LR(\pi)$ is an epimorphism, this implies that $\varepsilon_{\tilde{C}}$ is equal to π up to an isomorphism. Therefore, we get that $R(\varepsilon_{\tilde{C}})$ is equal to $R(\pi)$ up to an isomorphism but that in turn was equal to ξ up to an isomorphism. Therefore, there is an isomorphism

$$\Psi(\tilde{C}) = (R(\tilde{C}), R(\varepsilon_{\tilde{C}})) \cong (C, \xi),$$

and Ψ is essentially surjective.   □

6.7 Comonads

As usual, we do not spell out the dual notion of a monad, which is a comonad, in detail.

Definition 6.7.1 A *comonad* in a category $\mathcal{D}$ consists of an endofunctor $G: \mathcal{D} \to \mathcal{D}$, together with two natural transformations, $\varepsilon: G \Rightarrow \mathrm{Id}$ and $\delta: G \Rightarrow G \circ G$, such that the following diagrams commute:

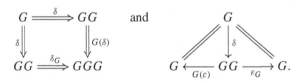

For a comonad G, the natural transformation ε is called the *counit* of the monad and δ is called the *comultiplication*. Comonads on a category $\mathcal{D}$ form a category with the obvious notion of morphisms.

Dual to 6.1.3, we can construct comonads from every pair of adjoint functors.

Theorem 6.7.2 Let $\mathcal{C} \underset{R}{\overset{L}{\rightleftarrows}} \mathcal{D}$ be an adjoint pair of functors. Then, the endofunctor $T = L \circ R: \mathcal{D} \to \mathcal{D}$ is a comonad on $\mathcal{D}$. The transformation $\delta: LR \Rightarrow LRLR$ is given by $L\eta_R$, where η is the unit of the adjunction and the transformation $\varepsilon: LR \Rightarrow \mathrm{Id}$ is the counit of the adjunction.

Exercise 6.7.3 Prove the theorem.

7

Abelian Categories

If you want to do homological algebra, you need projective and injective resolutions of objects, and for this, you need to be able to form kernels and cokernels of morphisms and you need to be able to add and subtract morphisms.

7.1 Preadditive Categories

Definition 7.1.1 A *preaddititve category* is a category $\mathcal{A}$, such that for every pair of objects A_1, A_2, there is an abelian group of morphisms from A_1 to A_2 and the composition of morphisms is a bilinear map.

A preadditive category with only one object is nothing but a ring. The endomorphisms of that object are an abelian group, and the composition of morphisms defines the multiplicative structure. Thus, a preadditive category can be thought of as a *ring with many objects*. A group with many objects in this sense is a groupoid, so one might call a preadditive category a *ringoid*.

Example 7.1.2 Let $\mathcal{C}$ be a category. Then, there is a *preadditive category associated with $\mathcal{C}$*, which we denote by $\mathbb{Z}[\mathcal{C}]$. Its objects are the objects of $\mathcal{C}$, and the morphisms are given by $\mathbb{Z}[\mathcal{C}](C_1, C_2) := \mathbb{Z}\{\mathcal{C}(C_1, C_2)\}$, where $\mathbb{Z}\{S\}$ denotes the free abelian group generated by a set S.

There are many other examples, such as the category of R-modules and R-linear maps for any unital and associative ring R, in particular, the category of abelian groups, is preadditive. The category of groups and group homomorphisms, Gr, however, is *not* preadditive.

Definition 7.1.3 Let $\mathcal{A}$ and $\mathcal{A}'$ be preadditive categories. A functor $F\colon \mathcal{A} \to \mathcal{A}'$ is *additive* if for any two objects A_1, A_2 of $\mathcal{A}$, the map $F\colon \mathcal{A}(A_1, A_2) \to \mathcal{A}'(F(A_1), F(A_2))$ is a group homomorphism.

Exercise 7.1.4 Let Preadd denote the category of all small preadditive categories and additive functors and let U denote the forgetful functor from Preadd to cat. Is $\mathbb{Z}[-]$ left adjoint to U?

Recall from Definition 1.5.1 that a zero object is an object that is terminal and initial. If a category $\mathcal{C}$ has a zero object, then for any two objects C_1 and C_2, there is a zero morphism, $0 \in \mathcal{C}(C_1, C_2)$, which is the composition of the unique morphism from C_1 to 0, followed by the unique morphism from 0 to C_2.

Remark 7.1.5 If $\mathcal{A}$ is a preadditive category and if $\mathcal{A}$ has a zero object, then for any pair of objects A_1, A_2 of $\mathcal{A}$, the zero morphism $0\colon A_1 \to A_2$ is the zero of the abelian group $\mathcal{A}(A_1, A_2)$, because the groups $\mathcal{A}(A_1, 0)$ and $\mathcal{A}(0, A_2)$ are trivial and the zero morphism is the composite of the two unique elements in these groups. Preadditive categories do not have to possess a zero object, but they always have zero morphisms.

Definition 7.1.6 Assume that a category $\mathcal{C}$ has zero morphisms. Then, the *kernel of a morphism* $f \in \mathcal{C}(C_1, C_2)$ is the equalizer of the morphisms $f, 0\colon C_1 \to C_2$. Dually, the *cokernel of a morphism* $f \in \mathcal{C}(C_1, C_2)$ is the coequalizer of the morphisms $f, 0\colon C_1 \to C_2$.

Note that every kernel is a monomorphism because it is an equalizer (see Remark 3.1.28), and every cokernel is an epimorphism (see Remark 3.1.17).

Remark 7.1.7 We defined kernels as equalizers and cokernels as coequalizers. In a preadditive category $\mathcal{A}$, actually *all* equalizers are kernels and *all* coequalizers are cokernels. We show this for kernels. The universal property of an equalizer diagram

$$A \overset{\varphi}{\longrightarrow} A_1 \underset{g}{\overset{f}{\rightrightarrows}} A_2$$

is that for every morphism $\alpha\colon A' \to A_1$ with $f \circ \alpha = g \circ \alpha$, there is a unique morphism ξ from A' to A with $\varphi \circ \xi = \alpha$.

As $\mathcal{A}$ is preadditive, the condition $f \circ \alpha = g \circ \alpha$ is equivalent to $(f - g) \circ \alpha = 0 = 0 \circ \alpha$. Hence, the equalizer diagram is equivalent to the equalizer diagram

$$A \overset{\varphi}{\longrightarrow} A_1 \underset{0}{\overset{f-g}{\rightrightarrows}} A_2,$$

and therefore, every equalizer is a kernel.

Example 7.1.8 If C is not a preadditive category, then it is in general *not* true that every monomorphism is a kernel. Consider for instance the category of groups, Gr. The trivial group is a zero object of Gr. Monomorphisms are just injective group homomorphisms, so, for instance, the inclusion of an arbitrary subgroup $i : H \hookrightarrow G$ is a monomorphism. However, if

$$N \xrightarrow{\varphi} G \underset{0}{\overset{f}{\rightrightarrows}} G'$$

is a kernel, then N can be identified with $N = \{g \in G, f(g) = e_{G'}\}$, and this is a *normal* subgroup of G.

Exercise 7.1.9 Let C be a category with a zero object. What is the kernel of a monomorphism $\alpha : C_1 \to C_2$? What is the kernel of the zero map $0 : C_1 \to C_2$?

In a preadditive category, initial and terminal objects have to agree if they exist.

Proposition 7.1.10 *Let $\mathcal{A}$ be a preadditive category. Then, the following are equivalent:*

- *There exists an initial object in $\mathcal{A}$.*
- *There exists a terminal object in $\mathcal{A}$.*
- *There exists a zero object in $\mathcal{A}$.*

Proof Of course, if we have a zero object, then we also have an initial and terminal object. We show that the existence of an initial object guarantees the existence of a zero object.

Let $\varnothing$ be an initial object of $\mathcal{A}$. Then, the abelian group $\mathcal{A}(\varnothing, \varnothing)$ consists of a single element, and hence, $1_\varnothing$ is the zero element of the group $\mathcal{A}(\varnothing, \varnothing)$. Let A be any object of $\mathcal{A}$. Then, the abelian group $\mathcal{A}(A, \varnothing)$ has at least a zero element. But for any $f \in \mathcal{A}(A, \varnothing)$, we get that f is equal to $1_\varnothing \circ f$. As the composition of morphisms is bilinear, this implies that f is the zero element of $\mathcal{A}(A, \varnothing)$, and hence, $\varnothing$ is also terminal and thus a zero object. $\square$

In the category of abelian groups, the product of two abelian groups A and B, $A \times B$, is isomorphic to (your favorite model of) the direct sum of A and B, $A \oplus B$. This is no bug but a feature.

Proposition 7.1.11 *Let $\mathcal{A}$ be a preadditive category and let A and B be two objects of $\mathcal{A}$. Then, the following are equivalent:*

(1) The coproduct of A and B exists in $\mathcal{A}$.
(2) The product of A and B exists in $\mathcal{A}$.

(3) There exists an object M in $\mathcal{A}$, together with morphisms in $\mathcal{A}$

$$p_A: M \to A, \quad p_B: M \to B, \quad i_A: A \to M \text{ and } i_B: B \to M,$$

such that

$$p_A \circ i_A = 1_A, \quad p_B \circ i_B = 1_B, \quad p_A \circ i_B = 0, \, p_B \circ i_A = 0 \text{ and}$$
$$i_A \circ p_A + i_B \circ p_B = 1_M.$$

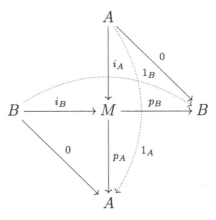

In addition, under these equivalent conditions, we have that i_A is a kernel of p_B, i_B is a kernel of p_A, and also p_A is a cokernel of i_B, and p_B is a cokernel of i_A.

Proof We show that the existence of binary products is equivalent to the existence of an object, as in (3). The proof of the other equivalence is dual to this proof.

Assume the product of A and B, $A \times B$, exists. Then, we have projection maps $p_A: A \times B \to A$ and $p_B: A \times B \to B$, and morphisms into $A \times B$ from any other object are determined by their composition with these projections. We define $i_A: A \to A \times B$ as the morphism with $p_A \circ i_A = 1_A$ and $p_B \circ i_A = 0$. Similarly, i_B is determined by $p_B \circ i_B = 1_B$ and $p_A \circ i_B = 0$. But then, we also get that

$$p_A \circ (i_A \circ p_A + i_B \circ p_B) = p_A, \quad p_B \circ (i_A \circ p_A + i_B \circ p_B) = p_B,$$

and hence, $(i_A \circ p_A + i_B \circ p_B)$ is equal to the identity on $A \times B$. So, (1) implies (3).

Conversely, assume that we have an object M as in (3). We show that M has the universal property of a product. Thus, let $f \in \mathcal{A}(A', A)$ and $g \in \mathcal{A}(A', B)$ be any pair of morphisms in $\mathcal{A}$. We define $\xi: A' \to M$ as

$$\xi = i_A \circ f + i_B \circ g.$$

Then, $p_A \circ \xi$ is, by definition, equal to f, and $p_B \circ \xi$ gives g.

If $\tau : A' \to M$ is any other morphism in $\mathcal{A}$ with the property that $p_A \circ \tau = f$ and $p_B \circ \tau = g$, then τ agrees with ξ because

$$
\begin{aligned}
\tau &= 1_M \circ \tau \\
&= (i_A \circ p_A + i_B \circ p_B) \circ \tau \\
&= i_A \circ f + i_B \circ g \\
&= \xi.
\end{aligned}
$$

Hence, (1) is equivalent to (3).

We still have to prove the claims about kernels and cokernels, and we show that the kernel of the projection p_B is i_A. The proof of the fact that the kernel of p_A is i_B is similar, and the proof about the identification of the cokernels is dual. The composition $p_B \circ i_A$ is trivial. If any other morphism $f : C \to M$ satisfies that $p_B \circ f$ is trivial, then we can factor f via i_A using $p_A \circ f$, because $p_A \circ i_A \circ p_A \circ f = p_A \circ f$ and $p_B \circ i_A \circ p_A \circ f = 0$. This determines $p_A \circ f$ as the unique morphism with $i_A \circ p_A \circ f = f$. The uniqueness is ensured because i_A is a monomorphism. $\square$

This result justifies the following definition:

Definition 7.1.12 Let $\mathcal{A}$ be a preadditive category. An object M, as in Proposition 7.1.11, is the *biproduct of A and B in $\mathcal{A}$*.

7.2 Additive Categories

Definition 7.2.1 A preadditive category $\mathcal{A}$ is an *additive category* if it has binary biproducts.

Lemma 7.2.2 *For any functor $F : \mathcal{A} \to \mathcal{B}$ between additive categories, the following are equivalent:*

- *The functor F is additive.*
- *The functor F preserves finite coproducts.*
- *The functor F preserves finite products.*
- *The functor F preserves binary biproducts.*

Proof If F is additive, then it preserves binary biproducts, because they are determined by the morphisms i_A, i_B and p_A, p_B and their additive relations.

Preservation of finite coproducts or products implies the preservation of binary biproducts.

If F preserves binary biproducts, then we have to show that it preserves the terminal object $0_{\mathcal{A}}$ (if that exists) in order for F to preserve finite products. As the target category $\mathcal{B}$ is preadditive, we have at least one element in $\mathcal{B}(B, F(0_{\mathcal{A}}))$ for all objects B of $\mathcal{B}$. As $0_{\mathcal{A}}$ is terminal in $\mathcal{A}$, the two morphisms $p_1 : 0_{\mathcal{A}} \times 0_{\mathcal{A}} \to 0_{\mathcal{A}}$ agree.   □

Example 7.2.3 The category of R-modules and R-linear maps is additive. However, the examples $\mathbb{Z}[\mathcal{C}]$ discussed in Example 7.1.2 will, in general, just be preadditive.

Exercise 7.2.4 Show that $\mathsf{Fun}(\mathcal{D}, \mathcal{A})$ is an additive category if $\mathcal{D}$ is small and if $\mathcal{A}$ is additive.

7.3 Abelian Categories

Definition 7.3.1 A preadditive category is an *abelian category* if it satisfies the following:

- There exists a zero object in $\mathcal{A}$.
- The category $\mathcal{A}$ has finite biproducts.
- Every morphisms $f \in \mathcal{A}(A, B)$ has a cokernel and a kernel.
- Every monomorphism is a kernel, and every epimorphism is a cokernel.

Remark 7.3.2 Note that the existence of kernels and cokernels and the existence of binary biproducts imply that $\mathcal{A}$ has all finite limits and colimits, because we can write an equalizer of two morphisms $f, g \in \mathcal{A}(A, B)$ as the kernel of $f - g$, and dually, the coequalizer of f and g is the cokernel of $f - g$. With 3.2.3, this ensures the existence of all finite limits and colimits.

One actually does not have to assume that $\mathcal{A}$ is preadditive. This follows from the remaining axioms. For a proof of this fact, see, for instance, [**Bo94-2**, 1.6.4].

Proposition 7.3.3 *Let $\mathcal{A}$ be an abelian category. A morphism $f \in \mathcal{A}(A, B)$ is an isomorphism if and only if it is an epi- and a monomorphism.*

Proof Let f be a monomorphism and an epimorphism. Then, we can write f as the kernel of a morphism $g : B \to C$. In particular, $g \circ f = 0$, but f is an epimorphism and $0 \circ f = 0$, and hence, g is the zero map. But every kernel of a zero map is an isomorphism (see Exercise 7.1.9).   □

Exercise 7.3.4 Let $f \in \mathcal{A}(A, B)$, where $\mathcal{A}$ is an abelian category. Prove that the following are equivalent:

- The morphism f is a monomorphism.
- The kernel of f is the zero object.
- For all objects C of $\mathcal{A}$ and for all morphisms $h \colon C \to A$, the triviality of the composition $f \circ h$ implies that h is the zero map.

Theorem 7.3.5 *Let $\mathcal{A}$ be an abelian category and let f be a morphism. Then, we can factor f as $f = i \circ p$, where p is an epimorphism and i is a monomorphism. Here, i is the kernel of the cokernel of f and p is the cokernel of the kernel of f.*

Proof   The proof is a diagram chase with morphisms.

For $f \colon A \to B$, we consider its kernel $j \colon \ker(f) \to A$ and the cokernel of j, $p \colon B \to \operatorname{coker}(j)$. As $f \circ j = 0$, we get a factorization of f through p, as $f = i \circ p$. We have to show that i is a monomorphism.

In Exercise 7.3.4, you proved that it suffices to show that for any morphism $h \colon C \to \operatorname{coker}(j)$ with $i \circ h = 0$, we obtain that $h = 0$.

If $i \circ h = 0$, then i factors through the cokernel of h, say $i = \xi \circ \pi$ with $\pi \colon \operatorname{coker}(j) \to \operatorname{coker}(h)$.

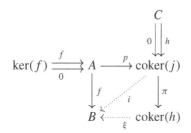

The composition $\pi \circ p$ is an epimorphism; hence, we can write it as a cokernel, so there is a $g \colon Z \to A$ with $\pi \circ p = \operatorname{coker}(g)$. The composition $f \circ g$ is equal to

$$i \circ p \circ g = \xi \circ \pi \circ p \circ g,$$

and this is trivial by the property of the cokernel. Therefore, we can factor g through the kernel of f and can express it as $g = j \circ \zeta$.

For a last factorization, we see that $p \circ g = p \circ j \circ \zeta$, but $p \circ j = 0$; hence, we can factor p through the cokernel of g as $s \circ (\pi \circ p)$. But p is an epimorphism, so this implies that $s \circ \pi = 1_{\operatorname{coker}(j)}$, and hence, π is a monomorphism.

As π is the cokernel of h by assumption, we have $\pi \circ h = 0$, but then, h was trivial to begin with. Therefore, i is a monomorphism.

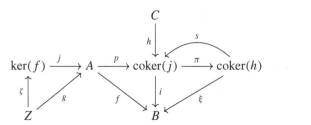

In the category of abelian groups, we can take p to be the surjection onto the image of f and i to be the inclusion of the image of f into the target abelian group.

Corollary 7.3.6 *Let f be a monomorphism in an abelian category, then f is the kernel of its cokernel. Dually, if f is an epimorphism, then it is the cokernel of its kernel.*

Proof Theorem 7.3.5, allows us to write f as $i \circ p$. If f is a monomorphism, then so is p, but then, p is an isomorphism, and hence, f is isomorphic to the kernel of its cokernel. Dually, for an epimorphism f, we see that, in the factorization, i has to be an isomorphism. □

For the concept of functor homology, the following fact is crucial.

Proposition 7.3.7 *Let $\mathcal{D}$ be a small category and let $\mathcal{A}$ be abelian. Then, the functor category $\mathsf{Fun}(\mathcal{D}, \mathcal{A})$ is abelian.*

Proof Colimits and limits in $\mathsf{Fun}(\mathcal{D}, \mathcal{A})$ are built pointwise (see Proposition 3.3.1); therefore, $\mathsf{Fun}(\mathcal{D}, \mathcal{A})$ has a zero object, finite biproducts, kernels, and cokernels.

We know from Corollary 3.3.2 that $f \in \mathsf{Fun}(\mathcal{D}, \mathcal{A})(F, G)$ is a monomorphism if and only if for every object D of $\mathcal{D}$, the component $f_D \colon F(D) \to G(D)$ is a monomorphism. For the cokernel of $f, g \colon G \Rightarrow H$, we get that g_D is the cokernel of f_D for all objects D of $\mathcal{D}$, and hence, f_D is the kernel of g_D. Therefore, f is the kernel of g.

Identifying epimorphisms as cokernels is done by a dual argument. □

8

Symmetric Monoidal Categories

Often, categories have an extra structure that is given by a product-like structure. Typical examples are cartesian products of sets or tensor products of vector spaces.

8.1 Monoidal Categories

Definition 8.1.1 A *strict monoidal category* $(\mathcal{C}, \otimes, e)$ is a category $\mathcal{C}$, together with a functor

$$\otimes : \mathcal{C} \times \mathcal{C} \to \mathcal{C}$$

and an object e of $\mathcal{C}$ satisfying:

$$\otimes \circ (\otimes \times \mathrm{Id}) = \otimes \circ (\mathrm{Id} \times \otimes) : \mathcal{C} \times \mathcal{C} \times \mathcal{C} \to \mathcal{C}, \quad \otimes \circ (e, \mathrm{Id}) = \otimes \circ (\mathrm{Id}, e) = \mathrm{Id},$$

where (e, Id) is the functor

$$(e, \mathrm{Id}) : \mathcal{C} \to \mathcal{C} \times \mathcal{C}$$

sending an object C of $\mathcal{C}$ to (e, C) and a morphism $f \in \mathcal{C}(C, C')$ to $(1_e, f)$. The functor (Id, e) is defined similarly.

You can think of $\otimes$ as a strictly associative and unital product. We actually want equality of functors in the preceding definition, and this implies, in particular, that for all objects C_1, C_2, C_3 of $\mathcal{C}$, we get

$$C_1 \otimes (C_2 \otimes C_3) = (C_1 \otimes C_2) \otimes C_3.$$

Also,

$$f \otimes (g \otimes h) = (f \otimes g) \otimes h$$

for all morphisms f, g, h in $\mathcal{C}$.

If we unravel the condition that $\otimes$ is a functor, then for all objects C_1, C_2 of $\mathcal{C}$, there is an object $C_1 \otimes C_2$ of $\mathcal{C}$, and for the identity morphisms and for the composition of morphisms, we obtain

$$1_{C_1} \otimes 1_{C_2} = 1_{C_1 \otimes C_2}, (f' \circ f) \otimes (g' \circ g) = (f' \otimes g') \circ (f \otimes g).$$

In particular, for $f : C_1 \to C_1'$ and $g : C_2 \to C_2'$, we can rewrite $f \otimes g$ as

$$f \otimes g = (f \otimes 1_{C_2'}) \circ (1_{C_1} \otimes g) = (1_{C_1'} \otimes g) \circ (f \otimes 1_{C_2}).$$

Examples 8.1.2

- Let M be a monoid with unit 1_M. Then, we can consider M as a discrete category, with the elements of M as objects. This is a strict monoidal category if we use the multiplication in M as $\otimes$, and use 1_M as the unit.
- Let $\mathcal{D}$ be a small category and let $\mathcal{C} = \mathsf{Fun}(\mathcal{D}, \mathcal{D})$ be the category of endofunctors of $\mathcal{D}$. Then, $\mathcal{C}$ is a strict monoidal category, with the composition of functors as multiplication and with the identity functor as a unit.

Example 8.1.3 If we change perspective and consider the category $\mathcal{C}_M$ with one object and with the monoid M as the set of endomorphisms, and if we assume that there is a strict monoidal structure on $\mathcal{C}_M$, then we can deduce that the monoid is actually commutative. This is the *Eckmann–Hilton argument* [**EH6162**, Theorem 5.4.2]. The proof is easy. We have the monoid multiplication and the strict monoidal structure. Thus, M is a monoid with respect to the multiplication in M and with respect to $\otimes$. Naturality of the monoidal structure gives an interchange law:

$$(m_1 \otimes m_2)(m_3 \otimes m_4) = (m_1 m_3) \otimes (m_2 m_4). \tag{8.1.1}$$

We first see that the unit of the monoid structure of M, 1_M, agrees with the unit of the monoidal structure, e:

$$1_M = 1_M 1_M = (e \otimes 1_M)(1_M \otimes e) = (e 1_M) \otimes (e 1_M) = e \otimes e = e.$$

Inserting the unit in the two middle terms gives

$$m_1 m_4 = (m_1 \otimes 1_M)(1_M \otimes m_4) = m_1 \otimes m_4,$$

and thus, both monoid structures agree.

Placing the unit at the two outer spots yields

$$m_2 m_3 = (1_M \otimes m_2)(m_3 \otimes 1_M) = m_3 \otimes m_2 = m_3 m_2,$$

and thus, we get the commutativity of the monoid structure.

Eckmann and Hilton used this result to show, for instance, that the fundamental group of an H-space is commutative.

Many relevant examples do *not* satisfy the strict assumptions of Definition 8.1.1. In the example of the tensor product of vector spaces, the associativity condition and unit condition don't hold on the nose but are satisfied up to canonical isomorphisms. You can use this example as a blueprint for the definition of a monoidal category, where one considers associativity and unit constraints up to coherent isomorphisms.

Definition 8.1.4 A *monoidal category* is a category $\mathcal{C}$, together with a functor $\otimes \colon \mathcal{C} \times \mathcal{C} \to \mathcal{C}$, an object e of $\mathcal{C}$, and natural isomorphisms α, λ, ρ as follows:

$$\alpha_{C_1,C_2,C_3} \colon C_1 \otimes (C_2 \otimes C_3) \cong (C_1 \otimes C_2) \otimes C_3, \quad \text{for all } C_1, C_2, C_3, \text{ and}$$

$$\lambda_C \colon e \otimes C \cong C, \quad \rho \colon C \otimes e \cong C, \quad \text{for all } C.$$

In addition, we have three coherence conditions:

(1) The natural isomorphism α satisfies the pentagon axiom; that is, the diagram

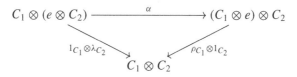

$$(8.1.2)$$

commutes for all objects C_1, C_2, C_3, C_4 of $\mathcal{C}$.

(2) The natural isomorphisms λ and ρ satisfy a triangle axiom; that is, the diagram

$$
\begin{array}{ccc}
C_1 \otimes (e \otimes C_2) & \xrightarrow{\quad\alpha\quad} & (C_1 \otimes e) \otimes C_2 \\
& {}_{1_{C_1} \otimes \lambda_{C_2}} \searrow \quad \swarrow {}_{\rho_{C_1} \otimes 1_{C_2}} & \\
& C_1 \otimes C_2 &
\end{array}
$$

commutes for all objects C_1, C_2 of $\mathcal{C}$.

The naturality conditions mean that the structural isomorphisms hold for all objects of $\mathcal{C}$ and that they are compatible with all morphisms in $\mathcal{C}$.

When monoidal categories were first introduced, longer lists of coherence conditions were demanded to hold (see [**ML63**]). Kelly [**K64**] noted that the coherence conditions of Definition 8.1.4 suffice. In the following exercises, you can prove that yourself (or you can cheat by consulting [**K64**]).

Exercise 8.1.5 Prove that the functors $e \otimes (-) \colon \mathcal{C} \to \mathcal{C}$ and $(-) \otimes e \colon \mathcal{C} \to \mathcal{C}$ are equivalences of categories.

Exercise 8.1.6 Show that

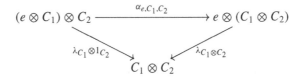

$$(e \otimes C_1) \otimes C_2 \xrightarrow{\alpha_{e,C_1,C_2}} e \otimes (C_1 \otimes C_2)$$

with $\lambda_{C_1} \otimes 1_{C_2}$ and $\lambda_{C_1 \otimes C_2}$ mapping to $C_1 \otimes C_2$

and

$$(C_1 \otimes C_2) \otimes e \xrightarrow{\alpha_{C_1,C_2,e}} C_1 \otimes (C_2 \otimes e)$$

with $\varrho_{C_1 \otimes C_2}$ and $1_{C_1} \otimes \varrho_{C_2}$ mapping to $C_1 \otimes C_2$

commute for all C_1, C_2 in $\mathcal{C}$.

Exercise 8.1.7 Use the preceding facts to prove that for e, the isomorphisms λ and ρ agree:

$$\lambda_e = \rho_e \colon e \otimes e \to e.$$

Remark 8.1.8 A highly nontrivial result (see [**ML98**, VII.2] and [**ML63**]) is that the conditions in Definition 8.1.4 ensure that every other meaningful coherence diagram commutes.

We will see many examples of monoidal categories later, and you have encountered some already. For instance:

Proposition 8.1.9 *If a category $\mathcal{C}$ possesses finite products and if $*$ denotes the terminal object, then $(\mathcal{C}, \times, *)$ is a monoidal category. Dually, if a category $\mathcal{C}$ possesses finite coproducts and if $\varnothing$ denotes the initial object, then $(\mathcal{C}, \sqcup, \varnothing)$ is a monoidal category.*

Proof You showed the unit axioms in these cases in 3.1.26 and 3.1.14. The associativity follows in a similar manner from the universal property of (co)products. □

Examples 8.1.10

- An instance of Proposition 8.1.9 is the category Sets, together with the usual product of sets as the bifunctor $\otimes$ and with a one-point set as the unit object.
- A second monoidal structure on the category Sets where we take the disjoint union of sets as $\otimes$ and the empty set as a unit object.

- Both monoidal structures on the preceding category Sets also give rise to a monoidal structure on the category of topological spaces. For two spaces X and Y, their product is a topological space with the product topology, and every topological space with just one point serves as a unit object. Similarly, the disjoint union of X and Y, $X \sqcup Y$, carries a natural topology, and the empty topological space is a unit for this monoidal structure.

- The category vect_K of K-vector spaces, together with the direct sum of K-vector spaces, is a monoidal category, where the zero vector space is the unit object. We can also take the category vect_K with the tensor product of K-vector spaces and the vector space K as a unit object.

- Let Fin be the skeleton of the category of finite sets with objects $\mathbf{n} = \{1, \ldots, n\}$ for $n \geq 0$ with $\mathbf{0} = \varnothing$. As a monoidal structure, we can take

$$\mathbf{n} \oplus \mathbf{m} := \mathbf{n} + \mathbf{m}$$

corresponding to the disjoint union of sets. Here, $e = \mathbf{0}$ is a unit. On the level of morphisms, we get for $f : \mathbf{n} \to \mathbf{n}'$ and $g : \mathbf{m} \to \mathbf{m}'$ the induced $f \oplus g : \mathbf{n} + \mathbf{m} \to \mathbf{n}' + \mathbf{m}'$, by defining

$$(f \oplus g)(i) = \begin{cases} f(i), & \text{for } 1 \leq i \leq n, \\ n' + g(i - n), & \text{for } n < i \leq n + m. \end{cases}$$

- For a commutative ring with unit k, we denote by $\mathrm{Ch}(k)$ the category of unbounded chain complexes of k-modules. This has a monoidal structure given by the tensor product of chain complexes. For two chain complexes C_* and C'_*, we define the chain complex $C_* \otimes C'_*$ as

$$(C_* \otimes C'_*)_n = \bigoplus_{p+q=n} C_p \otimes_k C'_q.$$

The differential on the tensor product, $d_\otimes$, is defined as

$$d_\otimes(c \otimes c') = d_C(c) \otimes c' + (-1)^p c \otimes d_{C'}(c')$$

if $c \in C_p$ and $c' \in C'_q$.

The unit of this symmetric monoidal structure is the zero sphere, $\mathbb{S}^0(k)$. This is the chain complex, which is trivial in all chain degrees but zero, and $\mathbb{S}^0(k)_0 = k$. All differentials in this chain complex are necessarily trivial.

- Let R be an associative ring with unit. Then, the category of R-bimodules is a monoidal category. The tensor product of two R-bimodules M and N over R, $M \otimes_R N$, is again an R-bimodule, because we only used the right R-module structure on M and the left R-module structure on N to form the tensor product.

Exercise 8.1.11 Consider the category Fin, and prove that the assignment

$$\mathbf{n} \otimes \mathbf{m} := \mathbf{nm}$$

also gives rise to a monoidal structure on Fin.

Every monoidal category $(\mathcal{C}, \otimes, e)$ possesses a preferred functor to the category of sets.

Definition 8.1.12 The functor

$$\mathcal{C}(e, -)\colon \mathcal{C} \to \mathsf{Sets}$$

is called the *underlying set functor*.

Remark 8.1.13 Quite often, the functor $\mathcal{C}(e, -)$ actually deserves its name. For instance, if we consider the category of k-modules for some commutative ring k with unit, then

$$k\text{-mod}(k, -)\colon k\text{-mod} \to \mathsf{Sets}$$

sends an k-module M to $k\text{-mod}(k, M) \cong M$, but on the latter, we only consider the set M and forget its module structure.

However, if we take the monoidal category of all small groupoids, with the product as the monoidal structure, then a morphism from the groupoid with one object and the identity morphism of that object to a groupoid G just picks out an object g of G, together with its identity morphism, so the underlying set functor applied to G gives the objects of G with trivial morphisms. Kelly [**K82**, p. 23] discusses a nice variety of examples.

Monoids are sets together with an associative and unital composition. This concept is available in every monoidal category.

Definition 8.1.14 Let $(\mathcal{C}, \otimes, e)$ be a monoidal category. A *monoid in* $\mathcal{C}$ is an object M of $\mathcal{C}$, together with a morphism $\mu \in \mathcal{C}(M \otimes M, M)$ and a morphism $\eta \in \mathcal{C}(e, M)$, which satisfy

$$\mu \circ (\mu \otimes 1_M) \circ \alpha = \mu \circ (1_M \otimes \mu) :$$

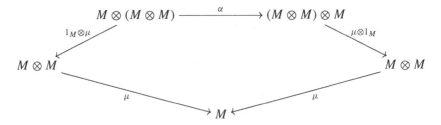

and

$$\mu \circ (\eta \otimes 1_M) = \lambda, \quad \mu \circ (1_M \otimes \eta) = \rho :$$

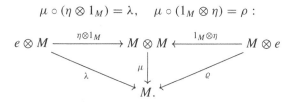

Examples 8.1.15

- In every monoidal category $(\mathcal{C}, \otimes, e)$, the unit e is a monoid.
- Monoids are monoids in the monoidal category $(\text{Sets}, \times, \{*\})$.
- Monoids in the category of endofunctors are nothing but monads, and we just spent a whole chapter, Chapter 6, discussing them.
- Let k be a commutative ring with unit. Monoids in the category of k-modules and k-linear maps are k-algebras.
- A differential graded k-algebra A_* is a monoid in $(\text{Ch}(k), \otimes, \mathbb{S}^0(k))$. Hence, the multiplication morphism μ is a chain map and has components

$$\mu_n : \bigoplus_{p+q=n} A_p \otimes_k A_q \to A_n,$$

and we have a unit map $\eta : \mathbb{S}^0(k) \to A_*$; thus, η is trivial in all chain degrees but zero. The fact that the multiplication is a morphism of chain complexes means that the differential on A_* satisfies the following *Leibniz rule*:

$$d_\otimes(a \otimes a') = d_{A_*}(a) \otimes a' + (-1)^p a \otimes d_{A_*}(a'), \quad a \in A_p, a' \in A_q.$$

In particular, A_0 is a k-algebra.

Fix a prime p. As a concrete example, one can consider the chain algebra A_* over the integers, which is generated by a generator in degree one, e, with $de = p$ and the relation $e^4 = 0$. Hence, e is not a cycle. As we do not require any graded commutativity, e^2 is not zero, and we obtain that

$$d(e^2) = pe - pe = 0;$$

hence, e^2 is a cycle and generates a nontrivial homology class. The homology of A_* is an exterior algebra on a generator in degree two over $\mathbb{F}_p$. This example features prominently in Dugger and Shipley's work on topologically equivalent differential graded algebras that are not quasi-isomorphic [**DS07**].

Definition 8.1.16 Let $(M_1.\mu_1, \eta_1)$, (M_2, μ_2, η_2) be two monoids in a monoidal category $\mathcal{C}$. A *morphism of monoids* from M_1 to M_2 is an $f \in \mathcal{C}(M_1, M_2)$, such that

$$\mu_2 \circ (f \otimes f) = f \circ \mu_1$$

and $f \circ \eta_1 = \eta_2$.

Monoids in $\mathcal{C}$, together with morphisms of monoids, form the category of monoids in $\mathcal{C}$.

Definition 8.1.17 Dually, a *comonoid* C in a monoidal category $(\mathcal{C}, \otimes, e)$ is an object C of $\mathcal{C}$ with a *comultiplication map* Δ, $\Delta \in \mathcal{C}(C, C \otimes C)$, which is coassociative and counital; that is, the comultiplication renders the diagram

commutative, and there is a counit morphism $\varepsilon_C \colon C \to e$ with the property that

$$(\varepsilon_C \otimes 1_C) \circ \Delta = \lambda^{-1} \text{ and } (1_C \otimes \varepsilon_C) \circ \Delta = \varrho^{-1}.$$

An alternative way of expressing the axioms is that a comonoid is a monoid in the opposite category of $\mathcal{C}$. The unit of every monoidal category $(C, \otimes, e)$ is a comonoid, but there are more interesting examples.

Example 8.1.18 A comonoid in the category of chain complexes over a commutative ring k, $\mathrm{Ch}(k)$, is a differential graded coalgebra C_*. The counit is a chain map $\varepsilon_{C_*} \colon C_* \to \mathbb{S}^0(k)$; in particular, ε_{C_*} is trivial in all degrees but zero. The comultiplication has components

$$\Delta_n \colon C_n \to \bigoplus_{p+q=n} C_p \otimes_k C_q.$$

The degree-zero part of C_*, C_0 is a coalgebra over k. Let c be an element of C_n. As for ordinary coalgebras, the counit condition forces the diagonal to be of the form

$$\Delta(c) = c \otimes 1_k + 1_k \otimes c + \sum_{p+q=n} c_p^{(1)} \otimes c_q^{(2)},$$

with $c_p^{(1)}$ in degree p and $c_q^{(2)}$ of degree q. As Δ is a morphism of chain complexes, it has to satisfy the co-Leibniz rule:

$$\Delta(dc) = \sum dc_p^{(1)} \otimes c_q^{(2)} + (-1)^p c_p^{(1)} \otimes dc_q^{(2)}.$$

Let X be a topological space and let α be a generator of the singular chains of degree n, that is, a continuous map $\alpha \colon \Delta^n \to X$. The *Alexander–Whitney map* sends α to

$$\mathrm{AW}(\alpha) = \sum_{p+q=n} v^p(\alpha) \otimes h^q(\alpha).$$

Here, $v^p(\alpha) = \partial_{p+1} \circ \cdots \circ \partial_n(\alpha)$ and $h^q(\alpha) = \partial_0 \circ \cdots \circ \partial_0(\alpha)$. This induces a coalgebra structure on the singular chain complex of a topological space X.

If we consider singular chains with coefficients in a field or if we happen to have flatness on corresponding level of homology groups, then this coproduct on chain level passes to a coproduct on the level of homology, and then, we get a comonoid structure in the category of graded abelian groups.

Exercise 8.1.19 Formulate and prove the dual of the Eckmann–Hilton argument for comonoids.

Definition 8.1.20 Let $(C_1, \Delta_1, \varepsilon_1)$ and $(C_2, \Delta_2, \varepsilon_2)$ be comonoids in a monoidal category $(\mathcal{C}, \otimes, e)$, then an $f \in \mathcal{C}(C_1, C_2)$ is a *morphism of comonoids*, if

$$(f \otimes f) \circ \Delta_1 = \Delta_2 \circ f \text{ and } \varepsilon_2 \circ f = \varepsilon_1.$$

Comonoids in $\mathcal{C}$, together with their morphisms, form a category.

Example 8.1.21 Let G be a group and let k be a commutative unital ring. The *group algebra*, $k[G]$, has as elements finite k-linear combinations of the group elements. Additively, it is the free k-module generated by the set G, and hence, an element of $k[G]$ is of the form $\sum_{i=1}^{n} \lambda_i g_i$ with $\lambda_i \in k$ and $g_i \in G$.

We define a multiplication on $k[G]$ by setting $\mu(g_1 \otimes_R g_2) = g_1 g_2$, where the latter term means the multiplication of g_1 and g_2 in the group G. We can extend μ bilinearly to all of $k[G]$. The neutral element $e \in G$ defines the unit $\eta: k \to k[G]$ by $1_k \mapsto 1_k e = e \in k[G]$.

For instance, if we denote the cyclic group of order n, C_n, multiplicatively as

$$C_n = \{1 = t^0 = t^n, t, \ldots, t^{n-1}\},$$

then it is easy to see that $\mathbb{Z}[C_n]$ has nontrivial zero divisors. For instance, both elements $1 - t$ and $1 + t + \cdots + t^{n-1}$ are nontrivial in $\mathbb{Z}[C_n]$ but

$$(1 - t)(1 + t + \cdots + t^{n-1}) = 1 + t + \cdots + t^{n-1} - (t + \cdots + t^{n-1} + t^n) = 0.$$

For general G, we can define a coalgebra structure on $k[G]$ by defining $\Delta(g)$ as $g \otimes g$ and by extending this definition k-linearly to all of $k[G]$.

8.2 Symmetric Monoidal Categories

In some of the earlier examples, we actually had more structure present than we made explicit. For instance, in the example of the category Fin, we have a natural isomorphism from $\mathbf{n} + \mathbf{m} = \mathbf{n} \oplus \mathbf{m}$ to $\mathbf{m} + \mathbf{n} = \mathbf{m} \oplus \mathbf{n}$. In this case, the isomorphism is given by the shuffle permutation $\chi : \mathbf{n} + \mathbf{m} \to \mathbf{m} + \mathbf{n}$,

$$\chi(i) = \begin{cases} m + i, & \text{if } 1 \leq i \leq n, \\ i - n, & \text{if } n < i \leq n + m. \end{cases}$$

Definition 8.2.1 A *symmetric monoidal category* consists of a monoidal category $(\mathcal{C}, \otimes, e; \alpha, \lambda, \rho)$, together with a natural isomorphism τ in $\mathcal{C}$ with

$$\tau_{C_1,C_2} : C_1 \otimes C_2 \cong C_2 \otimes C_1$$

for all objects C_1, C_2 of $\mathcal{C}$, such that τ satisfies the following conditions:

(1) For all objects C_1, C_2 of $\mathcal{C}$, $\tau_{C_2,C_1} \circ \tau_{C_1,C_2} = 1_{C_1 \otimes C_2}$.
(2) For all objects C of $\mathcal{C}$, $\rho_C = \lambda_C \circ \tau_{C,e} : C \otimes e \cong C$.
(3) The natural isomorphism τ is compatible with α in the sense that for all objects C_1, C_2, C_3 of $\mathcal{C}$, the following hexagon diagram commutes:

All examples in 8.1.10 except for the last one are actually symmetric monoidal categories. In the example of unbounded chain complexes of k-modules, the isomorphism τ is given by

$$\tau_{C_*,C'_*} : C_* \otimes_k C'_* \to C'_* \otimes_k C_*, \quad \tau_{C_*,C'_*}(c \otimes c') = (-1)^{pq} c' \otimes c$$

for $c \in C_p$ and $c' \in C'_q$.

Remark 8.2.2 The definition of a (symmetric) monoidal category above, given in terms of a monoidal product that satisfies certain conditions such as associativity in a coherent way, has the advantage of being concrete, but for a generalization to higher category theory, this description is not feasible. We will describe a different approach later in Section 11.8 and sketch its generalization to ∞ categories.

Example 8.2.3 The examples from Proposition 8.1.9 actually carry symmetric monoidal structures coming from the product or coproduct. The natural symmetry isomorphisms $\tau_{C_1,C_2} : C_1 \times C_2 \to C_2 \times C_1$ and $\tau_{C_1,C_2} : C_1 \sqcup C_2 \to C_2 \sqcup C_1$ stem from the symmetric group Σ_2.

Exercise 8.2.4 Show that on the category of unbounded chain complexes, the morphism $\tau_{C_*,C_*'} : C_* \otimes C_*' \to C_*' \otimes C_*$ with $\tau(c \otimes c') = (-1)^{pq} c' \otimes c$ for $c \in C_p$, $c' \in C_q'$, gives rise to a symmetric monoidal structure. Why does $\tau(c \otimes c') = c' \otimes c$ not work?

There is a strict version of a symmetric monoidal category.

Definition 8.2.5 A strict monoidal category $(\mathcal{C}, \otimes, e)$ is a *permutative category* if there is a natural isomorphism τ with

$$\tau_{C_1,C_2} : C_1 \otimes C_2 \cong C_2 \otimes C_1,$$

with

$$\tau_{C_1,C_2} \circ \tau_{C_2,C_1} = 1_{C_2 \otimes C_1}$$

for all objects C_1, C_2 of $\mathcal{C}$, such that τ satisfies

$$\tau_{C_1,C_2 \otimes C_3} = (1_{C_2} \otimes \tau_{C_1,C_3}) \circ (\tau_{C_1,C_2} \otimes 1_{C_3})$$

$$
\begin{array}{ccc}
C_1 \otimes C_2 \otimes C_3 & \xrightarrow{\ \tau_{C_1,C_2 \otimes C_3}\ } & C_2 \otimes C_3 \otimes C_1 \\
& \searrow_{\tau_{C_1,C_2} \otimes 1_{C_3}} \qquad \nearrow_{1_{C_2} \otimes \tau_{C_1,C_3}} & \\
& C_2 \otimes C_1 \otimes C_3 &
\end{array}
$$

and $\tau_{C,e} = 1_C$ for all objects C of $\mathcal{C}$.

The skeleton of the category of finite sets is actually a permutative category, and so are many other small categories based on the category of finite sets, such as the category of finite sets and injections, $\mathcal{I}$, and the category of finite sets and bijections, Σ.

In a symmetric monoidal category, we can talk about commutative monoids and cocommutative comonoids.

Definition 8.2.6 Let $\mathcal{C}$ be a symmetric monoidal category.

- A *commutative monoid in* $\mathcal{C}$ is a monoid M in $\mathcal{C}$, such that

$$\mu \circ \tau_{M,M} = \mu.$$

Commutative monoids in $\mathcal{C}$ form a full subcategory of the category of monoids in $\mathcal{C}$.

- A *cocommutative comonoid in* $\mathcal{C}$ is a comonoid C in $\mathcal{C}$, such that $\tau_{C,C} \circ \Delta = \Delta$. Cocommutative comonoids in $\mathcal{C}$ form a full subcategory of the category of comonoids in $\mathcal{C}$.

Examples 8.2.7

- Of course, commutative monoids are commutative monoids in (Sets, ×, {*}).

- Commutative k-algebras are commutative monoids in the category of k-modules and k-linear maps; for instance, the polynomial algebra $k[X]$ is a commutative k-algebra.

- Commutative monoids in $(\mathrm{Ch}(k), \otimes, \mathbb{S}^0(k))$ are commutative differential graded algebras. These are differential graded algebras A_*, such that

$$\mu(a' \otimes a) = (-1)^{pq} \mu(a \otimes a'), \text{ for } a \in A_p, a' \in A_q.$$

For instance, for $k = \mathbb{Z}$, we can consider the tensor product $A_* = \mathbb{Z}[X_2] \otimes_{\mathbb{Z}} \Lambda_{\mathbb{Z}}(X_1)$ of a polynomial algebra on a generator X_2 in degree two and an exterior algebra on a generator in degree one with $d(X_2) = X_1$. As the differential has to obey the Leibniz rule, this determines d on all of A_*. Note that, for instance,

$$d(X_2 X_1) = d(X_2)X_1 + X_2 d(X_1) = X_1^2 + 0 = 0;$$

hence, $X_2 X_1$ is a cycle. The only generator in degree four is X_2^2, and

$$d(X_2^2) = X_1 X_2 + X_2 X_1 = 2X_1 X_2,$$

and hence, we get $H_3(A_*) \cong \mathbb{Z}/2\mathbb{Z}$.

- Dually, cocommutative comonoids in the category of k-modules (here k is again a commutative ring with unit) are cocommutative coalgebras.

  We saw the example of a group algebra earlier. As $\tau \circ \Delta(g) = \Delta(g)$, we obtain a cocommutative coalgebra structure on $k[G]$.

The last example is actually richer than it seems, because $k[G]$ is not just a k-algebra and a k-coalgebra, but both structures are compatible in the following sense.

Definition 8.2.8 Let $\mathcal{C}$ be a symmetric monoidal category and let H be an object of $\mathcal{C}$. Then, H is a *bimonoid in $\mathcal{C}$* if it has a monoid structure (H, μ, η) and a comonoid structure (H, Δ, ε), such that Δ and ε are morphisms of monoids.

Remark 8.2.9 Let us describe what the last conditions amount to. The comonoid structure map Δ is a morphism from H to $H \otimes H$. We can endow $H \otimes H$ with a monoid structure by using the twist of the symmetric monoidal structure τ and by using the following composite as a multiplication:

$$H \otimes H \otimes H \otimes H \xrightarrow{1_H \otimes \tau_{H,H} \otimes 1_H} H \otimes H \otimes H \otimes H \xrightarrow{\mu \otimes \mu} H \otimes H.$$

The condition on Δ to being a morphism of monoids then requires the following diagram to commute:

$$
\begin{array}{ccc}
H \otimes H & \xrightarrow{\quad\quad\quad\quad\mu\quad\quad\quad\quad} & H \\
{\scriptstyle \Delta\otimes\Delta}\downarrow & & \downarrow{\scriptstyle \Delta} \\
H \otimes H \otimes H \otimes H \xrightarrow{1_H\otimes\tau_{H,H}\otimes 1_H} H \otimes H \otimes H \otimes H \xrightarrow{\mu\otimes\mu} & & H \otimes H
\end{array}
$$

and a compatibility condition with respect to the unit of H; that is,

$$
\begin{array}{ccc}
e & \xrightarrow{\quad\eta_H\quad} & H \\
{\scriptstyle \cong}\downarrow & & \downarrow{\scriptstyle \Delta} \\
e \otimes e & \xrightarrow{\eta_H\otimes\eta_H} & H \otimes H
\end{array}
$$

commutes.

The compatibility constraints for the counit ε are $\varepsilon \circ \eta = 1_H$ and the commutativity of

$$
\begin{array}{ccc}
H \otimes H & \xrightarrow{\mu} & H \\
{\scriptstyle \varepsilon\otimes\varepsilon}\downarrow & & \downarrow{\scriptstyle \varepsilon} \\
e \otimes e & \xrightarrow{\cong} & e.
\end{array}
$$

These conditions are equivalent to requiring that μ and η are morphisms of comonoids.

8.3 Monoidal Functors

If we consider functors between (symmetric) monoidal categories, then it is natural to ask that they should respect the monoidal structures. There are several options as to what extent the structures should be preserved.

Definition 8.3.1 Let $(\mathcal{C}, \otimes, e_{\mathcal{C}})$ and $(\mathcal{D}, \Box, e_{\mathcal{D}})$ be two monoidal categories.

- A functor $F : (\mathcal{C}, \otimes, e_{\mathcal{C}}) \to (\mathcal{D}, \Box, e_{\mathcal{D}})$ is a *lax monoidal functor* if for each pair of objects C_1, C_2 of $\mathcal{C}$, there is a morphism

$$\varphi_{C_1,C_2} : F(C_1)\Box F(C_2) \to F(C_1 \otimes C_2)$$

in $\mathcal{D}$, which is natural in C_1 and C_2, and there is a morphism

$$\eta : e_{\mathcal{D}} \to F(e_{\mathcal{C}})$$

in $\mathcal{D}$.

These morphisms fit into commutative diagrams for all objects C_1, C_2, C_3 and C:

$$F(C_1)\Box(F(C_2)\Box F(C_3)) \xrightarrow{\alpha_\mathcal{D}} (F(C_1)\Box F(C_2))\Box F(C_3)$$

$$1_{F(C_1)}\Box\varphi_{C_2,C_3} \downarrow \qquad\qquad\qquad \downarrow \varphi_{C_1,C_2}\Box 1_{F(C_3)}$$

$$F(C_1)\Box(F(C_2\otimes C_3)) \qquad (F(C_1\otimes C_2))\Box F(C_3)$$

$$\varphi_{C_1,C_2\otimes C_3} \downarrow \qquad\qquad\qquad \downarrow \varphi_{C_1\otimes C_2,C_3}$$

$$F(C_1\otimes(C_2\otimes C_3)) \xrightarrow{F(\alpha_C)} F((C_1\otimes C_2)\otimes C_3),$$

$$\begin{array}{ccc}
F(C)\Box e' \xrightarrow{\rho^\mathcal{D}_{F(C)}} F(C) & \text{and} & e'\Box F(C) \xrightarrow{\lambda^\mathcal{D}_{F(C)}} F(C) \\
1_{F(C)}\Box\eta \downarrow \quad \uparrow F(\rho^\mathcal{C}_C) & & \eta\Box 1_{F(C)} \downarrow \quad \uparrow F(\lambda^\mathcal{C}_C) \\
F(C)\Box F(e) \xrightarrow{\varphi_{C,e}} F(C\otimes e), & & F(e)\Box F(C) \xrightarrow{\varphi_{e,C}} F(e\otimes C).
\end{array}$$

- A lax monoidal functor $F\colon \mathcal{C} \to \mathcal{D}$ is *strong monoidal* if the structure morphisms φ and η are natural isomorphisms.
- A lax monoidal functor F is *strictly monoidal* if φ and η are identities.

If the categories involved are symmetric monoidal, then one can impose an additional symmetry condition.

Definition 8.3.2 Let $\mathcal{C}$ and $\mathcal{D}$ be symmetric monoidal categories. A functor $F\colon \mathcal{C} \to \mathcal{D}$ is *lax symmetric monoidal* if it is a lax monoidal functor, such that the diagrams

$$\begin{array}{ccc}
F(C_1)\Box F(C_2) & \xrightarrow{\tau^\mathcal{D}_{F(C_1),F(C_2)}} & F(C_2)\Box F(C_1) \\
\varphi_{C_1,C_2} \downarrow & & \downarrow \varphi_{C_2,C_1} \\
F(C_1\otimes C_2) & \xrightarrow{F(\tau^\mathcal{C}_{C_1,C_2})} & F(C_2\otimes C_1)
\end{array}$$

commute for all objects C_1, C_2 of $\mathcal{C}$. If these diagrams commute and F is strong (strictly) monoidal, then F is called *strong (strictly) symmetric monoidal*.

Examples 8.3.3
- The functor

$$\mathbb{Z}\{-\}\colon \mathsf{Sets} \to \mathsf{Ab}$$

that sends a set S to the free abelian group generated by S, $\mathbb{Z}\{S\}$ is a strong symmetric monoidal functor. There is a natural isomorphism

$$\mathbb{Z}\{S\} \otimes \mathbb{Z}\{T\} \to \mathbb{Z}\{S \times T\}$$

for all sets S and T, which sends a generator $s \otimes t$ to the generator $(s, t) \in \mathbb{Z}\{S \times T\}$. This natural isomorphism is compatible with the associativity and commutativity isomorphisms. In addition, we have $\mathbb{Z}\{*\} \cong \mathbb{Z}$.

- The forgetful functor

$$U : (\mathsf{Ab}, \otimes, \mathbb{Z}) \to (\mathsf{Sets}, \times, \{*\})$$

  is *not* strong monoidal.
- If we take the symmetric monoidal structure on Ab given by the product of groups, then the forgetful functor to sets is strong symmetric monoidal because it has a left adjoint, so it has to preserve products.

Often, one wants to replace a symmetric monoidal category by a permutative one or a monoidal category by a strict monoidal one. This is possible with the following strictification result:

Proposition 8.3.4

(1) For every monoidal category $(\mathcal{C}, \otimes, e)$, there is a strict monoidal category $(\mathsf{Str}(\mathcal{C}), \boxtimes, [])$ and strong monoidal functors $F : \mathsf{Str}(\mathcal{C}) \to \mathcal{C}$, $G : \mathcal{C} \to \mathsf{Str}(\mathcal{C})$, such that $F \circ G = \mathrm{Id}$ and $G \circ F \cong \mathrm{Id}$.

(2) If $\mathcal{C}$ is symmetric monoidal, then $\mathsf{Str}(\mathcal{C})$ is permutative and the functors involved are strong symmetric monoidal.

Proof

(1) The objects of $\mathsf{Str}(\mathcal{C})$ are words of finite length $(C_1, \ldots, C_n)$ with $n \geq 0$, such that the C_is are objects of $\mathcal{C}$. For $n = 0$, we have the empty word that we denote by $[]$. We define the morphisms in $\mathsf{Str}(\mathcal{C})$ with the help of the functor F. On objects, F assigns to the object $(C_1, \ldots, C_n)$ in $\mathsf{Str}(\mathcal{C})$ the object $(\cdots (C_1 \otimes C_2) \otimes \cdots) \otimes C_n$ of $\mathcal{C}$, and it sends the empty word $[]$ to the unit e of $\mathcal{C}$.

The set of morphisms in $\mathsf{Str}(\mathcal{C})$ from a word $(C_1, \ldots, C_n)$ to $(C'_1, \ldots, C'_m)$ is defined as

$$\mathsf{Str}(\mathcal{C})((C_1, \ldots, C_n), (C'_1, \ldots, C'_m)) := \mathcal{C}(F((C_1, \ldots, C_n)),$$

$$\times F((C'_1, \ldots, C'_m))). \quad (8.3.1)$$

By construction, $\mathsf{Str}(\mathcal{C})$ is a category.

In the following, we abbreviate $(C_1, \ldots, C_n)$ as $\underline{C}$. We define the monoidal structure on $\mathsf{Str}(\mathcal{C})$ by concatenation of words; that is, on objects, we define

$$\underline{C} \boxtimes \underline{C'} := (\underline{C}, \underline{C'}) = (C_1, \ldots, C_n, C'_1, \ldots, C'_m).$$

Concatenation with the empty word is a strict unit. For two morphisms $f\colon \underline{C}_1 \to \underline{C}_2$ and $g\colon \underline{C}'_1 \to \underline{C}'_2$, we define

$$f \boxtimes g \colon \underline{C}_1 \boxtimes \underline{C}'_1 \to \underline{C}_2 \boxtimes \underline{C}'_2$$

as the composite

$$
\begin{array}{ccccc}
F(\underline{C}_1 \boxtimes \underline{C}'_1) & = & F((\underline{C}_1, \underline{C}'_1)) & \xrightarrow{\ \alpha\ } & F(\underline{C}_1) \otimes F(\underline{C}'_1) \\
{\scriptstyle f \boxtimes g}\Big\downarrow & & & & \Big\downarrow{\scriptstyle f \otimes g} \\
F(\underline{C}_2 \boxtimes \underline{C}'_2) & = & F((\underline{C}_2, \underline{C}'_2)) & \xrightarrow{\ \alpha\ } & F(\underline{C}_2) \otimes F(\underline{C}'_2).
\end{array}
$$

Here, α is the unique associativity isomorphism in $\mathcal{C}$ comparing the two ways of setting parentheses. This definition is compatible with the composition of morphisms, and the coherence theorem [**ML98**, VII.2] ensures that for three morphisms f, g, h in $\mathrm{Str}(\mathcal{C})$, we have that

$$f \boxtimes (g \boxtimes h) = (f \boxtimes g) \boxtimes h.$$

With this structure, $\mathrm{Str}(\mathcal{C})$ is a strict monoidal category and $F\colon \mathrm{Str}(\mathcal{C})$ is a strong monoidal functor with $F[] = e$.

We define $G\colon \mathcal{C} \to \mathrm{Str}(\mathcal{C})$ on objects by sending an object C of $\mathcal{C}$ to the word of length one on C, (C) and by sending an $f \in \mathcal{C}(C, C')$ to $f \in \mathrm{Str}(\mathcal{C})((C), (C'))$. This, in fact, defines a functor, and we claim that G is strong monoidal. For any two objects C, C' of $\mathcal{C}$, we define the structure map

$$G(C) \boxtimes G(C') = (C) \boxtimes (C') = (C, C') \to G(C \otimes C') = (C \otimes C')$$

as the identity of $C \boxtimes C'$ (which corresponds to the identity of $C \otimes C'$ via (8.3.1)). For the unit, note that

$$\mathrm{Str}(\mathcal{C})([], (e)) = \mathcal{C}(F[], F(e)) = \mathcal{C}(e, e),$$

and we use the morphism $[] \to G(e) = (e)$ in $\mathrm{Str}(\mathcal{C})$, given by the identity on e. Hence, G is, in fact, strong monoidal.

The composition $F \circ G$ is the identity functor. Evaluating $G \circ F$ on an object $(C_1, \ldots, C_n)$ of $\mathrm{Str}(\mathcal{C})$ gives the one-letter word $(((\cdots (C_1 \otimes C_2) \otimes C_3) \otimes \cdots) \otimes C_n)$. The identity morphism on $(((\cdots (C_1 \otimes C_2) \otimes C_3) \otimes \cdots) \otimes C_n)$ is an isomorphism between $(F \circ G)(C_1, \ldots, C_n)$ and $(C_1, \ldots, C_n)$ in $\mathrm{Str}(\mathcal{C})$.

(2) If $\mathcal{C}$ is a symmetric monoidal category, then the symmetry isomorphism

$$\tau_{\underline{C}, \underline{C}'} \colon \underline{C} \boxtimes \underline{C}' \to \underline{C}' \boxtimes \underline{C}$$

in $\mathsf{Str}(\underline{C} \boxtimes \underline{C}', \underline{C}' \boxtimes \underline{C}) = C(F((\underline{C}, \underline{C}')), F((\underline{C}', \underline{C})))$ is defined as the composite

$$\tau_{\underline{C}, \underline{C}'} = \alpha^{-1} \circ \tau_{F(\underline{C}), F(\underline{C}')} \circ \alpha,$$

where α is again a uniquely determined associativity isomorphism. The fact that $\tau_{F(\underline{C}), F(\underline{C}')}$ satisfies compatibility conditions with respect to the unit and with respect to the associativity isomorphisms in C ensures that $\tau_{\underline{C}, \underline{C}'}$ satisfies the requirements for a symmetry isomorphism in a permutative category. Both functors F and G are compatible with these symmetry isomorphisms and are strong symmetric monoidal functors in this case. $\square$

Remark 8.3.5 Fix a permutation $\sigma \in \Sigma_n$. In a permutative category C, the symmetry isomorphism τ gives rise to a natural transformation from the n-fold tensor product $\otimes^n$ to the composite functor on C^n that first uses the left action of Σ_n on C^n (by the permutation of factors in the product) and then applies the functor $\otimes^n : C^n \to C$. We call this natural transformation τ_σ. A product of permutations ξ and σ, $\xi \circ \sigma$, is sent to $\tau_{\xi\sigma}$. Therefore, for every object C of a permutative category, there is a well-defined action of Σ_n on $C^{\otimes n}$. The coherence axioms of a symmetric monoidal category ensure that the same is true for any symmetric monoidal category.

Note that there is a right Σ_n-action on the translation category $\mathcal{E}_{\Sigma_n}$.

Lemma 8.3.6 [**May74**, §4] *For every permutative category C and every $n \geq 0$, there is a Σ_n-equivariant functor*

$$\tau_n : \mathcal{E}_{\Sigma_n} \times C^n \to C$$

that sends an object $(\sigma; C_1, \ldots, C_n)$ to $C_{\sigma^{-1}(1)} \otimes \cdots \otimes C_{\sigma^{-1}(n)}$ and a morphism $(\xi\sigma^{-1} : \sigma \to \xi; f_1, \ldots, f_n)$ with $f_i \in C(C_i, C_i')$, to the composition

$$\tau_{\xi\sigma^{-1}} \circ (f_{\sigma^{-1}(1)} \otimes \cdots \otimes f_{\sigma^{-1}(n)}) : C_{\sigma^{-1}(1)} \otimes \cdots \otimes C_{\sigma^{-1}(n)} \to C'_{\xi^{-1}(1)} \otimes \cdots \otimes C'_{\xi^{-1}(n)}.$$

Proof For proving that τ_n is a functor, it suffices to observe that for permutations $\xi, \sigma \in \Sigma_n$ and morphisms $g_i \in C(C_i', C_i'')$, it is true that

$$(g_{\xi^{-1}(1)} \otimes \cdots \otimes g_{\xi^{-1}(n)}) \circ \tau_{\xi\sigma^{-1}} = \tau_{\xi\sigma^{-1}} \circ (g_{\sigma^{-1}(1)} \otimes \cdots \otimes g_{\sigma^{-1}(n)}).$$

This shows that τ_n respects the composition of morphisms. It respects identity morphisms, and we have that

$$\tau_n(\sigma\xi, C_{\xi^{-1}(1)}, \ldots, C_{\xi^{-1}(n)}) = \tau(\sigma, C_1, \ldots, C_n). \qquad \square$$

Dual to the concept of monoidal functors is the following notion:

Definition 8.3.7 Let $(\mathcal{C}, \otimes, e_\mathcal{C})$ and $(\mathcal{D}, \square, e_\mathcal{D})$ be two monoidal categories.

- A functor $F\colon (\mathcal{C}, \otimes, e_\mathcal{C}) \to (\mathcal{D}, \square, e_\mathcal{D})$ is a *lax comonoidal functor* if for each pair of objects C_1, C_2 of $\mathcal{C}$, there is a morphism

$$\psi_{C_1, C_2}\colon F(C_1 \otimes C_2) \to F(C_1)\square F(C_2)$$

in $\mathcal{D}$, which is natural in C_1 and C_2, and there is a morphism

$$\varepsilon\colon F(e_\mathcal{C}) \to e_\mathcal{D}$$

in $\mathcal{D}$.

These morphisms satisfy coherence conditions dual to the ones in Definition 8.3.1.

- A lax comonoidal functor $F\colon \mathcal{C} \to \mathcal{D}$ is *strong comonoidal* if the structure morphisms ψ and ε are natural isomorphisms.
- A lax comonoidal functor F is *strictly comonoidal* if ψ and ε are identities.
- If $\mathcal{C}$ and $\mathcal{D}$ are symmetric monoidal categories, then a functor $F\colon \mathcal{C} \to \mathcal{D}$ is *lax symmetric comonoidal* if it is a lax monoidal functor, such that the diagram

$$
\begin{array}{ccc}
F(C_1 \otimes C_2) & \xrightarrow{F(\tau^\mathcal{C}_{C_1, C_2})} & F(C_2 \otimes C_1) \\
\psi_{C_1, C_2}\downarrow & & \downarrow\psi_{C_2, C_1} \\
F(C_1)\square F(C_2) & \xrightarrow{\tau^\mathcal{D}_{F(C_1), F(C_2)}} & F(C_2)\square F(C_1)
\end{array}
$$

commutes for all objects C_1, C_2 of $\mathcal{C}$.

If these diagrams commute and if F is strong (strictly) comonoidal, then F is called *strong (strictly) symmetric comonoidal*.

Note that a strong monoidal functor is automatically strong comonoidal and vice versa.

If you want that natural transformations between lax (symmetric) monoidal functor preserve structures, then the following condition is needed:

Definition 8.3.8 Let F and G be two lax (symmetric) monoidal functors with structure maps φ and ψ. A natural transformation $\gamma\colon F \Rightarrow G$ is *monoidal* if the diagrams

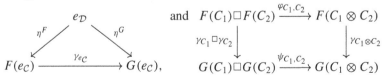

commute for all objects C_1, C_2 of $\mathcal{C}$.

8.4 Closed Symmetric Monoidal Categories

Definition 8.4.1 A symmetric monoidal category $\mathcal{C}$ is called *closed* if for all objects C of $\mathcal{C}$, the functor $(-) \otimes C$ possesses a right adjoint. We denote the right adjoint by $(-)^C$.

Making explicit what the adjunction property means gives us natural bijections

$$\mathcal{C}(A \otimes C, B) \cong \mathcal{C}(A, B^C)$$

for all objects A, B, C in $\mathcal{C}$. In particular, for $A = B^C$, we obtain evaluation morphisms

$$\mathrm{ev}\colon B^C \otimes C \to B.$$

Exercise 8.4.2 Use the defining adjunction to show that for every object C of $\mathcal{C}$, there is a canonical isomorphism from C^e to C if $\mathcal{C}$ is closed symmetric monoidal. For later reference, we call this isomorphism i_C, that is, $i_C\colon C^e \cong C$.

Exercise 8.4.3 What is a suitable notion of a functor between closed symmetric monoidal categories that preserves the structure?

Examples 8.4.4

- An example of a closed symmetric category is, of course, the category Sets, with the cartesian product of sets as the symmetric monoidal structure. Then, for sets S and T, the object S^T consists of the set of functions from T to S. The bijection

$$\mathrm{Sets}(U \times T, S) \cong \mathrm{Sets}(U, S^T)$$

is then the usual *exponential law*

$$S^{U \times T} \cong (S^T)^U,$$

with the explicit bijection sending a $g\colon U \times T \to S$ to the map $u \mapsto (t \mapsto g(u, t))$ for all $u \in U$ and $t \in T$.
- Let k be a commutative ring with unit. Then, the category of (left) k-modules is a closed symmetric monoidal category. We take the tensor product over k as the symmetric monoidal structure. The set of morphisms in this category carries a k-module structure, and the right adjoint to the functor $(-) \otimes M\colon k\text{-mod} \to k\text{-mod}$ is the functor $k\text{-mod}(M, -)$.
- Let k be a commutative ring with unit. Consider the category $\mathrm{Ch}(k)$ of unbounded chain complexes of k-modules. We saw that this category carries a symmetric monoidal structure, with the tensor product of chain complexes

as monoidal product. For two chain complexes C_* and C'_*, we define the chain complex $\text{HOM}(C_*, C'_*)$ as

$$\text{HOM}(C_*, C'_*)_n = \prod_{\ell \in \mathbb{Z}} k\text{-mod}(C_\ell, C'_{\ell+n});$$

that is, the nth chain group consists of $\mathbb{Z}$-indexed families of k-linear maps $f_\ell \colon C_\ell \to C'_{\ell+n}$. The differential, ∂, on $\text{HOM}(C_*, C'_*)$ is determined by $(\partial f)_\ell = d_{C'} \circ f_\ell - (-1)^n f_{\ell-1} \circ d_C$.

An element in $\text{HOM}(C_*, C'_*)_0$ consists of a family of k-linear maps $f_\ell \colon C_\ell \to C'_\ell$ for $\ell \in \mathbb{Z}$. If this family constitutes a zero cycle, that is, if $\partial f = 0$, then this is equivalent to the condition that $d_{C'} \circ f_\ell = f_{\ell-1} \circ d_C$ for all ℓ, and this means that the family $(f_\ell)_{\ell \in \mathbb{Z}}$ is a chain map.

- Pedicchio and Solimini showed [**PS86**, 2.4] that there is a unique closed symmetric monoidal structure on the category Top. You are aware that there are rather scary topological spaces out there, so it might not come as a surprise that this symmetric monoidal structure does not have the properties that you would like to have. For instance, the space of continuous maps carries the so-called pointwise topology and *not* the compact open topology in that structure. We will discuss as alternatives, the k-topology and compactly generated spaces, in Section 8.5 below.

The first example was special, because there, the monoidal structure was given by the categorical product.

Definition 8.4.5 A category $\mathcal{C}$ is *cartesian closed* if it has finite products and if the symmetric monoidal structure $(\mathcal{C}, \times, *)$ is closed.

The category Sets is cartesian closed, as we saw earlier. Let G be a group and let GSets be the category whose objects are sets with a left G-action and whose morphisms are equivariant maps, that is,

$$G\text{Sets}(S, T) = \{f \in \text{Sets}(S, T) \mid f(g.s) = g.f(s), \text{ for all } g \in G, s \in S\}.$$

This category is also cartesian closed. The product of sets induces a product on the category of G-sets for two G-sets S and T; we define the G-set S^T as $\text{Sets}(T, S)$, where the G-action is given by

$$(g.f)(t) = g.f(g^{-1}.t)$$

for $g \in G$, $f \in \text{Sets}(T, S)$ and $t \in T$.

Let $(\mathcal{C}, \otimes, e, \tau)$ be a symmetric monoidal category. An important class of objects in $\mathcal{C}$ comprises those objects C that have an inverse with respect to $\otimes$, that is, for which there is an object C' of $\mathcal{C}$, such that

$$C \otimes C' \cong e.$$

Definition 8.4.6 The *Picard groupoid of* C, Picard(C), is the category whose objects are the invertible objects of C and whose morphisms are isomorphisms between invertible objects.

Note that Picard(C) is again a symmetric monoidal category, with the structure inherited from C.

Example 8.4.7 Let k be a commutative ring with unit, then the category of k-modules, k-mod, is symmetric monoidal. The objects of Picard(k-mod) are invertible modules, that is, k-modules M, such that there is a k-module N with $M \otimes_k N \cong k$. Then, M is finitely generated projective of rank one, and the inverse of M is isomorphic to the dual of M, $N \cong k$-mod(M, k). In fact, the last property is true in any closed symmetric monoidal category [**HoPS97**, Proposition A.2.8].

Remark 8.4.8 Assume that C is closed symmetric monoidal. If the isomorphism classes of objects of Picard(C) form a set, then one can build the *Picard group of* C, Pic(C), which is the group whose elements are the isomorphism classes of objects, and the multiplication is given by

$$[C_1] \otimes [C_2] := [C_1 \otimes C_2].$$

This multiplication is well-defined and has $[e]$ as a neutral element, and the inverse of $[C]$ is $[C]^{-1} = [e^C]$.

8.5 Compactly Generated Spaces

We will give a brief introduction to the compactly generated topology and to the k-ification of spaces. These notions will allow us to define decent mapping spaces and well-behaved products, so that the categories of compactly generated spaces and the category of k-spaces carry closed symmetric monoidal structures. You know, for instance, that a product of two CW complexes won't be a CW complex in general, unless one of the factors is locally compact. Steenrod [**Ste67**] introduced the notion of compactly generated spaces, but he insisted on a full Hausdorff condition. Later, McCord [**McCo69**] introduced a slightly milder version with a weak Hausdorff condition and rather nice behavior. Vogt's approach [**V71**] axiomatizes the process of building convenient categories of topological spaces. Overviews can be found in [**May99**] and in Riehl's excellent summary [**Rie14**]. We mostly follow tom Dieck's book [**tD08**, 7.9] for proofs of useful facts about k-spaces and weak Hausdorff spaces.

We do *not* use the French convention; in our world, compact subspaces are *not* assumed to be Hausdorff.

Definition 8.5.1

- Let $(X, \mathcal{T})$ be a topological space and $A \subset X$ be a subspace. We call A *compactly closed, or k-closed*, if for all compact Hausdorff spaces K and every continuous map $f \colon K \to X$, the preimage $f^{-1}(A)$ is closed in K.
- A subset $O \subset X$ is *k-open* if for all compact Hausdorff spaces K and every continuous map $f \colon K \to X$, the preimage $f^{-1}(O)$ is open in K.
- The space X is a *k-space* if every compactly closed subspace A of X is closed.

We can force a space to become a k-space.

Definition 8.5.2 Let X be a topological space. The *k-ification of X, kX*, has the same underlying set as X, but we define a subset $A \subset X$ to be closed if and only if it is compactly closed.

Remark 8.5.3 Note that any open (closed) subset of X is automatically k-open (k-closed), and hence, the topology consisting of the k-open subsets of X is finer than the topology $\mathcal{T}$ of X. Therefore, the identity map $\iota \colon kX \to X$ is continuous. If $f \colon K \to X$ is a map from a compact Hausdorff space to X, then the map f is continuous when considered as a map to kX. If $g \colon X \to Y$ is a continuous map, then the same map is continous as a map from kX to kY; in fact, the assignment $k(-)$ is a functor from Top to Top, and it is idempotent.

Definition 8.5.4 We denote by k-Top the full subcategory of Top consisting of k-spaces.

The k-ification is a functor $k \colon$ Top $\to k$-Top. We also have an inclusion functor $i \colon k$-Top $\to$ Top.

Lemma 8.5.5 *The inclusion functor is left adjoint to the functor k; that is, k-Top is a reflective subcategory of* Top.

Proof The identity map $\iota \colon kY \to Y$ is continuous, and for every map of k-spaces $f \colon X \to kY$, we define $\mathrm{ad}(f) \colon iX \to Y$ as $\iota \circ f$. If X is a k-space, Y is any topological space, and $g \colon X \to Y$ is continous, then $kg \colon X = kX \to kY$ is continuous. This shows that the assignment $f \mapsto \mathrm{ad}(f)$ is surjective. Composing with ι is injective, and hence, we get a binatural bijection. $\square$

We also obtain that continuity of maps out of k-spaces can be tested with the help of compact Hausdorff spaces.

Lemma 8.5.6 *The following are equivalent:*

- *A space X is a k-space.*
- *A map of sets $g: X \to Y$ is continuous if and only if its precomposition $g \circ f: K \to Y$ is continuous for every continuous map $f: K \to X$ from a compact Hausdorff space into X.*

Proof Let X be a k-space and let $O \subset Y$ be an open subset. It suffices to show that $g^{-1}(O)$ is k-open in X. Take a continuous test map $f: K \to X$ with K compact Hausdorff, such that $g \circ f$ is continuous. Then, $f^{-1}(g^{-1}(O))$ is open, and hence, $g^{-1}(O)$ is k-open.

In order to show that X is a k-space, it suffices to show that the identity map $X \to kX$ is continuous, because then, $X = kX$ as topological spaces. By assumption, this map is continuous if and only if the precomposition with an arbitrary test map $f: K \to X$ is continuous. An open subset of kX is precisely a k-open subset, and thus, the claim follows. $\qquad\square$

One defines a symmetric monoidal product $(-) \times_k (-)$ on k-Top by setting

$$X \times_k Y := k(X \times Y), \qquad (8.5.1)$$

where $X \times Y$ is the product set of X and Y with the product topology.

Similarly, one can define arbitrary products. Let $(X_j)_{j \in J}$ be a family of k-spaces. Then, we define their product as $k(\prod_{j \in J} X_j)$. The following result tells us that this is justified.

Lemma 8.5.7 *The space $k(\prod_{j \in J} X_j)$, together with the projection maps $\varrho_j: k(\prod_{j \in J} X_j) \to k(X_j) = X_j$, is the product of the k-spaces X_j in the category k-Top.*

Proof As ι is left adjoint to k, the set of morphisms k-$\mathsf{Top}(Y, k(\prod_{j \in J} X_j))$ is in binatural bijection with $\mathsf{Top}(iY, \prod_{j \in J} X_j)$. Using the universal property of the product in Top, we get that

$$\mathsf{Top}(iY, \prod_{j \in J} X_j) \cong \prod_{j \in J} \mathsf{Top}(iY, X_j),$$

and using adjunction again, the claim follows. $\qquad\square$

We cite the following result from [**tD08**]:

Theorem 8.5.8 ([**tD08**], 7.9.12]) *The product in Top of a k-space with a locally compact space Hausdorff space is a k-space.*

Definition 8.5.9 A topological space X is *weak Hausdorff* if for all continuous maps $f \colon K \to X$ with K a compact Hausdorff space, the image $f(K)$ is closed in X.

Of course, Hausdorff spaces are weak Hausdorff. A point is compact, and hence, points are closed in every weak Hausdorff space. Thus, the weak Hausdorff condition implies T_1. As for Hausdorff spaces, one can characterize the weak Hausdorff condition via the diagonal $\Delta_X \subset X \times X$.

Lemma 8.5.10 [**tD08**, Proposition 7.9.14] *A k-space X is weak Hausdorff if and only if $\Delta_X \subset X \times_k X$ is closed.*

The weak Hausdorff condition allows for an internal description of compactly closed subsets.

Lemma 8.5.11 *Let X be a weak Hausdorff space. A subset $A \subset X$ is k-closed if and only if for all compact Hausdorff subsets $K \subset X$, the intersection $A \cap K$ is closed in K.*

Proof If $K \subset X$ is a compact Hausdorff subset of X, then the inclusion map $K \hookrightarrow X$ is among the maps that test k-closedness; hence, if A is k-closed, then $A \cap K$ is closed in K.

Let $f \colon L \to X$ be a continuous map from a compact Hausdorff space L and let $A \subset X$ be a subset of X, such that $A \cap K$ is closed in K for all compact Hausdorff $K \subset X$. As X is a weak Hausdorff space, the image $f(L)$ is closed in X, and hence, it is a compact Hausdorff subspace of X, and by assumption, $A \cap f(L)$ is closed. But then, $f^{-1}(A) = f^{-1}(A \cap f(L))$ is closed in L. $\square$

Remark 8.5.12 Note that the weak Hausdorff condition is preserved under k-ification.

Definition 8.5.13 A topological space X is *compactly generated* if it is a weak Hausdorff k-space. We denote the category of compactly generated spaces by cg.

We can now define an internal hom-object in k-Top. Let $\underline{\mathsf{Top}}(X, Y)$ denote the set of continuous maps from X to Y with the compact–open topology. For any continuous map $f \colon k(X \times Y) \to Z$, there is a map of sets $\psi(f) \colon X \to k\underline{\mathsf{Top}}(Y, Z)$.

Lemma 8.5.14 *If X and Y are k-spaces and if $f \colon k(X \times Y) \to Z$ is continuous, then $\psi(f)$ is continuous.*

Proof We first show a reduction to the case where X is Hausdorff and compact. Let $g \colon K \to X$ be an arbitrary continuous map from a compact Hausdorff space K to X. Thanks to Lemma 8.5.6, we have to show that $\psi(f) \circ g$ is continuous. But the composite $K \xrightarrow{g} X \xrightarrow{\psi(f)} k\underline{\mathsf{Top}}(Y, Z)$ is the adjoint of $k(K \times Y) \xrightarrow{g \times 1_Y} k(X \times Y) \xrightarrow{f} Z$. So, if we show the claim for $X = K$, then we are done. Thus, without loss of generality, we can assume that X is compact and Hausdorff. By Theorem 8.5.8 for such X, we obtain $X \times_k Y = X \times Y$ and the adjoint $\psi(f) \colon X \to \underline{\mathsf{Top}}(Y, Z)$ is continuous. As X is now assumed to be Hausdorff and compact, this also implies that $\psi(f) \colon X \to k\underline{\mathsf{Top}}(Y, Z)$ is continuous. $\qquad\square$

The underlying set of $k\underline{\mathsf{Top}}(X, Y)$ is the set of continuous maps from X to Y. Hence, on point-set level, we have evaluation maps $\mathrm{ev}_{X,Y} \colon k\underline{\mathsf{Top}}(X, Y) \times_k X \to Y$ that are given by sending a pair (g, x) to $g(x)$.

Lemma 8.5.15 *For every k-space X, the evaluation map*

$$ev_{X,Y} \colon k\underline{\mathsf{Top}}(X, Y) \times_k X \to Y$$

is continuous.

Proof   Let $f \colon K \to k\underline{\mathsf{Top}}(X, Y) \times_k X$ be a continuous map, where K is a compact Hausdorff space. We have to show that $\mathrm{ev}_{X,Y} \circ f$ is continuous. The test map f has two continuous components, $f = (f_1, f_2)$, where $f_1 \colon K \to k\underline{\mathsf{Top}}(X, Y)$ and $f_2 \colon K \to X$. The map f_1 is continuous if and only if the composite $f_1' := \iota \circ f_1 \colon K \to \underline{\mathsf{Top}}(X, Y)$ is continuous. This map is in turn continuous if and only if its adjoint map $\mathrm{ad}(f_1') \colon K \times X \to Y$ is also continuous, and – by two-fold use of the exponential law for usual mapping spaces – the continuity of the latter is equivalent to the continuity of the adjoint map $\varphi_1 \colon X \to \underline{\mathsf{Top}}(K, Y)$. But as X is a k-space, we can test continuity of φ_1 via test maps from compact Hausdorff spaces by 8.5.6. Let $h \colon L \to X$ be a continuous map and let L be a compact Hausdorff space. Then, $\varphi_1 \circ h \colon L \to \underline{\mathsf{Top}}(K, Y)$ is the adjoint morphism to $\underline{\mathsf{Top}}(h, Y) \circ f_1'$, and this is a composition of continuous maps.

The composite $\mathrm{ev}_{X,Y} \circ f$ is equal to

$$\mathrm{ad}(f_1') \circ (1_K, f_2) \colon K \to K \times X \to Y,$$

and we showed that this map is a composition of continuous maps. $\qquad\square$

We can now show the crucial ingredient for the fact that the space of morphisms $k\underline{\mathsf{Top}}(X, Y)$ is an internal hom-object in k-Top that turns the symmetric monoidal product $(-) \times_k (-)$ on k-Top into a closed structure.

Theorem 8.5.16 *For k-spaces X, Y, and Z, there is a natural homeomorphism*

$$k\underline{\mathsf{Top}}(X, k\underline{\mathsf{Top}}(Y, Z)) \cong k\underline{\mathsf{Top}}(X \times_k Y, Z).$$

Proof   First, we note that the usual exponential bijection gives a bijection

$$\phi \colon \mathsf{Top}(X, k\underline{\mathsf{Top}}(Y, Z)) \to k\underline{\mathsf{Top}}(X \times_k Y, Z) : \psi$$

for all k-spaces X, Y, and Z, because we have seen in Lemma 8.5.14 that for every continuous f, $\psi(f)$ is continuous. For every $g \in k\underline{\mathsf{Top}}(X, k\underline{\mathsf{Top}}(Y, Z))$, we can express $\phi(g)$ as the composition

$$X \times_k Y \xrightarrow{\;g \times_k 1_Y\;} k\underline{\mathsf{Top}}(Y, Z) \times_k Y \xrightarrow{\;\mathrm{ev}_{Y,Z}\;} Z. \qquad (8.5.2)$$

Thus, $\phi(g)$ is a composition of continuous maps and hence is continuous.

As ϕ and ψ are inverse to each other, it remains to show that both maps are continuous.

We first determine what $\phi(\phi)$ is. By (8.5.2), we know that we can rewrite $\phi(\phi)$ as

$$\phi(\phi) = \mathrm{ev}_{X \times_k Y, Z} \circ (\phi \times_k 1_{X \times_k Y}).$$

The diagram

$$
\begin{array}{ccc}
k\underline{\mathsf{Top}}(X, k\underline{\mathsf{Top}}(Y, Z)) \times_k X \times_k Y & \xrightarrow{\;\mathrm{ev}_{X,k\underline{\mathsf{Top}}(Y,Z)} \times_k 1_Y\;} & k\underline{\mathsf{Top}}(Y, Z) \times_k Y \\[4pt]
{\scriptstyle \phi \times_k 1_{X \times_k Y}} \big\downarrow & & \big\downarrow {\scriptstyle \mathrm{ev}_{Y,Z}} \\[4pt]
k\underline{\mathsf{Top}}(X \times_k Y, Z) \times_k X \times_k Y & \xrightarrow{\;\mathrm{ev}_{X \times_k Y, Z}\;} & Z
\end{array}
$$

is commutative, and therefore, $\phi(\phi)$ is equal to

$$\mathrm{ev}_{Y,Z} \circ (\mathrm{ev}_{X,k\underline{\mathsf{Top}}(Y,Z)} \times_k 1_Y).$$

The latter is continuous, and thus, so is $\phi = \psi(\phi(\phi))$.

We know from (8.5.2) that $\phi(\psi)$ is equal to $\mathrm{ev}_{X,k\underline{\mathsf{Top}}(Y,Z)} \circ (\psi \times_k 1_X)$, but checking what this composite does tells us that it sends a pair (f, x) with $f \in k\underline{\mathsf{Top}}(X \times_k Y, Z)$ and an $x \in X$ to $f(x, -) \in k\underline{\mathsf{Top}}(Y, Z)$. This agrees with the composite

$$\psi(\mathrm{ev}_{X \times_k Y, Z}) \colon k\underline{\mathsf{Top}}(X \times_k Y, Z) \times_k X \to k\underline{\mathsf{Top}}(Y, Z).$$

This shows that $\phi(\psi)$ – and hence ψ – is the image of the continuous map $\mathrm{ev}_{X \times_k Y, Z}$ under ψ and thus continuous. $\square$

Remark 8.5.17 Theorem 8.5.16 implies that the category of k-spaces is closed symmetric monoidal. For two spaces X, Y in cg, their k-product $X \times_k Y$ is again in cg because $X \times Y$ is in cg, and k-ification preserves the weak Hausdorff condition. One also gets that $k\underline{\mathsf{Top}}(X, Y)$ is in cg if X and Y are [**tD08**, Proposition 7.9.21] and that cg is a closed symmetric monoidal category.

8.6 Braided Monoidal Categories

In the definition of a symmetric monoidal category (Definition 8.2.1), we required that the natural isomorphism τ with $\tau_{C_1,C_2} : C_1 \otimes C_2 \to C_2 \otimes C_1$ squares to the identity: $\tau_{C_1,C_2} \circ \tau_{C_2,C_1} = 1_{C_2 \otimes C_1}$ for all objects C_1, C_2. But, sometimes, we might consider symmetries that are not of this form. Think of a braid on two strands:

If you concatenate another copy of that braid to it, then this doesn't untwist, but you get a double twist:

Still, twisting is an isomorphism in this situation because you can untwist using the overcrossing, with the right strand being on top

and then you can pull the strands straight to get two parallel strands.

We only give a rather terse account of braided monoidal categories. For more background, see [**ML98**, XI.1], [**K95**], and [**JoSt93**]

Definition 8.6.1 A *braided monoidal category* is a monoidal category $(\mathcal{C}, \otimes, e)$, together with a binatural isomorphism

$$\beta_{C_1,C_2} : C_1 \otimes C_2 \to C_2 \otimes C_1,$$

for all objects C_1, C_2 of $\mathcal{C}$, that satisfies the hexagon axiom; that is, the two hexagon diagrams

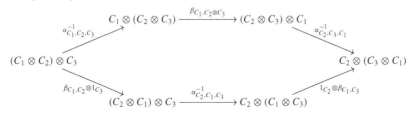

and

$$(C_1 \otimes C_2) \otimes C_3 \xrightarrow{\beta_{C_1 \otimes C_2, C_3}} C_3 \otimes (C_1 \otimes C_2)$$

with α_{C_1,C_2,C_3} from $C_1 \otimes (C_2 \otimes C_3)$ and α_{C_3,C_1,C_2} to $(C_3 \otimes C_1) \otimes C_2$, and $1_{C_1} \otimes \beta_{C_2,C_3}$ to $C_1 \otimes (C_3 \otimes C_2) \xrightarrow{\alpha_{C_1,C_3,C_2}} (C_1 \otimes C_3) \otimes C_2$ and $\beta_{C_1,C_3} \otimes 1_{C_2}$

commute for all objects C_1, C_2, and C_3 of $\mathcal{C}$.

We call β the *braiding* of the braided monoidal structure.

Remark 8.6.2

- Note that if β is a braiding, then β^{-1} can also be used as a braiding.
- If we started with a strict monoidal structure, that is, if the αs are identities, then the hexagon axioms simplify to the conditions that

$$\beta_{C_1, C_2 \otimes C_3} = (1_{C_2} \otimes \beta_{C_1,C_3}) \circ (\beta_{C_1,C_2} \otimes 1_{C_3}) \text{ and } \beta_{C_1 \otimes C_2, C_3}$$

$$= (\beta_{C_1,C_3} \otimes 1_{C_2}) \circ (1_{C_1} \otimes \beta_{C_2,C_3}).$$

The unit e in the monoidal structure of $\mathcal{C}$ is compatible with the braiding in the following sense:

Lemma 8.6.3 *For all objects C of C, we have $\varrho_C = \lambda_C \circ \beta_{C,e}$*

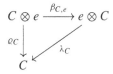

and dually $\lambda_C = \varrho_C \circ \beta_{e,C}$.

Proof   For the proof, we expand the first hexagon diagram, inserting unit isomorphisms. We start with two objects C and C' of C:

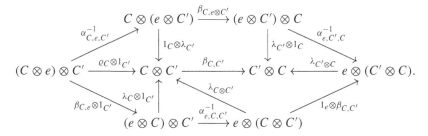

The triangle at the bottom left commutes because the remainder of the diagram is commutative and consists of isomorphisms. Using $C' = e$ and postcomposing with ϱ give the first claim. Using the second hexagonal diagram gives the second claim.   □

Remark 8.6.4 In addition, the preceding lemma implies $\beta_{e,C} = \beta_{C,e}^{-1}$ because

$$\varrho_C \circ \beta_{e,C} \circ \beta_{C,e} = \lambda_C \circ \beta_{C,e} = \varrho_C.$$

In the braid group on three strands, we see the following relation:

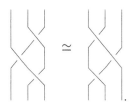

Its categorical analog holds.

Proposition 8.6.5 *For all objects C_1, C_2, C_3 in C, the following diagram commutes:*

$$
\begin{array}{ccccc}
(C_2 \otimes C_1) \otimes C_3 & \xleftarrow{\ \beta_{C_1,C_2} \otimes 1_{C_3}\ } & (C_1 \otimes C_2) \otimes C_3 & \xrightarrow{\ \alpha^{-1}_{C_1,C_2,C_3}\ } & C_1 \otimes (C_2 \otimes C_3) \\
{\scriptstyle \alpha^{-1}_{C_2,C_1,C_3}}\downarrow & & & & \downarrow{\scriptstyle 1_{C_1} \otimes \beta_{C_2,C_3}} \\
C_2 \otimes (C_1 \otimes C_3) & & & & C_1 \otimes (C_3 \otimes C_2) \\
{\scriptstyle 1_{C_2} \otimes \beta_{C_1,C_3}}\downarrow & & & & \downarrow{\scriptstyle \alpha_{C_2,C_1,C_3}} \\
C_2 \otimes (C_3 \otimes C_1) & & & & (C_1 \otimes C_3) \otimes C_2 \\
{\scriptstyle \alpha_{C_2,C_1,C_3}}\downarrow & & & & \downarrow{\scriptstyle \beta_{C_1,C_3} \otimes 1_{C_2}} \\
(C_2 \otimes C_3) \otimes C_1 & & & & (C_3 \otimes C_1) \otimes C_2 \\
{\scriptstyle \beta_{C_2,C_3} \otimes 1_{C_1}}\downarrow & & & & \downarrow{\scriptstyle \alpha^{-1}_{C_3,C_1,C_2}} \\
(C_3 \otimes C_2) \otimes C_1 & \xrightarrow{\ \alpha^{-1}_{C_3,C_2,C_1}\ } & C_3 \otimes (C_2 \otimes C_1) & \xleftarrow{\ 1_{C_3} \otimes \beta_{C_1,C_2}\ } & C_3 \otimes (C_1 \otimes C_2).
\end{array}
$$

In particular, if C is a strict monoidal category, then we get the categorical Yang–Baxter equation:

$$(1_{C_3} \otimes \beta_{C_1,C_2}) \circ (\beta_{C_1,C_3} \otimes 1_{C_2}) \circ (1_{C_1} \otimes \beta_{C_2,C_3})$$

$$= (\beta_{C_2,C_3} \otimes 1_{C_1}) \circ (1_{C_2} \otimes \beta_{C_1,C_3}) \circ (\beta_{C_1,C_2} \otimes 1_{C_3}).$$

Proof We note that the preceding diagram can be glued together from two hexagonal diagrams and a square.

The hexagons are precisely the ones from the hexagon axiom, and the square commutes because of the naturality of β. $\qquad\square$

Example 8.6.6 The collection of braid groups, Br_n, $n \geq 0$, gives rise to a braided monoidal category $\mathcal{B}$, whose underlying category is a groupoid. The objects of $\mathcal{B}$ are the natural numbers, including zero and

$$\mathcal{B}(\mathbf{n}, \mathbf{m}) = \begin{cases} \text{Br}_n, & \text{if } n = m, \\ \varnothing, & \text{if } n \neq m. \end{cases}$$

Here, we follow the convention that $\text{Br}_0 = \text{Br}_1$ is the trivial group. We define the monoidal structure on $\mathcal{B}$ via $\mathbf{n} \oplus \mathbf{m} = \mathbf{n} + \mathbf{m}$, which corresponds to the disjoint union of sets. The object $\mathbf{0}$ is a strict unit and $(\mathcal{B}, \oplus, \mathbf{0})$ is a strict monoidal category. For the braiding

$$\beta_{n,m} : \mathbf{n} \oplus \mathbf{m} \to \mathbf{m} \oplus \mathbf{n},$$

we use the braid that moves the first n strands to the spots $m + 1, \ldots, m + n$, and moves the last m strands to the spots $1, \ldots, m$, where the last m strands move over the first n strands.

Remark 8.6.7 Similar to the Σ_n-action on $C^{\otimes n}$ for every object C of a symmetric monoidal category (see Remark 8.3.5), any object C in a braided monoidal category gives rise to an object $C^{\otimes n}$ with a Br_n-action [**ML98**, XI.5, Theorem 2].

Example 8.6.8 One can combine the category $\mathcal{B}$ with the category of order-preserving injections, such that morphisms from $\mathbf{n}$ to $\mathbf{m}$ are braided injections from $\mathbf{n}$ to $\mathbf{m}$. This category features prominently in [**ScSo16**], and we will come back to it later (see Section 14.6.6), as it helps to model two-fold based loop spaces.

Some bialgebras H possess a universal R-matrix and are called *braided bialgebras*. A universal R-matrix is an invertible element $R \in H \otimes H$ that satisfies

$$\tau \circ \Delta(h) = R\Delta(h)R^{-1}, (\Delta \otimes 1_H)(R) = R_{13}R_{23}, \text{ and } (1_H \otimes \Delta)(R) = R_{13}R_{12}.$$

Here, the elements R_{13}, R_{23}, R_{13}, and R_{12} live in $H \otimes H \otimes H$, and they are stretched versions of R; for instance, if $R = \sum_i h_i^1 \otimes h_i^2$, then

$$R_{13} = \sum_i h_i^1 \otimes 1 \otimes h_i^2.$$

Example 8.6.9 Let H be a bialgebra. Then, the category of H-modules is a braided monoidal category if and only if H is braided. See [**K95**, XIII] for details.

9

Enriched Categories

In general, we require that for any pair of objects in a category, there is a set of morphisms from one object to the other. You know several examples where morphism sets carry more structure, for instance, if A and B are abelian groups, then the set of homomorphisms $\mathrm{Ab}(A, B)$ is itself an abelian group. In other examples, we do not just take the set of morphisms and observe that it has extra structure but we also consider morphism objects in a category different from sets. We have seen the example of the chain complex $\mathrm{HOM}(C_*, C'_*)$, where chain maps correspond to the zero cycles in $\mathrm{HOM}(C_*, C'_*)$. This was already mentioned as an example of a closed symmetric monoidal category. The concept of enriched categories allows for hom-objects in a general closed symmetric monoidal category. Good accounts on enriched category theory are [**K82, Day70b, Day70a**], [**Rie14**, I.3], [**Du70**], and [**Bo94-2**, Chapter 6]. In contrast to [**Lu09**], we only study enrichments in closed *symmetric* monoidal categories.

9.1 Basic Notions

Definition 9.1.1 Let $(\mathcal{V}, \odot, e)$ be a closed symmetric monoidal category. A *category enriched in* $\mathcal{V}$ or a $\mathcal{V}$-*category* $\mathcal{C}$ consists of a class of objects, and for all pairs of objects C_1, C_2, there is an object in $\mathcal{V}$, $\underline{\mathcal{C}}(C_1, C_2)$. There is a *composition morphism* in $\mathcal{V}$ for all objects C_1, C_2, C_3 of $\mathcal{C}$

$$m \colon \underline{\mathcal{C}}(C_1, C_2) \odot \underline{\mathcal{C}}(C_2, C_3) \to \underline{\mathcal{C}}(C_1, C_3),$$

and there is a *unit morphism* in $\mathcal{V}$, $\eta_C \colon e \to \underline{\mathcal{C}}(C, C)$ for each object C of $\mathcal{C}$. These morphisms satisfy a unit condition and an associativity condition. We require that the following diagrams commute for all objects C_1, C_2, C_3, C_4 of $\mathcal{C}$:

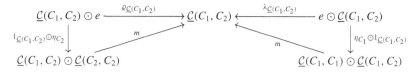

and

$$(\underline{\mathcal{C}}(C_1, C_2) \odot \underline{\mathcal{C}}(C_2, C_3)) \odot \underline{\mathcal{C}}(C_3, C_4) \xrightarrow{\alpha^{-1}_{\underline{\mathcal{C}}(C_1,C_2),\underline{\mathcal{C}}(C_2,C_3),\underline{\mathcal{C}}(C_3,C_4)}} \underline{\mathcal{C}}(C_1, C_2) \odot (\underline{\mathcal{C}}(C_2, C_3) \odot \underline{\mathcal{C}}(C_3, C_4))$$

with the left vertical map $m \odot 1_{\underline{\mathcal{C}}(C_3,C_4)}$ to $\underline{\mathcal{C}}(C_1, C_3) \odot \underline{\mathcal{C}}(C_3, C_4)$, the right vertical map $1_{\underline{\mathcal{C}}(C_1,C_2)} \odot m$ to $\underline{\mathcal{C}}(C_1, C_2) \odot \underline{\mathcal{C}}(C_2, C_4)$, and both m maps going to $\underline{\mathcal{C}}(C_1, C_4)$.

Examples 9.1.2

- Every category (in our sense) is a category enriched in the closed symmetric monoidal category of sets.
- We learned in Definition 7.1.1 that a preaddititve category is a category enriched in the category of abelian groups. For every pair of objects C_1, C_2, there is an abelian group of morphisms from C_1 to C_2, and the composition of morphisms is a bilinear map.
- Every closed symmetric monoidal category $\mathcal{V}$ is enriched in itself. In order to easy readability, we will write $\underline{\mathcal{V}}(V_1, V_2)$ for $V_2^{V_1}$ in this chapter.
- The category of complex vector spaces $\mathcal{V}_{\mathbb{C}}$ from [**BDR04**] has as objects the natural numbers, including zero, and

$$\mathcal{V}_{\mathbb{C}}(n, m) = \begin{cases} \varnothing, & \text{if } n \neq m, \\ U(n), & \text{for } n = m. \end{cases}$$

Here, $U(n)$ is the unitary group with its natural topology. Composition is induced by matrix multiplication, and this is a continuous map. The unit maps $\eta_n: \{*\} \to U(n)$ send the one-point space to the unit matrix in $U(n)$. Hence, $\mathcal{V}_{\mathbb{C}}$ is enriched in k-Top.

Definition 9.1.3 Let $(\mathcal{V}, \odot, e)$ be a closed symmetric monoidal cocomplete category and let $\mathcal{D}$ be a small category. The *standard enrichment of $\mathcal{D}$ in $\mathcal{V}$* is

$$\underline{\mathcal{D}}(D_1, D_2) = \bigsqcup_{\mathcal{D}(D_1,D_2)} e.$$

Note that the assumption that $\mathcal{V}$ is closed ensures that the coproduct distributes over the $\odot$-product in $\mathcal{V}$.

We saw an example of this construction already in Example 7.1.2. If we consider the closed symmetric monoidal category of abelian groups $(\text{Ab}, \otimes, \mathbb{Z})$, then the construction $\mathbb{Z}[\mathcal{C}]$ is a special case of the definition given earlier. For the category k-mod, you obtain $k\{\mathcal{D}(D_1, D_2)\}$ as the standard enriched morphism object.

Definition 9.1.4 Let $\mathcal{C}$ and $\mathcal{C}'$ be two $\mathcal{V}$-enriched categories. A $\mathcal{V}$-*functor* $F: \mathcal{C} \to \mathcal{C}'$ assigns to every object C of $\mathcal{C}$ an object $F(C)$ of $\mathcal{C}'$, and for every pair of objects C_1, C_2 of $\mathcal{C}$, it induces a morphism in $\mathcal{V}$

$$F = F_{C_1, C_2}: \underline{\mathcal{C}}(C_1, C_2) \to \underline{\mathcal{C}}'(F(C_1), F(C_2)),$$

such that these morphisms are compatible with the composition morphisms m of $\mathcal{C}$ and m' of $\mathcal{C}'$ and the unit morphisms η of $\mathcal{C}$ and η' of $\mathcal{C}'$, that is, the diagram

$$
\begin{array}{ccc}
\underline{\mathcal{C}}(C_1, C_2) \odot \underline{\mathcal{C}}(C_2, C_3) & \xrightarrow{\ m\ } & \underline{\mathcal{C}}(C_1, C_3) \\
\Big\downarrow{\scriptstyle F \odot F} & & \Big\downarrow{\scriptstyle F} \\
\underline{\mathcal{C}}'(F(C_1), F(C_2)) \odot \underline{\mathcal{C}}'(F(C_2), F(C_3)) & \xrightarrow{\ m'\ } & \underline{\mathcal{C}}'(F(C_1), F(C_3))
\end{array}
$$

commutes for all objects C_1, C_2, C_3 of $\mathcal{C}$ and

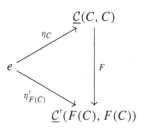

commutes for all objects C of $\mathcal{C}$.

Remark 9.1.5 Assume that $F: \mathcal{C} \to \mathcal{V}$ is a $\mathcal{V}$-functor. In this case, we can combine the structure morphism

$$F = F_{C_1, C_2}: \underline{\mathcal{C}}(C_1, C_2) \to \underline{\mathcal{V}}(F(C_1), F(C_2)),$$

with the evaluation map, ev, coming from the closed structure in $\mathcal{V}$ in order to obtain a morphism, which we call ev by slight abuse of notation.

$$ev: \underline{\mathcal{C}}(C_1, C_2) \odot F(C_1) \xrightarrow{F \odot 1_{F(C_1)}} \underline{\mathcal{V}}(F(C_1), F(C_2)) \odot F(C_1) \xrightarrow{ev} F(C_2).$$
$$\tag{9.1.1}$$

Example 9.1.6 Assume that $\mathcal{C}$ and $\mathcal{C}'$ are preadditive categories, that is, $\mathcal{C}$ and $\mathcal{C}'$ are enriched in abelian groups. An Ab-functor $F: \mathcal{C} \to \mathcal{C}'$ is nothing but an additive functor. The abelian group $\underline{\mathcal{C}}(C_1, C_2)$ has an underlying set, and an

Ab-functor induces a functor $\mathcal{C} \to \mathcal{C}'$. The fact that F induces a homomorphism $F \colon \underline{\mathcal{C}}(C_1, C_2) \to \underline{\mathcal{C}'}(F(C_1), F(C_2))$ is precisely the condition of additivity, as in Section 7.1.3.

Exercise 9.1.7 Take a small category $\mathcal{D}$ with the standard enrichment, as in Definition 9.1.3, in a closed symmetric monoidal cocomplete category $\mathcal{V}$, and show that $\mathcal{V}$-functors correspond to ordinary functors.

Example 9.1.8 Let $\mathcal{C}$ be a $\mathcal{V}$-enriched category and let C be an object of $\mathcal{C}$. The assignment $C' \mapsto \underline{\mathcal{C}}(C, C')$ defines a $\mathcal{V}$-functor

$$\underline{\mathcal{C}}(C, -) \colon \mathcal{C} \to \mathcal{V}.$$

Let C_1 and C_2 be two objects of $\mathcal{C}$. We define the morphism in $\mathcal{V}$

$$\underline{\mathcal{C}}(C, -)_{C_1, C_2} \colon \underline{\mathcal{C}}(C_1, C_2) \to \underline{\mathcal{V}}(\underline{\mathcal{C}}(C, C_1), \underline{\mathcal{C}}(C, C_2))$$

as the adjoint of the composition morphism

$$m \colon \underline{\mathcal{C}}(C, C_1) \odot \underline{\mathcal{C}}(C_1, C_2) \to \underline{\mathcal{C}}(C, C_2)).$$

Dually, we can form the contravariant morphism functor

$$\underline{\mathcal{C}}(-, C) \colon \mathcal{C}^o \to \mathcal{V},$$

with

$$\underline{\mathcal{C}}(-, C)_{C_1, C_2} \colon \underline{\mathcal{C}}(C_1, C_2) \to \underline{\mathcal{V}}(\underline{\mathcal{C}}(C_2, C), \underline{\mathcal{C}}(C_1, C))$$

as the adjoint of

$$m \colon \underline{\mathcal{C}}(C_1, C_2) \odot \underline{\mathcal{C}}(C_2, C) \to \underline{\mathcal{C}}(C_1, C)).$$

Definition 9.1.9 Let $F, G \colon \mathcal{C} \to \mathcal{C}'$ be two $\mathcal{V}$-functors. A $\mathcal{V}$-*natural transformation* $\xi \colon F \Rightarrow G$ consists of morphisms in $\mathcal{V}$

$$\xi_C \colon e \to \underline{\mathcal{C}'}(F(C), G(C))$$

for every object C of $\mathcal{C}$. These morphisms satisfy a naturality condition. For all objects C_1, C_2 of $\mathcal{C}$, the following diagram commutes:

$$(9.1.2)$$

that is,

$$m' \circ (\xi_{C_1} \odot G) \circ \lambda^{-1} = m' \circ (F \odot \xi_{C_2}) \circ \varrho^{-1}.$$

Remark 9.1.10 If $F, G \colon \mathcal{C} \to \mathcal{V}$ are two $\mathcal{V}$-functors and $\xi \colon F \Rightarrow G$ is a $\mathcal{V}$-natural transformation, then the morphisms $\xi_C \colon e \to \underline{\mathcal{V}}(F(C), G(C))$ are adjoint to morphisms from $e \odot F(C)$ to $G(C)$, and we identify them with morphisms

$$\xi' \colon F(C) \to G(C).$$

Using these adjoint structure maps, we can rewrite condition (9.1.2) in the familiar form by demanding that

$$
\begin{array}{ccc}
\underline{\mathcal{C}}(C_1, C_2) & \xrightarrow{\hspace{2cm} F \hspace{2cm}} & \underline{\mathcal{V}}(F(C_1), F(C_2)) \\
{\scriptstyle G} \downarrow & & \downarrow {\scriptstyle \underline{\mathcal{V}}(1_{F(C_1)}, \xi'_{C_2})} \\
\underline{\mathcal{V}}(G(C_1), G(C_2)) & \xrightarrow[\underline{\mathcal{V}}(\xi'_{C_1}, 1_{G(C_2)})]{} & \underline{\mathcal{V}}(F(C_1), G(C_2))
\end{array}
$$

commutes for all objects C_1 and C_2 of $\mathcal{C}$.

Remark 9.1.11 If we fix a symmetric monoidal category $\mathcal{V}$, then the collection of small $\mathcal{V}$-enriched categories, $\mathrm{cat}_{\mathcal{V}}$, forms a 2-category, whose 1-morphisms are $\mathcal{V}$-functors and whose 2-morphisms are $\mathcal{V}$-natural transformations.

Exercise 9.1.12 Let $F \colon \mathcal{V} \to \mathcal{W}$ be a lax symmetric monoidal functor between closed symmetric monoidal categories and let $\mathcal{C}$ be a $\mathcal{V}$-enriched category. Use F to define a $\mathcal{W}$-enrichment on $\mathcal{C}$.

9.2 Underlying Category of an Enriched Category

Consider the category of unbounded chain complexes over a commutative ring k, with its internal morphism object $\mathrm{HOM}(-, -)$ and the tensor product of chain complexes as part of the monoidal structure. Chain maps $f \colon C_* \to C'_*$ between two chain complexes C_* and C'_* correspond to zero cycles in $\mathrm{HOM}(C_*, C'_*)$, and these, in turn, correspond to chain maps from the chain complex $\mathbb{S}^0(k)$ (which is k in chain degree zero and zero elsewhere) to $\mathrm{HOM}(C_*, C'_*)$. So, one can recover the ordinary category of chain complexes from the enriched one.

This can be done in general.

Definition 9.2.1 Let C be a category enriched in a closed symmetric monoidal category V. The *underlying category of* C, C_0 has as objects the objects of C and

$$C_0(C_1, C_2) = V(e, \underline{C}(C_1, C_2)).$$

Theorem 9.2.2 *The underlying category of an enriched category is a category.*

Proof We define the identity morphism 1_C as $\eta_C : e \to V(e, \underline{C}(C, C))$ and the composition of morphisms $f \in V(e, \underline{C}(C_1, C_2))$ and $g \in V(e, \underline{C}(C_2, C_3))$, $g \circ f$ as

$$e \cong e \odot e \xrightarrow{\quad f \odot g \quad} \underline{C}(C_1, C_2) \odot \underline{C}(C_2, C_3) \xrightarrow{\quad m \quad} \underline{C}(C_1, C_3);$$

hence, we have to show that

(1) $g \circ 1_{C_2} = g$;
(2) $1_{C_2} \circ f = f$; and
(3) $h \circ (g \circ f) = (h \circ g) \circ f$ for all f, g, as above, and $h \in V(e, \underline{C}(C_3, C_4))$.

For (1), note that the diagram

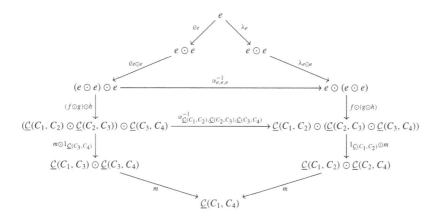

commutes because of the unit condition on η_{C_2} and $\lambda_{\underline{C}(C_2, C_3)}$ in any V-enriched category. The proof of (2) is dual, using the condition for η and ϱ.

For (3), observe that the hat of the diagram

commutes because of the unit condition in a monoidal category. The middle square commutes because of the naturality of α, and the bottom pentagon commutes because of the associativity condition of m in a $\mathcal{V}$-enriched category. $\qquad\square$

Exercise 9.2.3 Give an alternative proof of Theorem 9.2.2 by showing that the functor $\mathcal{V}(e, -) \colon \mathcal{V} \to$ Sets is lax symmetric monoidal and by using Exercise 9.1.12.

Corollary 9.2.4 *Let $f \in C_0(C_1, C_2)$. Then, f induces a morphism in $\mathcal{V}$:*

$$\underline{\mathcal{C}}(1_C, f) = f_* \colon \underline{\mathcal{C}}(C, C_1) \to \underline{\mathcal{C}}(C, C_2).$$

This assignment satisfies $(1_C)_* = 1_{\underline{\mathcal{C}}(C,C_1)}$ and $(g \circ f)_* = g_* \circ f_*$ for every $g \in C_0(C_2, C_3)$.

Proof We define $\underline{\mathcal{C}}(1_C, f) = f_*$ as

$$\underline{\mathcal{C}}(C, C_1) \cong \underline{\mathcal{C}}(C, C_1) \odot e \xrightarrow{1_{\underline{\mathcal{C}}(C,C_1)} \odot f} \underline{\mathcal{C}}(C, C_1) \odot \underline{\mathcal{C}}(C_1, C_2) \xrightarrow{m} \underline{\mathcal{C}}(C, C_2).$$

It is straightforward to check the properties. $\qquad\square$

Remark 9.2.5 Similarly, $f \in C_0(C_1, C_2)$ induces a map $\underline{\mathcal{C}}(f, 1_C) = f^* \colon \underline{\mathcal{C}}(C_2, C) \to \underline{\mathcal{C}}(C_1, C)$ with $(g \circ f)^* = f^* \circ g^*$ and $(1_C)^* = 1_{\underline{\mathcal{C}}(C,C_1)}$.

Proposition 9.2.6 *If $\mathcal{V}$ is a closed symmetric monoidal category, which we view as a self-enriched category, then the underlying category of $\mathcal{V}$ is isomorphic to $\mathcal{V}$ itself.*

Proof For two objects V_1, V_2 of $\mathcal{V}$, the adjunction

$$\mathcal{V}(e, \underline{\mathcal{V}}(V_1, V_2)) \cong \mathcal{V}(e \odot V_1, V_2) \cong \mathcal{V}(V_1, V_2) \tag{9.2.1}$$

tells us that we get a bijection between the morphism set of the underlying category of $\mathcal{V}$ and $\mathcal{V}$. For $f \in \mathcal{V}(e, \underline{\mathcal{V}}(V_1, V_2))$, we denote its adjoint in $\mathcal{V}(V_1, V_2)$ by $\phi(f)$.

We have to show that the composition of morphisms is compatible with these bijections and that the morphisms $\eta_V \colon e \to \underline{\mathcal{V}}(V, V)$ correspond to the identity morphism 1_V on the object V in the category $\mathcal{V}$ under the adjunction in (9.2.1). For the latter, the adjoint of η_V is $\lambda_V \colon e \odot V \cong V$ and the evaluation map is adjoint to $1_{\underline{\mathcal{V}}(V,V)}$; hence, under the bijection from (9.2.1), $\phi(\eta_V)$ is 1_V.

As m is adjoint to

$$\underline{\mathcal{V}}(V_1, V_2) \odot \underline{\mathcal{V}}(V_2, V_3) \odot V_1 \xrightarrow{1_{\underline{\mathcal{V}}(V_1, V_2)} \odot \tau_{\underline{\mathcal{V}}(V_2, V_3), V_1}} \underline{\mathcal{V}}(V_1, V_2) \odot V_1 \odot \underline{\mathcal{V}}(V_2, V_3)$$

$$V_3 \xleftarrow{\quad ev \quad} \underline{\mathcal{V}}(V_2, V_3) \odot V_2 \xleftarrow{\tau_{V_2, \underline{\mathcal{V}}(V_2, V_3)}} V_2 \odot \underline{\mathcal{V}}(V_2, V_3),$$

with $ev \odot 1_{\underline{\mathcal{V}}(V_2, V_3)}$ on the right vertical arrow,

the adjunction maps the composition to

$$V_1 \xrightarrow{(\lambda_{e \odot V_1}) \circ \lambda_{V_1}} e \odot e \odot V_1 \xrightarrow{f \odot g \odot 1_{V_1}} \underline{\mathcal{V}}(V_1, V_2) \odot \underline{\mathcal{V}}(V_2, V_3) \odot V_1$$

$$\downarrow{1_{\underline{\mathcal{V}}(V_1, V_2)} \odot \tau_{\underline{\mathcal{V}}(V_2, V_3), V_1}}$$

$$\underline{\mathcal{V}}(V_1, V_2) \odot V_1 \odot \underline{\mathcal{V}}(V_2, V_3)$$

$$\downarrow{ev \odot 1_{\underline{\mathcal{V}}(V_2, V_3)}}$$

$$V_3 \xleftarrow{\quad ev \quad} \underline{\mathcal{V}}(V_2, V_3) \odot V_2 \xleftarrow{\tau_{V_2, \underline{\mathcal{V}}(V_2, V_3)}} V_2 \odot \underline{\mathcal{V}}(V_2, V_3).$$

The following diagram

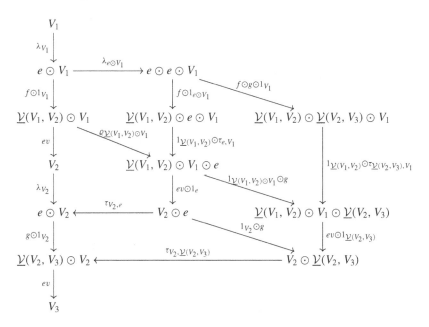

commutes, and therefore, $\phi(g) \circ \phi(f)$ on the left-hand side of the diagram is the same as $\phi(m(f \odot g))$ on the right-hand side. Hence, (9.2.1) is compatible with compositions. $\qquad\square$

9.3 Enriched Yoneda Lemma

We first prove a set version of a $\mathcal{V}$-enriched variant of the Yoneda lemma.

Proposition 9.3.1 *Let C be an object of the $\mathcal{V}$-enriched category $\mathcal{C}$ and let $F: \mathcal{C} \to \mathcal{V}$ be a $\mathcal{V}$-functor. Then, there is a bijection between the set of $\mathcal{V}$-natural transformations from the functor $\underline{\mathcal{C}}(C, -)$ to F and the underlying set of $F(C)$, $\mathcal{V}(e, F(C))$.*

Proof Let ξ be a $\mathcal{V}$-natural transformation with components

$$\xi_{C'}: \underline{\mathcal{C}}(C, C') \to F(C').$$

We send ξ to the morphism

$$f_\xi: e \xrightarrow{\eta_C} \underline{\mathcal{C}}(C, C) \xrightarrow{\xi_C} F(C).$$

Conversely, given $f \in \mathcal{V}(e, F(C))$, we define $\xi_{C'}^{f}$ as

$$\underline{\mathcal{C}}(C, C') \xrightarrow{F} \underline{\mathcal{V}}(F(C), F(C')) \xrightarrow{\underline{\mathcal{V}}(f, 1_{F(C')})} \underline{\mathcal{V}}(e, F(C')) \xrightarrow{i_{F(C')}} F(C'),$$

where $i_{F(C')}: \underline{\mathcal{V}}(e, F(C')) \to F(C')$ is the isomorphism that uses the closed monoidal structure in $\mathcal{V}$ (see Exercise 8.4.2).

We have to show that $\xi_{C'}^{f}$ is $\mathcal{V}$-natural in C' and that the two maps are inverse to each other. The $\mathcal{V}$-naturality follows from the binaturality of $\underline{\mathcal{C}}(-, -)$.

If we start with $f \in \mathcal{V}(e, F(C))$, then $f_{\xi_C^{f}}$ is

$$e \xrightarrow{\eta_C} \underline{\mathcal{C}}(C, C) \xrightarrow{F} \underline{\mathcal{V}}(F(C), F(C)) \xrightarrow{\underline{\mathcal{V}}(f, 1_{F(C)})} \underline{\mathcal{V}}(e, F(C)) \xrightarrow{i_{F(C)}} F(C).$$

But $F \circ \eta_C = \eta_{F(C)}$ and the latter is adjoint to the isomorphism $e \odot F(C) \cong F(C)$. Hence, $f^* \circ \eta_{F(C)} = \underline{\mathcal{V}}(f, 1_{F(C)}) \circ \eta_{F(C)}$ is adjoint to the morphism

$$f \circ \lambda = f \circ \varrho: e \odot e \to F(C),$$

and the application of i cancels the first unit isomorphism, and thus, $f_{\xi_C^{f}} = f$.

Conversely, given ξ, we have to check that $\xi^{f_\xi} = \xi$. We can fit $\xi^{f_\xi} = i_{F(C')} \circ \underline{\mathcal{V}}(f_\xi, 1_{F(C')}) \circ F$ into the following commutative diagram:

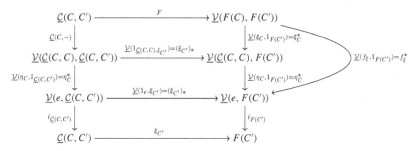

As $\underline{C}(C, -) \colon \underline{C}(C, C') \to \underline{V}(\underline{C}(C, C), \underline{C}(C, C'))$ is adjoint to m, the composite $\underline{V}(\eta_C, 1_{\underline{C}(C,C')}) \circ \underline{C}(C, -)$ is adjoint to $e \odot \underline{C}(C, C') \cong \underline{C}(C, C')$ and postcomposing with i corresponds to getting rid of this unit isomorphism; hence, the composition downward on the left-hand side of the diagram is the identity. $\square$

Of course, we want to improve this version of an enriched Yoneda lemma to an isomorphism in $\mathcal{V}$.

Definition 9.3.2 Let C and C' be enriched over $\mathcal{V}$. The category $C \odot C'$ has as objects pairs (C, C'), where C is an object of C and C' is an object of C'. We set

$$(\underline{C} \odot \underline{C}')((C_1, C_1'), (C_2, C_2')) := \underline{C}(C_1, C_2) \odot \underline{C}'(C_1', C_2').$$

The composition morphism $m_\odot$ is defined as $(m \odot m') \circ (2, 3)$, where $(2, 3) \in \Sigma_4$ denotes the permutation that switches the second and third $\odot$-factors in

$$\underline{C}(C_1, C_2) \odot \underline{C}'(C_1', C_2') \odot \underline{C}(C_2, C_3) \odot \underline{C}'(C_2', C_3')$$

for objects C_1, C_2, C_3 of C and C_1', C_2', C_3' of C'. Here, m and m' are the composition morphisms of C and C'.

With this definition, $C \odot C'$ is again a $\mathcal{V}$-category.

Exercise 9.3.3 Let $\mathcal{V}$ be a symmetric monoidal category. Show that the category of small $\mathcal{V}$-categories and $\mathcal{V}$-functors, $\mathrm{cat}_\mathcal{V}$, is a symmetric monoidal category with $\odot \colon \mathrm{cat}_\mathcal{V} \times \mathrm{cat}_\mathcal{V} \to \mathrm{cat}_\mathcal{V}$ from Definition 9.3.2 as monoidal structure.

We extend the notion of dinatural families and ends to the enriched context.

Definition 9.3.4 Let C be a $\mathcal{V}$-category, let V be an object of $\mathcal{V}$, and let $G \colon C^o \odot C \to \mathcal{V}$ be a $\mathcal{V}$-functor. A family of morphisms $(\partial_C \in \mathcal{V}(V, G(C, C)))_C$ over all objects C in C is a $\mathcal{V}$-*dinatural family* if for all objects C and C' of C, the diagram

$$
\begin{array}{ccc}
V & \xrightarrow{\ \ \partial_C\ \ } & G(C, C) \\
{\scriptstyle \partial_{C'}}\big\downarrow & & \big\downarrow{\scriptstyle \gamma_{C,C'}} \\
G(C', C') & \xrightarrow[\ \ \beta_{C',C}\ \ ]{} & \underline{\mathcal{V}}(\underline{C}(C, C'), G(C, C'))
\end{array}
$$

commutes.

Here, $\beta_{C',C}$ is adjoint to the evaluation map

$$G(C', C') \odot \underline{C}(C, C') \to G(C, C'),$$

using the contravariant component of G, and dually, $\gamma_{C,C'}$ is adjoint to

$$G(C, C) \odot \underline{C}(C, C') \to G(C, C')$$

and uses the covariant component of G.

We want to replace the *set* of $\mathcal{V}$-natural transformations in the weak enriched Yoneda lemma with an object in $\mathcal{V}$. We do this by forming ends in $\mathcal{V}$.

Definition 9.3.5 Let $\mathcal{C}$ be a $\mathcal{V}$-category, let $G: \mathcal{C}^o \odot \mathcal{C} \to \mathcal{V}$ be a $\mathcal{V}$-functor, and let $(\partial_C \in \mathcal{V}(V, G(C, C)))_C$ be a $\mathcal{V}$-dinatural family. Then, $(\partial_C)_C$ is a $\mathcal{V}$-*end of G* if for all $\mathcal{V}$-dinatural families $(\mu_C \in \mathcal{V}(V', G(C, C)))_C$, there is a unique $f \in \mathcal{V}(V', V)$ with $\mu_C = \partial_C \circ f$ for all C.

Using the notation introduced in Section 4.6, we denote the object V by $\int_{\mathcal{C}} G(C, C)$, and then, $\partial_{C'}: \int_{\mathcal{C}} G(C, C) \to G(C', C')$. The usual disclaimer applies. Enriched ends do not have to exist. As before, they do exist if $\mathcal{V}$ is complete.

Proposition 9.3.6 (Enriched Yoneda lemma) *Assume that $\mathcal{V}$ is a closed symmetric monoidal complete category and that $\mathcal{C}$ is a $\mathcal{V}$-enriched category. Let $F: \mathcal{C} \to \mathcal{V}$ be a $\mathcal{V}$-functor and let C be an object of $\mathcal{C}$. Then, $F(C)$ is isomorphic to the $\mathcal{V}$-end*

$$\int_{\mathcal{C}} \underline{\mathcal{V}}(\mathcal{C}(C, C'), F(C')).$$

Proof   Let $\nu_{C'}: F(C) \to \underline{\mathcal{V}}(\underline{\mathcal{C}}(C, C'), F(C'))$ be the adjoint of

$$F: \underline{\mathcal{C}}(C, C') \to \underline{\mathcal{V}}(F(C), F(C')),$$

and assume that $(\mu_{C'}: V \to \underline{\mathcal{V}}(\underline{\mathcal{C}}(C, C'), F(C')))_{C'}$ is an arbitrary $\mathcal{V}$-dinatural family. The adjoint of $\mu_{C'}$ is

$$\mu'_{C'}: \underline{\mathcal{C}}(C, C') \to \underline{\mathcal{V}}(V, F(C')).$$

This family is a natural transformation from $\underline{\mathcal{C}}(C, -)$ to $\underline{\mathcal{V}}(V, F(-))$, and hence, the set version of the enriched Yoneda lemma 9.3.1 ensures the existence of a unique morphism $f \in \mathcal{V}(V, F(C))$, such that we can write $\mu'_{C'}$ as the composite

$$\underline{\mathcal{C}}(C, C') \xrightarrow{\ F\ } \underline{\mathcal{V}}(F(C), F(C')) \xrightarrow{\ \underline{\mathcal{V}}(f, 1_{F(C')})\ } \underline{\mathcal{V}}(V, F(C')).$$

As $F: \underline{\mathcal{C}}(C, C') \to \underline{\mathcal{V}}(F(C), F(C'))$ is adjoint to $\nu_{C'}$, this implies that $\nu_{C'} \circ f = \mu_{C'}$. $\square$

If $\mathcal{A}$ is a preadditive category and if A is an object of $\mathcal{A}$, then Proposition 9.3.6 has the following form:

Corollary 9.3.7 *The natural transformations from the representable functor* $\mathcal{A}(A, -)$ *to an additive functor* $F \colon \mathcal{A} \to$ Ab *form an abelian group and are isomorphic to* $F(A)$.

There is also a multilinear variant of the additive Yoneda lemma. I used a version of this in [**Ri03**].

Proposition 9.3.8 *Let* $A_1, \ldots, A_n$ *be objects of a preadditive category* $\mathcal{A}$ *and let* $F \colon \mathcal{A}^n \to$ Ab *be a functor that is additive in every component. Then, the natural transformations from the functor* $(B_1, \ldots, B_n) \mapsto \mathcal{C}(A_1, B_1) \otimes \cdots \otimes \mathcal{C}(A_n, B_n)$ *to* F *form an abelian group, and this group is isomorphic to* $F(A_1, \ldots, A_n)$.

Proof We define a morphism $Y(F, (A_1, \ldots, A_n))$ by sending a natural transformation

$$\eta \colon \mathcal{C}(A_1, -) \otimes \cdots \otimes \mathcal{C}(A_n, -) \Rightarrow F$$

to $\eta_{(A_1, \ldots, A_n)}(1_{A_1} \otimes \cdots \otimes 1_{A_n}) \in F(A_1, \ldots, A_n)$. For any $a \in F(A_1, \ldots, A_n)$ and every $f_1 \otimes \cdots \otimes f_n \in \mathrm{Ab}(A_1, B_1) \otimes \cdots \otimes \mathrm{Ab}(A_n, B_n)$, we set

$$\tau(F, A_1, \ldots, A_n)_a(f_1 \otimes \cdots \otimes f_n) := F(f_1, \ldots, f_n)(a) \in F(B_1, \ldots, B_n).$$

As F is multiadditive, this assignment is well-defined. One shows, as in the proof of the "classical" Yoneda lemma (Theorem 2.2.2), that $Y(F, (A_1, \ldots, A_n))$ is inverse to $\tau(F, A_1, \ldots, A_n)$. $\square$

9.3.1 Enriched Co-Yoneda Lemma

The enriched Yoneda lemma is a statement about ends involving a representable functor. The enriched co-Yoneda lemma is concerned with enriched coends involving a representable functor.

Exercise 9.3.9 Dualize the definition of a $\mathcal{V}$-end to a $\mathcal{V}$-coend.

Proposition 9.3.10 (co-Yoneda lemma) *Let* $\mathcal{D}$ *be a small category enriched in* $\mathcal{V}$. *If* $\mathcal{V}$ *has all small colimits, then for every* $\mathcal{V}$-*functor* $F \colon \mathcal{D}^o \to \mathcal{V}$, *the* $\mathcal{V}$-*coend* $\int^{\mathcal{D}} F(D) \odot \underline{\mathcal{D}}(D_1, D)$ *exists, and there is a natural isomorphism in* $\mathcal{V}$:

$$\left(\int^{\mathcal{D}} F(D) \odot \underline{\mathcal{D}}(D_1, D) \right) \cong F(D_1).$$

Proof The existence of the $\mathcal{V}$-coend is clear, and the remainder of the proof is just an enriched analog of the proof of Theorem 5.4.8. $\square$

9.4 Cotensored and Tensored Categories

In applications, one usually encounters tensors and cotensors of enriched categories.

Definition 9.4.1 Let $(\mathcal{V}, \odot, e)$ be a closed symmetric monoidal category and let $\mathcal{C}$ be a category enriched in $\mathcal{V}$. Let C be an object of $\mathcal{C}$ and let V be an object of $\mathcal{V}$.

- The *tensor of C with V exists* if there is an object $C \otimes V$ of $\mathcal{C}$, together with isomorphisms in $\mathcal{V}$

$$\underline{\mathcal{C}}(C \otimes V, C') \cong \underline{\mathcal{V}}(V, \underline{\mathcal{C}}(C, C'))$$

  that are natural in C'.
- The *cotensor of C' with V exists* if there is an object $(C')^V$ of $\mathcal{C}$, together with isomorphisms in $\mathcal{V}$

$$\underline{\mathcal{C}}(C, (C')^V) \cong \underline{\mathcal{V}}(V, \underline{\mathcal{C}}(C, C'))$$

  that are natural in C.
- The category $\mathcal{C}$ is *tensored in $\mathcal{V}$* if the tensor of C with V exists for all objects C of $\mathcal{C}$ and V of $\mathcal{V}$.
- The category $\mathcal{C}$ is *cotensored in $\mathcal{V}$* if the cotensor of C with V exists for all objects C of $\mathcal{C}$ and V of $\mathcal{V}$.

Example 9.4.2 If $\mathcal{V}$ is a closed symmetric monoidal category, then $\mathcal{V}$ is tensored and cotensored over itself.

Example 9.4.3 Let $\mathcal{D}$ be a small category and let $(\mathcal{C}, \otimes, e_{\mathcal{C}})$ be a closed symmetric monoidal category. Then, the category $\mathrm{Fun}(\mathcal{D}, \mathcal{C})$ is tensored and cotensored over $\mathcal{C}$. For a functor $F \colon \mathcal{D} \to \mathcal{C}$ and an object C of $\mathcal{C}$, we define the tensor of F with C as

$$(F \otimes C)(D) = F(D) \otimes C$$

and the cotensor as

$$(F^C)(D) = F(D)^C,$$

where $F(D)^C$ is the internal morphism object of C and $F(D)$ in $\mathcal{C}$.

If $\mathcal{C}$ is (co)tensored over $\mathcal{V}$, then there are some canonical rules for calculating iterated tensors and cotensors.

Proposition 9.4.4 *Assume that C is tensored over V. Then*

(1) $C \otimes e \cong C$ for all objects C of C.
(2) For all objects V_1, V_2 of V and for all objects C of C,

$$(C \otimes V_1) \otimes V_2 \cong C \otimes (V_1 \odot V_2) \cong C \otimes (V_2 \odot V_1) \cong (C \otimes V_2) \otimes V_1. \quad (9.4.1)$$

Proof   The first claim follows because for all objects C' of C, we have

$$\underline{C}(C \otimes e, C') \cong \underline{V}(e, \underline{C}(C, C')),$$

but we saw already in Exercise 8.4.2 that $\underline{V}(e, \underline{C}(C, C')) \cong \underline{C}(C, C')$; hence, $C \otimes e \cong C$.

We will only show the first isomorphism in (2); the others follow from the symmetric monoidal structure of V. Again, we consider an arbitrary object C' of C and obtain

$$\begin{aligned}
\underline{C}((C \otimes V_1) \otimes V_2, C') &\cong \underline{V}(V_2, \underline{C}(C \otimes V_1, C')) \\
&\cong \underline{V}(V_2, \underline{V}(V_1, \underline{C}(C, C'))) \\
&\cong \underline{V}(V_1 \odot V_2, \underline{C}(C, C')) \\
&\cong \underline{C}(C \otimes (V_1 \odot V_2), C').
\end{aligned}$$

Here, the first, second, and last isomorphisms are given by the defining isomorphisms of tensors as in Definition 9.4.1 and the third isomorphism uses that V is closed symmetric monoidal. Hence, $(C \otimes V_1) \otimes V_2 \cong C \otimes (V_1 \odot V_2)$.   □

Exercise 9.4.5 Formulate and prove the dual rules for calculating iterated cotensors.

Proposition 9.4.6 *If C is tensored over V, then $(-) \otimes V : C \to C$ is a V-functor. Dually, if C is cotensored over V, then $(-)^V : C \to C$ is a V-functor.*

Proof   Note that there is an isomorphism in V:

$$\underline{C}(C_1 \otimes \underline{C}(C_1, C_2), C_2) \cong \underline{V}(\underline{C}(C_1, C_2), \underline{C}(C_1, C_2)),$$

and therefore, we get a bijection on the underlying sets of morphisms:

$$\begin{aligned}
C_0(C_1 \otimes \underline{C}(C_1, C_2), C_2) &\cong V(e, \underline{V}(\underline{C}(C_1, C_2), \underline{C}(C_1, C_2))) \\
&\cong V(\underline{C}(C_1, C_2), \underline{C}(C_1, C_2)).
\end{aligned}$$

Let ε denote the morphism in $\mathcal{C}_0(C_1 \otimes \underline{\mathcal{C}}(C_1, C_2), C_2)$ that corresponds to $1_{\underline{\mathcal{C}}(C_1,C_2)}$ under the isomorphism. We consider the following composition of morphisms in $\mathcal{V}$:

$$e \xrightarrow{\eta_{C_2 \otimes V}} \underline{\mathcal{C}}(C_2 \otimes V, C_2 \otimes V) \xrightarrow{(\varepsilon \otimes 1_V)^*} \underline{\mathcal{C}}((C_1 \otimes \underline{\mathcal{C}}(C_1, C_2)) \otimes V, C_2 \otimes V)$$

$$\downarrow \cong$$

$$\underline{\mathcal{C}}((C_1 \otimes V) \otimes \underline{\mathcal{C}}(C_1, C_2), C_2 \otimes V)$$

$$\downarrow \cong$$

$$\underline{\mathcal{V}}(\underline{\mathcal{C}}(C_1, C_2), \underline{\mathcal{C}}(C_1 \otimes V, C_2 \otimes V)).$$

The first isomorphism uses the associativity property of tensors from Proposition 9.4.4, and the second isomorphism is the defining isomorphism of tensors.

The compositon corresponds to a morphism in $\mathcal{V}$ from $\underline{\mathcal{C}}(C_1, C_2)$ to $\underline{\mathcal{C}}(C_1 \otimes V, C_2 \otimes V)$, which turns $(-) \otimes V$ into a $\mathcal{V}$ functor.

The proof for $(-)^V \colon \mathcal{C} \to \mathcal{C}$ being a $\mathcal{V}$-functor is dual and uses the isomorphism $\underline{\mathcal{C}}(C_2^V, C_2^V) \cong \underline{\mathcal{V}}(V, \underline{\mathcal{C}}(C_2^V, C_2))$. $\square$

Example 9.4.7 If $\mathcal{C}$ is a category that has small coproducts, then $\mathcal{C}$ is tensored over the category of sets, Sets. For an object C of $\mathcal{C}$ and any set X, we define

$$C \otimes X := \bigsqcup_{x \in X} C,$$

so $C \otimes X$ is the X-copower of C in $\mathcal{C}$.

Dually, if $\mathcal{C}$ is a category with small products, then it is cotensored over Sets with

$$C^X := \prod_{x \in X} C,$$

so C^X is the X-power of C in $\mathcal{C}$.

9.5 Categories Enriched in Categories

The category of all small categories, cat, is closed symmetric monoidal, with the product of categories as monoidal structure, and it is bicomplete. What does it mean for a category $\mathcal{C}$ to be enriched in cat? By definition, we get for all pairs of objects C_1, C_2 of $\mathcal{C}$, a category of morphism $\underline{\mathcal{C}}(C_1, C_2)$. We think of an object f of $\underline{\mathcal{C}}(C_1, C_2)$ as a morphism from C_1 to C_2. For two such objects f and g, we have morphisms in the category $\underline{\mathcal{C}}(C_1, C_2)$ from f to g.

It is common to call the objects of $\underline{C}(C_1, C_2)$ 1-*morphisms* and to call the morphisms in $\underline{C}(C_1, C_2)$ 2-*morphisms*. We will draw them as 2-cells

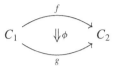

if $\phi \in \underline{C}(C_1, C_2)(f, g)$.

The fact that $\underline{C}(C_1, C_2)$ is a category ensures that we have a strictly associative composition of morphisms. If $\phi \colon f \Rightarrow g$ and $\phi' \colon g \Rightarrow h$, then the composition is $\phi' \circ \phi \colon f \Rightarrow h$, and we have strict associativity of threefold compositions, and the identity morphisms $1_f \in \underline{C}(C_1, C_2)(f, f)$ acts as a strict unit.

The composition rules in enriched categories ensure that the composition

$$\underline{C}(C_1, C_2) \times \underline{C}(C_2, C_3) \to \underline{C}(C_1, C_3)$$

is also strictly associative and unital.

Definition 9.5.1 A *strict 2-category* is a category C enriched in cat.

We saw an example of such a category in Remark 2.1.6 when we studied cat. Let C and C' be two objects of cat, that is, two small categories. Then, there is a category of morphisms from C to C' whose objects are functors from C to C' and whose morphisms are natural transformations between such functors. As we have the interchange laws for the vertical and horizontal compositions of natural transformations, as in (2.1.1), this establishes cat as a category enriched in cat, that is, a strict 2-category.

Example 9.5.2 Let $(C, \otimes, e)$ be a strict monoidal category (recall Definition 8.1.1). There is an associated strict 2-category ΣC with a single object $*$, whose 1-morphisms are the objects of C and whose 2-morphisms are the morphisms of C. Conversely, if D is a strict 2-category with one object, then D gives rise to a strict monoidal category.

Remark 9.5.3 Every strict 2-category can be viewed as a category enriched in simplicial sets (see Definition 10.2.1). You will learn about the nerve of a small category in Definition 11.1.1. This is a simplicial set, and the assignment of sending a small category to its nerve is strong symmetric monoidal, so an enrichment in cat gives rise to an enrichment in simplicial sets.

9.6 Bicategories

For many purposes, such a level of strictness, as in a strict 2-category, is too much to ask for. The notion of bicategories goes back to Bénabou [**Ben67**]. In bicategories, several equalities in the definition of a strict 2-category are replaced by natural isomorphisms.

Definition 9.6.1 [**Ben67**] A *bicategory* $\mathcal{B}$ consists of the following:

- A class of objects of $\mathcal{B}$.
- For each pair of objects B_1, B_2 of $\mathcal{B}$, there is a category $\mathcal{B}(B_1, B_2)$. As in the setting of strict 2-categories, it is common to call the objects of $\mathcal{B}$ 0-cells the objects of $\mathcal{B}(B_1, B_2)$ 1-cells and the morphisms in $\mathcal{B}(B_1, B_2)$ 2-cells.
- For each triple of objects B_1, B_2, B_3 of $\mathcal{B}$, there is a *composition functor*

$$c_{B_1,B_2,B_3} : \mathcal{B}(B_1, B_2) \times \mathcal{B}(B_2, B_3) \to \mathcal{B}(B_1, B_3).$$

- For each object B of $\mathcal{B}$, there is an object I_B of $\mathcal{B}(B, B)$, called the *identity arrow of B*. Its identity morphism 1_{I_B} in the category $\mathcal{B}(B, B)$ is denoted by i_B and is called the *identity 2-cell of B*. We will identify I_B with the corresponding functor $I_B : [0] \to \mathcal{B}(B, B)$.
- For every quadruple of objects B_1, B_2, B_3, B_4 of $\mathcal{B}$, there is a natural isomorphism α_{B_1,B_2,B_3,B_4} between the functors

$$c_{B_1,B_2,B_4} \circ (\mathrm{Id} \times c_{B_2,B_3,B_4}) \text{ and } c_{B_1,B_3,B_4} \circ (c_{B_1,B_2,B_3} \times \mathrm{Id}).$$

We call α the *associativity isomorphism*.
- For each pair of objects B_1, B_2 of $\mathcal{B}$, there are two natural isomorphisms λ_{B_1,B_2} and ϱ_{B_1,B_2}, where λ_{B_1,B_2} is a natural isomorphism between the functors $c_{B_1,B_1,B_2} \circ (I_{B_1} \times \mathrm{Id})$ and the canonical isomorphism of categories $[0] \times \mathcal{B}(B_1, B_2) \cong \mathcal{B}(B_1, B_2)$. Analogously, ϱ is a natural isomorphism between $c_{B_1,B_2,B_2} \circ (\mathrm{Id} \times I_{B_2})$ and the isomorphism $\mathcal{B}(B_1, B_2) \times [0] \cong \mathcal{B}(B_1, B_2)$.

These natural isomorphisms have to satisfy the following coherence axioms:

- The associativity isomorphisms are coherent, so that the diagram of natural transformations

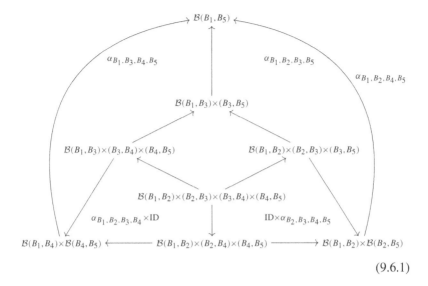

$$(9.6.1)$$

commutes for all B_i. Here, the α on the right-hand side denotes the outer 2-cell of the diagram.

- The natural isomorphisms λ and ϱ are compatible with α, so that the diagram of 2-cells commutes for all B_i:

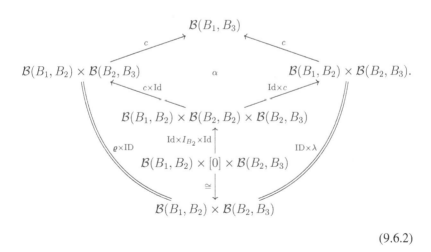

$$(9.6.2)$$

If you don't like diagram (9.6.1) because of its outer 2-cell, then you can draw the diagram on a cube.

Let $X_{i,i+1}$ be an object of $\mathcal{B}(B_i, B_{i+1})$. If we abbreviate $c(B_i, B_{i+1}, B_{i+2})$ $(X_{i,i+1}, X_{X_{i+1,i+2}})$ with $X_{i+1,i+2} \circ X_{i,i+1}$ and if we suppress the objects at 1 and α, then one can express (9.6.1) as a pentagon axiom.

$$
\begin{array}{ccc}
X_{4,5} \circ (X_{3,4} \circ (X_{2,3} \circ X_{1,2})) & \xrightarrow{\;1 \circ \alpha\;} & X_{4,5} \circ ((X_{3,4} \circ X_{2,3}) \circ X_{1,2}) \\
\Big\| \alpha & & \Big\| \alpha \\
(X_{4,5} \circ X_{3,4}) \circ (X_{2,3} \circ X_{1,2}) & & (X_{4,5} \circ (X_{3,4} \circ X_{2,3})) \circ X_{1,2} \\
& \searrow^{\alpha} \quad \swarrow^{\alpha \circ 1} & \\
& ((X_{4,5} \circ X_{3,4}) \circ X_{2,3}) \circ X_{1,2} &
\end{array}
$$

Similarly, (9.6.2) can be expressed as a triangle axiom.

Examples 9.6.2

- If $\mathcal{B}$ is a bicategory with a single object $*$, then the category of morphisms, $\mathcal{B}(*, *)$, forms a monoidal category (recall Definition 8.1.4). For two objects C_1, C_2 of the category $\mathcal{B}(*, *)$ set $C_1 \otimes C_2 := C_2 \circ C_1$. In these examples, the coherence condition from (9.6.1) reduces to the usual pentagon coherence condition for monoidal categories and (9.6.2) corresponds to the triangle condition for the coherence between the associativity isomorphism and the left and right units. Conversely, if C is a monoidal category, then it gives rise to a one-object bicategory, which one often denotes by ΣC.

- Every 2-category is a bicategory, where the αs, ϱs, and λs are identities.

- Let Bim be the bicategory that associates with two rings R_1, R_2 the category $\mathrm{Bim}(R_1, R_2)$, whose objects are the R_1-R_2-bimodules and whose morphisms are the morphisms of R_1-R_2-bimodules.

Exercise 9.6.3 Let C be a category that has pullbacks. Show that the category of spans in C forms a bicategory.

Definition 9.6.4 A *morphism (F, φ) of bicategories* from a bicategory $\mathcal{B}$ to a bicategory $\mathcal{B}'$ consists of the following:

- An assignment $B \mapsto F(B)$ on the level of objects.
- A family of functors

$$F(B_1, B_2) \colon \mathcal{B}(B_1, B_2) \to \mathcal{B}'(F(B_1), F(B_2)).$$

- For every object B of $\mathcal{B}$, there is a morphism in $\mathcal{B}'(F(B), F(B))$ (that is, a 2-cell):

$$\varphi \colon I_{F(B)} \Rightarrow F(I_B).$$

- For every triple of objects B_1, B_2, B_3 of $\mathcal{B}$, there is a natural transformation
$\varphi(B_1, B_2, B_3) \colon c \circ (F(B_1, B_2) \times F(B_2, B_3)) \Rightarrow F(B_1, B_3) \circ c$:

$$
\begin{array}{ccc}
\mathcal{B}(B_1, B_2) \times \mathcal{B}(B_2, B_3) & \xrightarrow{\quad c \quad} & \mathcal{B}(B_1, B_3) \\
{\scriptstyle F(B_1,B_2) \times F(B_2,B_3)} \downarrow & \varphi(B_1,B_2,B_3) & \downarrow {\scriptstyle F(B_1,B_3)} \\
\mathcal{B}'(F(B_1), F(B_2)) \times \mathcal{B}'(F(B_2), F(B_3)) & \xrightarrow{\quad c \quad} & \mathcal{B}'(F(B_1), F(B_3)).
\end{array}
$$

These functors and natural transformations have to satisfy the following properties:

(1) For any object $(X_{1,2}, X_{2,3}, X_{3,4})$ of $\mathcal{B}(B_1, B_2) \times \mathcal{B}(B_2, B_3) \times \mathcal{B}(B_3, B_4)$, the diagram

$$
\begin{array}{ccc}
F(B_3,B_4)(X_{3,4}) \circ (F(B_2,B_3)(X_{2,3}) \circ F(B_1,B_2)(X_{1,2})) & \xrightarrow{\ \alpha\ } & (F(B_3,B_4)(X_{3,4}) \circ F(B_2,B_3)(X_{2,3})) \circ F(B_1,B_2)(X_{1,2}) \\
{\scriptstyle 1 \circ \varphi(B_1,B_2,B_3)(X_{1,2},X_{2,3})} \downarrow & & \downarrow {\scriptstyle \varphi(B_2,B_3,B_4)(X_{2,3},X_{3,4}) \circ 1} \\
F(B_3,B_4)(X_{3,4}) \circ F(B_1,B_3)(X_{2,3} \circ X_{1,2}) & & F(B_2,B_4)(X_{3,4} \circ X_{2,3}) \circ F(B_1,B_2)(X_{1,2}) \\
{\scriptstyle \varphi(B_1,B_3,B_4)(X_{2,3} \circ X_{1,2},X_{3,4})} \downarrow & & \downarrow {\scriptstyle \varphi(B_1,B_2,B_4)(X_{1,2},X_{3,4} \circ X_{2,3})} \\
F(B_1,B_4)(X_{3,4} \circ (X_{2,3} \circ X_{1,2})) & \xrightarrow{\ F(B_1,B_4)(\alpha)\ } & F(B_1,B_4)((X_{3,4} \circ X_{2,3}) \circ X_{1,2})
\end{array}
$$

commutes.

(2) For every object X of $\mathcal{B}(B_1, B_2)$, the following unit diagrams commute:

$$
\begin{array}{ccc}
I_{F(B_2)} \circ F(B_1, B_2)(X) & \xrightarrow{\ \varphi(B_2) \circ 1\ } & F(B_2, B_2)(I_{B_2}) \circ F(B_1, B_2)(X) \\
{\scriptstyle \lambda} \downarrow & & \downarrow {\scriptstyle \varphi(B_1,B_2)(I_{B_2},X)} \\
F(B_1, B_2)(X) & \xleftarrow{\ F(\lambda)\ } & F(B_1, B_2)(I_{B_2} \circ X)
\end{array}
$$

and

$$
\begin{array}{ccc}
F(B_1, B_2)(X) \circ I_{F(B_1)} & \xrightarrow{\ 1 \circ \varphi(B_1)\ } & F(B_1, B_2)(X) \circ F(B_1, B_1)(I_{B_1}) \\
{\scriptstyle \varrho} \downarrow & & \downarrow {\scriptstyle \varphi(B_1,B_2)(X,I_{B_1})} \\
F(B_1, B_2)(X) & \xleftarrow{\ F(\varrho)\ } & F(B_1, B_2)(X \circ I_{B_1}).
\end{array}
$$

Remark 9.6.5 There is an abundance of different names in the setting of bicategories. As the notion of a morphism of bicategories reduces to the notion of a lax monoidal functor for the one-object bicategory, it is rather common to call such morphisms *lax functors* between bicategories. There is also the notion of a *pseudofunctor*; for these, one requires the natural transformations that are part of the data of a morphism, to be natural isomorphisms. Thomason [**T79**] uses the variant that reduces to lax comonoidal functors between monoidal categories and calls such functors *op-lax*, so this corresponds to reversing the direction of the structure maps φ shown earlier. The idea of not using functors but something weaker that is still coherent goes back to Grothendieck [**G-SGA1**, Exposé VI].

You can now continue and define an analog of natural transformations for morphisms between bicategories. These are called *modifications*. For more background on these concepts, see [**Le∞**, 1.3]. As every monoidal category can be strictified, as in Proposition 8.3.4, every bicategory is biequivalent to a strict 2-category [**Le∞**, 2.3].

9.7 Functor Categories

Let $\mathcal{D}$ be a small category and let $\mathcal{C}$ an arbitrary category. Recall that we denote by $\mathsf{Fun}(\mathcal{D}, \mathcal{C})$ the category of functors from $\mathcal{D}$ to $\mathcal{C}$ with natural transformations as morphisms.

Examples 9.7.1

- Let $\mathcal{I}$ denote the category of finite sets and injections with objects $\mathbf{n} = \{1, \ldots, n\}$ for $n \geq 0$, where $\mathbf{0}$ denotes the empty set. Morphisms in this category are injective functions. The category of functors from $\mathcal{I}$ to a category $\mathcal{C}$ plays an important role, for instance, in Bökstedt's definition of topological Hochschild homology [**Bö∞**] from the 1980s, in the setting of FI-modules [**CEF15**] and in the study of $\mathcal{I}$-spaces [**SaSc12**]. We will encounter this category again later.

- The category Σ has again the same objects as $\mathcal{I}$, but we only consider bijections as maps; hence,

$$\Sigma(\mathbf{n}, \mathbf{m}) = \begin{cases} \varnothing, & \text{if } n \neq m \\ \Sigma_n, & \text{if } n = m. \end{cases}$$

Hence, one can view Σ as the subcategory of $\mathcal{I}$ of isomorphisms, that is, $\Sigma = \mathrm{Iso}(\mathcal{I})$.

Let $\mathcal{C}$ be an arbitrary category. Then, the category $\mathsf{Fun}(\Sigma, \mathcal{C})$ is called the *category of symmetric sequences in $\mathcal{C}$*. If X is an object of $\mathsf{Fun}(\Sigma, \mathcal{C})$, then $X(\mathbf{n})$ is called the *nth level of X*. Symmetric sequences play an important role in mathematics. You can find more about symmetric sequences of module categories and their algebraic properties in [**Sto93**], and a comprehensive overview is in [**AM10**]. We will come back to them later when we discuss symmetric spectra.

- We denote by Γ the category of finite pointed sets and pointed maps with objects $[n] = \{0, \dots, n\}$ for $n \geq 0$, where 0 is the basepoint of $[n]$. Morphisms are functions of finite sets sending 0 to 0. Note that our use of Γ is the opposite to Segal's category Γ [**Se74**].

We saw in Proposition 3.3.1 that (co)completeness of C transfers to (co)completeness of $\mathsf{Fun}(\mathcal{D}, C)$. Enrichments can also be transferred.

Proposition 9.7.2 *Let $\mathcal{D}$ be a small category and let C be a category enriched in $\mathcal{V}$. If $\mathcal{V}$ has all small limits, then $\mathsf{Fun}(\mathcal{D}, C)$ is $\mathcal{V}$-enriched.*

Proof For two objects F, G of $\mathsf{Fun}(\mathcal{D}, C)$, we have to define an object in $\mathcal{V}$, $\underline{\mathsf{Fun}}(\mathcal{D}, C)(F, G)$. We construct this as the $\mathcal{V}$-end of the diagram

$$\underline{C}(F(-), G(-)) \colon \mathcal{D}^o \times \mathcal{D} \to \mathcal{V}. \qquad \square$$

9.8 Day Convolution Product

The Day convolution product is a simple procedure to transfer symmetric monoidal structures to the level of functor categories. This monoidal product features in several applications, and we present some of those.

Definition 9.8.1
- Let C be a symmetric monoidal category and let $\mathcal{D}$ be a small category. For two functors $F, G \in \mathsf{Fun}(\mathcal{D}, C)$, we can form their *external product* $F \boxtimes G \in \mathsf{Fun}(\mathcal{D} \times \mathcal{D}, C)$ by defining

$$(F \boxtimes G)(D_1, D_2) := F(D_1) \otimes G(D_2), \quad (F \boxtimes G)(f_1, f_2) := F(f_1) \otimes G(f_2)$$

for all objects D_1, D_2 of $\mathcal{D}$ and all morphisms f_1, f_2 in $\mathcal{D}$.
- Let $(\mathcal{D}, \sqcup, 0, \tau)$ and $(C, \otimes, e_C, c)$ be two symmetric monoidal categories, and assume that $\mathcal{D}$ is small and that C is cocomplete. The *Day convolution product* $F \square G$ of two functors $F, G \in \mathsf{Fun}(\mathcal{D}, C)$ is defined as the left Kan extension of the external product $F \boxtimes G$ along the functor $\sqcup \colon \mathcal{D} \times \mathcal{D} \to \mathcal{D}$:

In the following, we will work with *enriched functors and their Day convolution product*, and hence, the left Kan extension described previously takes place in an enriched setting. In the applications, we will often use the standard enrichment (see Definition 9.1.3) of a small category $\mathcal{D}$ in a cocomplete closed symmetric monoidal category $(\mathcal{C}, \otimes, e_C)$ with

$$\underline{\mathcal{D}}(D_1, D_2) = \bigsqcup_{\mathcal{D}(D_1, D_2)} e_C,$$

and then, $\mathcal{C}$-functors $F \colon \mathcal{D} \to \mathcal{C}$ correspond to ordinary functors from $\mathcal{D}$ to $\mathcal{C}$. By $\mathsf{Func}_\mathcal{C}(\mathcal{D}, \mathcal{C})$, we denote the category of $\mathcal{C}$-functors and $\mathcal{C}$-natural transformations.

Note that if $\mathcal{D}$ is a small symmetric monoidal category enriched in $\mathcal{C}$ and $\mathcal{C}$ is a cocomplete closed symmetric monoidal category, then we have a pointwise left Kan extension (see Definition 4.1.5), and hence, we can also express the functor $F \square G$ as the coend

$$\int^{\mathcal{D} \times \mathcal{D}} \underline{\mathcal{D}}(D_1 \sqcup D_2, -) \otimes (F \boxtimes G)(D_1, D_2)$$
$$= \int^{\mathcal{D} \times \mathcal{D}} \underline{\mathcal{D}}(D_1 \sqcup D_2, -) \otimes F(D_1) \otimes G(D_2).$$

An immediate reformulation is then of the following form:

Lemma 9.8.2 *If $\mathcal{D}$ is a small symmetric monoidal category enriched in $\mathcal{C}$ and if $\mathcal{C}$ is a cocomplete closed symmetric monoidal category, then there is an isomorphism*

$$\mathsf{Func}_\mathcal{C}(\mathcal{D}, \mathcal{C})(F \square G, H) \cong \mathsf{Func}_\mathcal{C}(\mathcal{D} \times \mathcal{D}, \mathcal{C})(F \boxtimes G, H \circ (- \sqcup -)).$$

With the coend interpretation of the Day convolution product, it is rather transparent that we get a symmetric monoidal structure on $\mathsf{Func}_\mathcal{C}(\mathcal{D}, \mathcal{C})$.

Proposition 9.8.3 *Assume that $(\mathcal{D}, \sqcup, 0, \tau)$ is a small symmetric monoidal category enriched in $\mathcal{C}$, where $\mathcal{C}$ is a cocomplete closed symmetric monoidal category. Then, the category of enriched functors from $\mathcal{D}$ to $\mathcal{C}$, $\mathsf{Func}_\mathcal{C}(\mathcal{D}, \mathcal{C})$ is a symmetric monoidal category with the Day convolution product. The unit for $(\mathsf{Func}_\mathcal{C}(\mathcal{D}, \mathcal{C}), \square)$, is given by $\underline{\mathcal{D}}(0, -)$, and the symmetry is induced by the ones in $\mathcal{D}$ and in $\mathcal{C}$.*

Proof   Writing down a full proof is a tedious task, and Day has done it in [**Day70b**]. We sketch the main idea.

The associativity of the Day convolution product follows from the associativity of the coend description given earlier, together with the associativity for the monoidal structures in $\mathcal{D}$ and $\mathcal{C}$.

$$(F \square G) \square H = \int^{\mathcal{D} \times \mathcal{D}} \underline{\mathcal{D}}(D_1 \sqcup D_4, -)$$

$$\otimes \left(\int^{\mathcal{D} \times \mathcal{D}} \underline{\mathcal{D}}(D_2 \sqcup D_3, D_1) \otimes F(D_2) \square G(D_3) \right) \otimes H(D_4)$$

$$\cong \int^{\mathcal{D} \times \mathcal{D} \times \mathcal{D}} \underline{\mathcal{D}}(D_2 \sqcup D_3 \sqcup D_4, -) \otimes (F(D_2) \otimes G(D_3)) \otimes H(D_4).$$

The unit condition follows in a similar manner, and we only show the left unit condition

$$\underline{\mathcal{D}}(0, -) \square F = \int^{\mathcal{D} \times \mathcal{D}} \underline{\mathcal{D}}(D_1 \sqcup D_2, -) \otimes \underline{\mathcal{D}}(0, D_1) \otimes F(D_2)$$

$$\cong \int^{\mathcal{D}} \underline{\mathcal{D}}(0 \sqcup D_2, -) \otimes F(D_2) \cong F.$$

For the symmetry isomorphism

$$\chi_{F,G} \colon F \square G \to G \square F,$$

we have to apply the symmetry in $\mathcal{C}$, c, in order to interchange the $\otimes$-factors $F(D_1) \otimes G(D_2)$, but we also have to apply the twist in $\mathcal{D}$, τ in order to get an isomorphism

$$\underline{\mathcal{D}}(\tau, 1) \otimes c_{F(D_1), G(D_2)} \colon \underline{\mathcal{D}}(D_1 \sqcup D_2, -) \otimes F(D_1) \otimes G(D_2)$$
$$\to \underline{\mathcal{D}}(D_2 \sqcup D_1, -) \otimes G(D_2) \otimes F(D_1) \qquad (9.8.1)$$

that combines to a binatural isomorphism $\chi_{F,G}$. $\qquad \square$

Corollary 9.8.4 *Under the assumptions of Proposition 9.8.3 and if $\mathcal{C}$ has small limits, then the symmetric monoidal structure on $\mathrm{Func}(\mathcal{D}, \mathcal{C})$ is closed.*

Proof We define an internal hom-object $\underline{\mathcal{C}}^{\mathcal{D}}(F, G) \colon \mathcal{D} \to \mathcal{C}$ as the $\mathcal{C}$-functor that sends an object D of $\mathcal{D}$ to the end

$$\int_{\mathcal{D}} \underline{\mathcal{C}}(F(D_1), G(D \sqcup D_1)).$$

For checking that this actually defines an adjoint to $\square$, we use

$$\mathrm{Func}(\mathcal{D}, \mathcal{C})(F \square G, H) \cong \mathrm{Func}_{\mathcal{C}}(\mathcal{D} \times \mathcal{D}, \mathcal{C})(F \boxtimes G, H \circ (- \sqcup -))$$

and the latter is isomorphic to

$$\int_{\mathcal{D} \times \mathcal{D}} \underline{\mathcal{C}}(F(D_1), \underline{\mathcal{C}}(G(D_2), H(D_1 \sqcup D_2)))$$

$$\cong \int_{\mathcal{D}} \underline{\mathcal{C}}(F(D_1), \underline{\mathcal{C}}^{\mathcal{D}}(G, H)(D_1))$$

$$\cong \mathrm{Func}_{\mathcal{C}}(\mathcal{D}, \mathcal{C})(F, \underline{\mathcal{C}}^{\mathcal{D}}(G, H)). \qquad \square$$

The fact that $\mathsf{Func}_{\mathcal{C}}(\mathcal{D}, \mathcal{C})$ is closed with respect to the Day convolution product enables us to calculate the product of two enriched representable functors.

Proposition 9.8.5 *Assume that $\mathcal{C}$ has small limits and that the assumptions of Proposition 9.8.3 are satisfied. Then, for any two objects D_1, D_2 of $\mathcal{D}$,*

$$\underline{\mathcal{D}}(D_1, -) \square \underline{\mathcal{D}}(D_2, -) \cong \underline{\mathcal{D}}(D_1 \sqcup D_2, -).$$

Proof   The adjunction property implies that

$$\mathsf{Func}_{\mathcal{C}}(\mathcal{D}, \mathcal{C})(\underline{\mathcal{D}}(D_1, -) \square \underline{\mathcal{D}}(D_2, -), H) \cong$$
$$\mathsf{Func}_{\mathcal{C}}(\mathcal{D}, \mathcal{C})(\underline{\mathcal{D}}(D_1, -), \underline{\mathcal{C}}^{\mathcal{D}}(\underline{\mathcal{D}}(D_2, -), H)),$$

but $\underline{\mathcal{C}}^{\mathcal{D}}(\underline{\mathcal{D}}(D_2, -), H) \cong \int_{\mathcal{D}} \mathcal{C}(\underline{\mathcal{D}}(D_2, D_3), H(- \sqcup D_3)) \cong H(- \sqcup D_2)$ and

$$\mathsf{Func}_{\mathcal{C}}(\mathcal{D}, \mathcal{C})(\underline{\mathcal{D}}(D_1, -), H(- \sqcup D_2)) \cong H(D_1 \sqcup D_2) \cong$$
$$\mathsf{Func}_{\mathcal{C}}(\mathcal{D}, \mathcal{C})(\underline{\mathcal{D}}(D_1 \sqcup D_2, -), H),$$

and hence, $\underline{\mathcal{D}}(D_1, -) \square \underline{\mathcal{D}}(D_2, -) \cong \underline{\mathcal{D}}(D_1 \sqcup D_2, -)$.   □

Examples 9.8.6

- We already studied the first example in Example 4.1.7. Let $(\mathcal{C}, \otimes, e_{\mathcal{C}})$ be a bicomplete closed symmetric monoidal category, let $\mathcal{D}$ be the category Σ, and assume that Σ is enriched in $\mathcal{C}$ in the standard way, that is,

$$\underline{\Sigma}(n, m) = \bigsqcup_{\Sigma(n,m)} e_{\mathcal{C}}.$$

The category of symmetric sequences in $\mathcal{C}$, $\mathcal{C}^{\Sigma}$ possesses a symmetric monoidal structure given by the Day convolution product. The external product of two symmetric sequences X, Y is

$$(X \boxtimes Y)(\mathbf{p}, \mathbf{q}) = X(\mathbf{p}) \otimes Y(\mathbf{q}).$$

The Day convolution product can be made explicit in this example, and we obtain

$$(X \square Y)(\mathbf{n}) = \bigsqcup_{p+q=n} \Sigma_n \times_{\Sigma_p \times \Sigma_q} X(\mathbf{p}) \otimes Y(\mathbf{q}).$$

Here, $\bigsqcup$ denotes the coproduct in $\mathcal{C}$, and $\Sigma_n \times \mathcal{C} = \bigsqcup_{\Sigma_n} \mathcal{C}$ carries the Σ_n-action that permutes the summands.

The unit for this structure is given by the symmetric sequence $u = \underline{\mathcal{D}}(0, -)$. This can be made explicit. We obtain $u(0) = e_{\mathcal{C}}$ and $u(n) = \varnothing$ for all positive n, where $\varnothing$ denotes the initial object of $\mathcal{C}$.

For the twist map

$$c_\square \colon X \square Y \to Y \square X,$$

the twist $c_\square$ on level n is defined as

$$c_\square(\mathbf{n}) \colon (X \square Y)(\mathbf{n}) \to (Y \square X)(\mathbf{n}),$$

where $c_\square(\mathbf{n})$ applies the twist map

$$c_\otimes \colon X(\mathbf{p}) \otimes Y(\mathbf{q}) \to Y(\mathbf{q}) \otimes X(\mathbf{p})$$

and uses the shuffle permutation $\chi_{p,q} \in \Sigma_n$ that exchanges the first p elements with the last q elements on the permutation coordinate in Σ_n.

Example 9.8.7 Let $\mathcal{I}$ be the category of finite sets and injections from Example 9.7.1. As $\mathcal{C}$, we take any symmetric monoidal cocomplete category, in which $\mathcal{I}$ is enriched in the standard way. In this example, the unit 0 is initial in $\mathcal{I}$, and hence,

$$\underline{\mathcal{I}}(0, -) \colon \mathcal{I} \to \mathcal{C}$$

is the constant functor with value $e_\mathcal{C}$.

A version of the following result can be found in [**MMSS01**, §20]. The category of lax (symmetric) monoidal functors has as objects lax (symmetric) monoidal functors and as morphisms monoidal natural transformations, as in Definition 8.3.8.

Proposition 9.8.8 *Let $\mathcal{D}$ be a small category symmetric monoidal category and let $\mathcal{C}$ be a symmetric monoidal category that is cocomplete. Then, there is an equivalence of categories between the category of lax (symmetric) monoidal functors and the category of (commutative) monoids in* $\mathsf{Func}(\mathcal{D}, \mathcal{C})$ *with respect to the Day convolution product.*

Proof The universal property of maps out of the Day convolution product from Lemma 9.8.2

$$\mathsf{Func}(\mathcal{D}, \mathcal{C})(F \square G, H) \cong \mathsf{Func}(\mathcal{D} \times \mathcal{D}, \mathcal{C})(F \boxtimes G, H \circ (- \sqcup -))$$

allows us to see that multiplications $m \colon F \square F \to F$ are in bijection with $\mathcal{C}$-natural transformations $F \boxtimes F \Rightarrow F \circ (- \sqcup -)$.

A functor F is a lax monoidal functor if an only if there is such a $\mathcal{C}$-natural transformation

$$\varphi \colon F \boxtimes F \Rightarrow F \circ ((-) \sqcup (-))$$

and a morphism $e_\mathcal{C} \to F(0)$ satisfying compatibility conditions. The associativity of the monoid structure is equivalent to the associativity condition on φ.

Starting with φ, we denote the resulting multiplication $F \square F \to F$ by m_φ. If $m : F \square F \to F$ is given, we call the resulting C-natural transformation φ_m.

It remains to check the equivalence of the unit condition on the monoid and on φ and the equivalence of the commutativity of the monoid structure and symmetry condition on φ.

Assume that F is lax monoidal with unit $\eta_F : e_C \to F(0)$. We need a C-natural transformation $\eta : \underline{D}(0, -) \Rightarrow F$, and we define the C-component of η as

$$\eta_C : \underline{D}(0, C) \cong \underline{D}(0, C) \otimes e_C \xrightarrow{1_{\underline{D}(0,C)} \otimes \eta_F} \underline{D}(0, C) \otimes F(0) \xrightarrow{ev} F(C).$$

Here, ev is the evaluation map from (9.1.1).

Conversely, if F is a monoid with unit $\eta : \underline{D}(0, -) \Rightarrow F$, then the unit of the enriched structure $e_C \to \underline{D}(0, 0)$ composed with η_0 gives a morphism $e_C \to F(0)$, and this gives a unit for the lax monoidal structure.

These assignments are inverse to each other. Starting with $\eta : \underline{D}(0, -) \Rightarrow F$ and defining first $\eta_F : e_C \to F(0)$ from it and then the natural transformation associated with it give the composition from the upper-left corner to the lower-left corner in the diagram

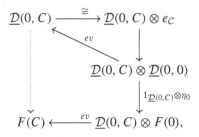

and as the upper triangle commutes, the dashed arrow is nothing but η_C.

Vice versa, starting with $\eta_F : e_C \to F(0)$ and defining $\eta : \underline{D}(0, -) \Rightarrow F$ from it and then restricting to the resulting morphism $e_C \to F(0)$ yields the composite

$$e_C \xrightarrow{\eta_F} \underline{D}(0, 0) \xrightarrow{\cong} \underline{D}(0, 0) \otimes e_C$$
$$\downarrow{1_{\underline{D}(0,0)} \otimes \eta_F}$$
$$\underline{D}(0, 0) \otimes F(0)$$
$$\downarrow{ev}$$
$$F(0),$$

and by the unitality condition on the action of enriched morphisms on enriched functors, this is equal to η_F.

Assume that the monoid F is abelian, that is, $m \circ \chi_{F,F} = m$ as elements in $\mathsf{Func}(\mathcal{D}, \mathcal{C})(F \Box F, F)$. Then, for $\varphi_m : F \boxtimes F \to F \circ (- \sqcup -)$, we get with (9.8.1) that

$$F(\tau_{D_1, D_2}) \circ (\varphi_m)_{D_1, D_2} = (\varphi_m)_{D_2, D_1} \circ c_{F(D_1), F(D_2)}$$

and that is precisely the symmetry of φ_m.

A morphism of monoids $\alpha : F \to G$ is a $\mathcal{C}$-natural transformation, which satisfies that for all objects D_1, D_2, the diagram

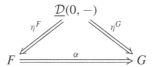

commutes and vice versa. It remains to check that the condition on the unit transfers as well.

If $\alpha : F \Rightarrow G$ satisfies that

$$
\begin{array}{ccc}
 & \underline{\mathcal{D}}(0, -) & \\
{}^{\eta^F} \nearrow & & \searrow {}^{\eta^G} \\
F & \underset{\alpha}{\Longrightarrow} & G
\end{array}
$$

commutes, then the induced unit condition for $e_{\mathcal{C}} \to \underline{\mathcal{D}}(0, 0) \to F(0)$ and $e_{\mathcal{C}} \to \underline{\mathcal{D}}(0, 0) \to G(0)$ is satisfied because

$$
\begin{array}{ccc}
 & & {}^{\eta_0^F} \to F(0) \\
e_{\mathcal{C}} & \xrightarrow{u} \underline{\mathcal{D}}(0, 0) & \downarrow {}^{\alpha_0} \\
 & & {}_{\eta_0^F} \to G(0)
\end{array}
$$

commutes. Starting with $\eta_F : e_{\mathcal{C}} \to F(0)$ gives the unit diagram

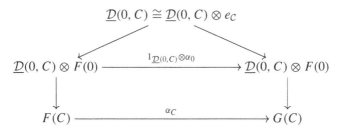

for the monoid morphism α, and this commutes because of the naturality condition of α and the unit property of η_F. $\qquad\square$

Examples 9.8.9 In all our examples, we use the standard enrichment, so $\mathrm{Fun}_C(\mathcal{D}, C) = \mathrm{Fun}(\mathcal{D}, C)$. In the following examples, E_∞-structures feature. We will define them in Definitions 12.3.6 and 12.4.1 for spaces and chain complexes. For now, just think of them as encoding homotopy commutativity up to all coherence constraints that you can think of.

- Commutative monoids in $\mathrm{Fun}(\mathcal{I}, \mathrm{Top})$ are called *commutative $\mathcal{I}$-space monoids* in [**SaSc12**], and with a suitable model structure, they correspond to topological spaces with an E_∞-structure, and hence, they allow us to replace a homotopy-coherent multiplication by a strictly commutative one.
- Similarly, if k is a commutative ring, then commutative monoids in $\mathrm{Fun}(\mathcal{I}, \mathrm{Ch}(k))$ are *commutative $\mathcal{I}$-chain complexes*, and they model unbounded chain complexes over k with an E_∞-structure and commutative algebra spectra over the Eilenberg–Mac Lane spectrum Hk [**RiS17**].
- For functors $F \in \mathrm{Fun}(\Gamma, s\mathrm{Sets}_*)$ with $F(0) = *$, commutative monoids are *commutative Γ-rings*. We will encounter them and their associated commutative ring spectra later again in Remark 14.3.15.

Remark 9.8.10 We will see later in the example of functors from the category of braided injections (see Definition 14.6.1) that a braided monoidal structure on the indexing category can be used to turn the Day convolution product into a braided monoidal structure. Similarly, if the indexing category is just monoidal, one obtains a monoidal Day convolution product.

Part II

From Categories to Homotopy Theory

10

Simplicial Objects

We focus on those aspects of simplicial objects that we will need later. The definition of simplicial sets goes back to Eilenberg and Zilber [**EZ50**] in their comparison of singular chains on a product of spaces, $S_*(X \times Y)$, and the tensor product of the singular chains, $S_*(X) \otimes S_*(Y)$. Kan used simplicial sets as a combinatorial model for homotopy theory [**K57, K58b**]. We leave out the discussion of model category structures on simplicial objects in suitable categories and refer to [**Q67**] and [**GJ09**] for that. Further comprehensive accounts of simplicial sets and related topics are [**GZ67**] and [**May67**].

10.1 The Simplicial Category

We consider the finite set $\{0, 1, \ldots, n\}$ with its natural ordering $0 < 1 < \cdots < n$ and call this ordered set $[n]$ for all $n \geq 0$.

Definition 10.1.1 The *simplicial category*, Δ, has as objects the ordered sets $[n]$, $n \geq 0$, and the morphisms in Δ are the order-preserving functions, that is, functions $f : [n] \to [m]$, such that $f(i) \leq f(j)$ for all $i < j$.

The category Δ is small. As the only order-preserving bijection of the set $[n]$ is the identity map, Δ has only trivial automorphism groups. Hence, if you take the associated category of isomorphisms, $\mathrm{Iso}(\Delta)$, then this is the discrete category on the objects $[n]$ for $n \geq 0$.

As an example, consider the set of morphisms $\Delta([2], [1])$. If an $f \in \Delta$ $([2], [1])$ is not surjective, then it has to map all elements in $[2]$ to either 0 or 1.

If it is surjective, it has to hit one value twice, and this can be either 0 or 1. Therefore, the set $\Delta([2], [1])$ has four elements:

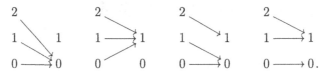

There are certain basic morphisms in Δ, and we will see in Lemma 10.1.4 that we can describe every morphism in Δ via these building blocks. Consider, for instance, the order-preserving map

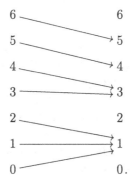

We can factor it as an order-preserving surjection, followed by an order-preserving injection:

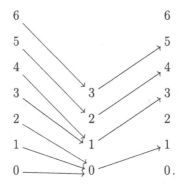

Definition 10.1.2

- For $0 \le i \le n$, let $\delta_i \colon [n-1] \to [n]$ be the order-preserving inclusion that misses the value i in the target:

$$\delta_i(j) = \begin{cases} j, & \text{for } 0 \le j \le i - 1, \\ j + 1, & \text{for } i \le j \le n - 1. \end{cases}$$

- For $0 \leq j \leq n$, let $\sigma_j \colon [n+1] \to [n]$ be the order-preserving surjective map that sends j and $j + 1$ to j:

$$\sigma_j(k) = \begin{cases} k, & \text{for } 0 \leq k \leq j, \\ k - 1, & \text{for } j < k \leq n + 1. \end{cases}$$

Compositions of these maps satisfy certain identities.

Lemma 10.1.3 *The following relations hold:*

$$\begin{aligned} \delta_j \circ \delta_i &= & \delta_i \circ \delta_{j-1}, \quad & i < j, \\ \sigma_j \circ \sigma_i &= & \sigma_i \circ \sigma_{j+1}, \quad & i \leq j, \quad \text{and} \end{aligned}$$

$$\sigma_j \circ \delta_i = \begin{cases} \delta_i \circ \sigma_{j-1}, & i < j, \\ 1_{[n]}, & i = j, j + 1, \\ \delta_{i-1} \circ \sigma_j, & i > j + 1. \end{cases}$$

In the example discussed earlier, you can then write the surjection as composites of σ_is and the injection as a composition of δ_js. That works in general.

Lemma 10.1.4 *Every morphism* $1_{[n]} \neq f \colon [n] \to [m]$ *in* Δ *can be written as a composition*

$$f = \delta_{i_1} \circ \cdots \circ \delta_{i_r} \circ \sigma_{j_1} \circ \cdots \circ \sigma_{j_s},$$

with $0 \leq i_r < \cdots < i_1 \leq m$ *and* $0 \leq j_1 < \cdots < j_s < n$, *where* $m = n - s + r$. *This decomposition is unique.*

Proof You can factor every f as an order-preserving surjection, followed by the order-preserving inclusion of the image into the target. You express the surjection as a composite of the maps σ_js and express the inclusion as a composite of δ_is. The simplicial identities from Lemma 10.1.3 guarantee that these decompositions are unique if we impose the condition on the indices i_k and j_k, as we did. $\qquad \square$

10.2 Simplicial and Cosimplicial Objects

Definition 10.2.1 Let $\mathcal{C}$ be an arbitrary category. A *simplicial object in* $\mathcal{C}$ is a contravariant functor from Δ to $\mathcal{C}$. A *cosimplicial object in* $\mathcal{C}$ is a covariant functor from Δ to $\mathcal{C}$.

How can we describe simplicial objects in an explicit manner? Assume that we have a functor $X \colon \Delta^{op} \to \mathcal{C}$. Then, for every object $[n] \in \Delta$, we have an object $X([n]) =: X_n$ in $\mathcal{C}$. As all morphisms in Δ can be described as a composite of δ_is and σ_js, it suffices to know what the maps $X(\delta_i) =: d_i : X_n \to X_{n-1}$ and $X(\sigma_j) =: s_j$ do. Hence, if you want to describe a simplicial object, then you have to understand the sequence of objects $X_0, X_1, \cdots$ and the morphisms d_i, s_j in $\mathcal{C}$. These maps satisfy the dual relations to the one in Lemma 10.1.3:

$$
\begin{aligned}
d_i \circ d_j &= d_{j-1} \circ d_i, & i < j, \\
s_i \circ s_j &= s_{j+1} \circ s_i, & i \le j, \quad \text{and} \\
d_i \circ s_j &= \begin{cases} s_{j-1} \circ d_i, & i < j, \\ 1_{[n]}, & i = j, j+1, \\ s_j \circ d_{i-1}, & i > j+1. \end{cases} & (10.2.1)
\end{aligned}
$$

Thus, a simplicial object can be visualized as a diagram of the form

$$
X_0 \rightrightarrows X_1 \mathrel{\substack{\longleftarrow \\ \longrightarrow \\ \longleftarrow}} X_2 \quad \ldots,
$$

where the morphisms $\leftarrow$ correspond to the d_is, whereas the morphisms $\rightarrow$ are given by the s_js. Note that on X_n, you have $n+1$ maps going out to the left and to the right.

Definition 10.2.2 The d_is are called *face maps* and the s_js are called *degeneracy maps*.

For a concrete category $\mathcal{C}$ with a faithful functor $U : \mathcal{C} \to$ Sets, the elements $x \in U(X_n)$ are the *n-simplices of X*. We will omit the functor U from the notation. Elements of the form $x = s_i y \in X_n$ for a $y \in X_{n-1}$ are called *degenerate n-simplices*.

We will see later that degenerate simplices are irrelevant for certain properties of a simplicial object.

Simplicial objects in a category $\mathcal{C}$ form a category where the morphisms are natural transformations of functors. We denote this category by $s\mathcal{C}$.

As $\mathcal{C}$ is an arbitrary category, we can consider simplicial R-modules, simplicial sets, simplicial rings, simplicial topological spaces, and many more. Simplicial sets are particularly important because they model topological spaces (see 10.6.1). Simplicial objects in an abelian category $\mathcal{A}$ model nonnegatively graded chain complexes over $\mathcal{A}$. The famous Dold–Kan correspondence (see Theorem 10.11.2) is an equivalence of categories between $s\mathcal{A}$ and $\mathsf{Ch}_{\ge 0}(\mathcal{A})$.

Definition 10.2.3 Let $\mathcal{C}$ be any category. For any object C of $\mathcal{C}$, we define the *constant simplicial object*, $c(C)$, with $c(C)_n = C$ and $d_i = s_i = 1_C$ for all i.

Exercise 10.2.4 Show that $c: \mathcal{C} \to s\mathcal{C}$ is a functor and that it is full and faithful.

Definition 10.2.5 Let $\Delta_n: \Delta^{op} \to$ Sets be the functor given by $[m] \mapsto \Delta([m], [n])$.

Remark 10.2.6 Often, the functor Δ_n is denoted by Δ^n, but as it covariantly depends on n, we stick to Δ_n.

Note that every map in $\Delta([2], [1])$ (and actually every map in $\Delta([n], [1])$ for $n \geq 2$) is degenerate. In $\Delta([1], [1])$, there is a nondegenerate element, the identity on $[1]$, whereas the map that sends both elements of $[1]$ to zero can be written as $s_0(\delta_1) = \delta_1 \circ \sigma_0$. Similarly, the map that sends all of $[1]$ to 1 is equal to $\delta_0 \circ \sigma_0$ and hence are also degenerate. There are two nondegenerate 0-simplices, namely $\delta_0 = d_0(1_{[1]})$ and $\delta_1 = d_1(1_{[1]}): [0] \to [1]$. We will see later that the representable functor Δ_1 is a simplicial model for an interval (with one 1-cell and two 0-cells corresponding to the nondegenerate simplices).

The Yoneda lemma (2.2.2) identifies the set X_n with the set of natural transformations from Δ_n to X for every simplicial set X:

$$X_n \cong s\text{Sets}(\Delta_n, X). \tag{10.2.2}$$

Definition 10.2.7 The *category of elements of a simplicial set* X, el(X), is the category $X\backslash \Delta^o$ associated with the functor $X: \Delta^o \to$ Sets. Explicitly, the objects of el(X) are the $x \in X_n$ for some n. The morphisms in el(X) from $x \in X_n$ to $y \in X_m$ are all $f \in \Delta([n], [m])$ with $X(f)(y) = x$.

Remark 10.2.8 The Yoneda lemma (10.2.2) gives an alternative description of el(X). Objects are morphisms of simplicial sets $x: \Delta_n \to X$, and a morphism from $x: \Delta_n \to X$ to $y: \Delta_m \to X$ is an $f \in \Delta([n], [m])$, such that the diagram

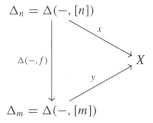

commutes.

The following result is a corollary of Density Theorem 5.4.3:

Proposition 10.2.9 *For every simplicial set X, there is an isomorphism of simplicial sets*

$$\text{colim}_{\text{el}(X)} \Delta_n \cong X. \tag{10.2.3}$$

The category of simplicial sets is closed symmetric monoidal.

Definition 10.2.10 For two simplicial sets X and X', their *product* is the simplicial set $X \times X'$ with $(X \times X')_n = X_n \times X'_n$ and coordinatewise structure maps.

The *simplicial set of morphisms from X to X'* is $s\underline{\text{Sets}}(X, X')$ with p-simplices:

$$s\underline{\text{Sets}}(X, X')_p = s\text{Sets}(X \times \Delta_p, X').$$

Note that Δ_p is covariant in $[p]$, and this determines the simplicial structure maps of $s\underline{\text{Sets}}(X, X')$.

Exercise 10.2.11 Show that the preceding definition of the product and the internal homomorphism object turn the category of simplicial sets into a closed symmetric monoidal category.

10.3 Interlude: Joyal's Category of Intervals

There is an alternative explicit description of the category Δ^o due to Joyal in terms of intervals [**Jo-a∞**]. I first learned about this approach to Δ^o from Martin Markl.

Definition 10.3.1 Let **I** denote the *category of proper intervals* whose objects are the objects $[n]$ of Δ for $n \geq 1$. Morphisms are order-preserving maps $f : [n] \to [m]$, such that $f(0) = 0$ and $f(n) = m$.

There are variants of this definition where one doesn't require n to be at least one. Then, the corresponding simplicial category will be the diagram category describing augmented simplicial objects.

Joyal describes an object $[n + 1]$ of **I** as the possible *Dedekind cuts* [**Jo-a∞**, p. 2] of the object $[n]$ of Δ. Let us mark a cut with the symbol "|". For instance, the Dedekind cuts of the object $[3]$ are

$$| \, 0123, \, 0 \, | \, 123, \, 01 \, | \, 23, \, 012 \, | \, 3 \text{ and } 0123 \, |;$$

thus, in particular, a cut in front of the smallest number 0 is allowed and so is a cut after the largest number, here 3, and therefore, the number of Dedekind cuts of $[n]$ is $n + 2$ and can be identified with the object $[n + 1]$.

We can identify the set of cuts of $[n]$ with the set $\Delta([n], [1])$. If $f \in \Delta([n], [1])$, then there is an i in $[n]$, such that $f(j) = 0$ for all $0 \leq j \leq i - 1$ and $f(j) = 1$ for all $j \geq i$. A morphism $g\colon [m] \to [n]$ induces a map $(-) \circ g\colon \Delta([n], [1])$. If the cut described by f was at 0, that is, $f(j) = 1$ for all j, then the cut described by $f \circ g$ is also at 0. Similarly, if f is the constant map with value 0, that is, it describes the cut at n, then $f \circ g$ is also the constant map with value 0, and hence, $f \circ g$ describes the cut at m. Thus, the precomposition with g induces a map of intervals and thus, we obtain the following:

Lemma 10.3.2 *There is a functor* $D\colon \Delta^o \to \mathbf{I}$, *such that* $D[n] = [n + 1]$.

For the following, it is useful to have a direct description of the effect of D on morphisms. This can be found, for instance, in [**Ou10**]. For $f \in \Delta([n], [m])$, the morphism $D(f) \in \mathbf{I}([m + 1], [n + 1])$ can be described as follows: $D(f)(i)$ is uniquely determined by the property that the preimage $f^{-1}\{i, i + 1, \ldots, m\}$ is equal to $\{D(f)(i), D(f)(i) + 1, \ldots, n\}$. The morphism $D(f)$ satisfies

$$D(f)(i) \leq j \Leftrightarrow i \leq f(j) \text{ for all } i \in [m], j \in [n]. \tag{10.3.1}$$

Note that $D(f)(0) = 0$ and $D(f)(m + 1) = n + 1$, because $D(f)$ is a map of intervals, and thus, (10.3.1) determines $D(f)$.

Example 10.3.3 Consider the following morphism $f\colon [4] \to [3]$ in Δ:

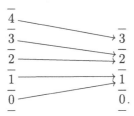

Here, we indicated the cuts of $[4]$ and $[3]$ in the picture. If you focus on the cuts, then for the dual morphism, you connect those cuts that are unobstructed by the arrows of f.

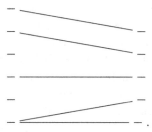

As we identify a category and an opposite category, you also have to reflect the picture, and hence, the dual morphism of f, $D(f)$, is

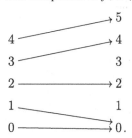

Thus, you can construct the dual morphism $D(f)$ by using the preimages of cuts under f.

Exercise 10.3.4 Draw $D(\delta_i^o)$ and $D(\sigma_j^o)$.

Joyal also describes an inverse to this functor. If $[n]$ is an interval, then we can consider the internal Dedekind cuts of $[n]$, for example, for $[n] = [3]$, the list is

$$0 \mid 123,\, 01 \mid 23 \text{ and } 012 \mid 3.$$

The number of internal cuts of $[n]$ is n and can be identified with the object $[n-1]$ of Δ. We can also identify it as the set of morphisms of intervals $\mathbf{I}([n], [1])$. Therefore, precomposition with a morphism of intervals $g \in \mathbf{I}([m], [n])$ gives a morphism in $\mathbf{I}([m], [1])$. Identifying the latter set again with the object $[m-1]$ of Δ, we get that the corresponding morphism $g^\vee : [n-1] \to [m-1]$ is order-preserving. Thus we get the following:

Lemma 10.3.5 *The assignment* $[n] \mapsto [n-1]$ *and* $\mathbf{I}([n], [m]) \ni g \mapsto g^\vee$ *defines a functor*

$$(-)^\vee : \mathbf{I}^o \to \Delta.$$

Again, we can describe the morphism $g^\vee$ explicitly in terms of g as a morphism from $[n-1]$ to $[m-1]$. The preimage $g^{-1}\{0, \ldots, j\}$ is $\{0, \ldots, g^\vee(j)\}$. Therefore,

$$i \le g^\vee(j) \Leftrightarrow g(i) \le j \text{ for all } i \in [n-1] \text{ and } j \in [m-1]. \qquad (10.3.2)$$

You can also construct $g^\vee$ from g by considering preimages of internal cuts under g.

Summarizing the results, we get a duality between the categories Δ^o and $\mathbf{I}$.

Theorem 10.3.6 *The functor D is an isomorphism of categories with inverse $(-)^\vee$.*

Proof It is clear that D and $(-)^\vee$ are inverse to each other on the level of objects. Applying (10.3.2) to the trivial relation that $g(i) \le g(i)$ for every $g \in \mathbf{I}([n], [m])$ and for all $i \in [n]$ yields $i \le g^\vee(g(i))$ for all $i \in [n]$. With (10.3.1), we obtain

$$D(g^\vee)(i) \le g(i) \text{ for all } i \in [n].$$

By the definition of $D(f)$, we have that $f(D(f)(i))$ is always greater than or equal to i, and that implies by (10.3.2) (with $g^\vee = f$) that

$$g(i) \le D(g^\vee)(i)$$

for all i, and thus, we get that $g = D(g^\vee)$.

Similarly, if $f \in \Delta([n], [m])$, then $f(j) \le f(j)$ for all $j \in [n]$ implies with (10.3.1) that $D(f)(f(j)) \le j$ for all j, and this, in turn, implies with (10.3.2) that

$$f(j) \le D(f)^\vee(j) \text{ for all } j \in [n].$$

As $g(g^\vee(i))$ is always less than or equal to i by definition of $g^\vee$, we get (with $g = D(f)$) by using (10.3.1) that

$$D(f)^\vee(i) \le f(i),$$

and therefore, $f = D(f)^\vee$. $\qquad\qquad\qquad\qquad\qquad\qquad\square$

Corollary 10.3.7 *The category of simplicial objects in a category $\mathcal{C}$ is isomorphic to the category of covariant functors from $\mathbf{I}$ to $\mathcal{C}$.*

Remark 10.3.8 If we denote by $\mathbf{I}^{\text{big}}$ the *category of arbitrary intervals*, that is, linearly ordered sets with a smallest and a biggest element, such that morphisms are order-preserving and preserve the smallest and biggest elements, then a morphism $f \in \mathbf{I}^{\text{big}}(D[n], [0, 1])$ from $D[n] = [n + 1]$ to the unit interval $[0, 1]$ is nothing but a point in the topological n-simplex Δ^n. By definition, $f(0) = 0$ and $f(n + 1) = 1$. The $(n + 1)$-tuple $(t_0 = f(1), t_1 = f(2) - f(1), \ldots, t_n = 1 - f(n))$ is an element of Δ^n.

10.4 Bar and Cobar Constructions

Let $(\mathcal{C}, \otimes, 1)$ be a monoidal category, not necessarily symmetric. Let (C, μ, η) be a monoid in $\mathcal{C}$, let N be a right C-module with structure map $\rho \colon N \otimes C \to N$, and let M be a left C-module with structure map $\lambda \colon C \otimes M \to M$.

We assign to $[n] \in \Delta$ the object $N \otimes C^{\otimes n} \otimes M$ and define

$$d_i \colon N \otimes C^{\otimes n} \otimes M \to N \otimes C^{\otimes n-1} \otimes M,$$

$$d_i = \begin{cases} N \otimes C^{\otimes i-1} \otimes \mu \otimes C^{\otimes n-i-1} \otimes M, & \text{if } 0 < i < n, \\ \rho \otimes C^{\otimes n-1} \otimes M, & \text{if } i = 0, \text{ and} \\ N \otimes C^{\otimes n-1} \otimes \lambda, & \text{for } i = n. \end{cases}$$

The degeneracies are defined by

$$s_i \colon N \otimes C^{\otimes n} \otimes M \to N \otimes C^{\otimes n+1} \otimes M, \quad s_i = N \otimes C^{\otimes i} \otimes \eta \otimes C^{\otimes n-i} \otimes M.$$

Proposition 10.4.1 *For every monoid C, every right C-module N, and left C-module M, the preceding notions define a simplicial object in $\mathcal{C}$.*

Proof The simplicial identities (10.2.1) are easy to check: If $|i - j| > 1$, then d_i and d_j don't interfere, and the equality $d_i \circ d_j = d_{j-1} \circ d_i$ for $i < j + 1$ is caused by an index shift due to the multiplication in C or a module action map. The relation $d_i \circ d_{i+1} = d_i \circ d_i$ corresponds to the associativity of μ for $0 < i < n$ and to the associativity of the C-module structures for $i = 0, n$. Inserting the unit in different places does not cause any relation but a reindexing, so $s_i \circ s_j = s_{j+1} \circ s_i$ holds for $i \leq j$. The relations $d_i \circ s_i = 1 = d_{i+1} \circ s_i$ follow from the unit condition of the monoid structure on C and the unit condition for the module structures. In all other cases, the d_is and s_js don't interfere, except for an index shift. $\qquad \square$

Definition 10.4.2 The simplicial object $B(N, C, M)$, $[n] \mapsto B(N, C, M)_n = N \otimes C^{\otimes n} \otimes M$ is the *two-sided bar construction on N, C, and M*.

Remark 10.4.3 One can dualize the two-sided bar construction. For a comonoid T in $\mathcal{C}$, a right T-comodule N, and a left T-comodule M, the assignment $[n] \mapsto N \otimes T^{\otimes n} \otimes M$ defines a cosimplicial object, called the *(two-sided) cobar construction*. This is often denoted by $\Omega(N, T, M)$.

10.4.1 Applications of Bar and Cobar Constructions

We present some sample applications of bar and cobar constructions. As these constructions are quite ubiquitous in homological algebra and algebraic topology, the following is just a tiny appetizer. We will encounter more examples of bar constructions later.

- If $\mathcal{C}$ is the category of abelian groups and $C = R$ is an associative ring, N is a right R-module, and M is a left R-module, then the definition of the two-sided bar construction recovers the one from homological algebra [**W94**, Example 8.7.2]. It is an important tool to generate resolutions and is used in many applications where you need to calculate Tor and Ext groups such as group (co)homology, Hochschild (co)homology, and others.
- Monads C on a category $\mathcal{C}$ give rise to an important class of examples [**May72**, §9]. Right C-modules are then functors $F: \mathcal{C} \to \mathcal{C}$ with a module structure map $\rho: F \circ C \to F$, which is compatible with the unit and the multiplication of the monad C. Similarly, we can define left C-modules. For a right C-module F and a left C-module G, we get a two-sided bar construction $B(F, C, G)$.

  We can modify the preceding construction to allow the left C-module to be a functor from some category $\mathcal{D}$ to $\mathcal{C}$ and the right C-module to be a functor from $\mathcal{C}$ to $\mathcal{D}$. Then, the two-sided bar construction is a simplicial object in $\mathcal{D}$.

  Consider a monad $C = RL$ associated with an adjoint pair of functors (L, R). Then, L is a right C-module with structure map

$$\rho = \varepsilon_L: LC = L(RL) = (LR)L \to L,$$

and any C-algebra has a left C-action. In particular, if D is a C-algebra, then $R(D)$ has a left C-action, and we obtain a two-sided bar construction:

$$B(L, C, R(D)). \tag{10.4.1}$$

- As an example of a monad, we consider the loop-suspension adjunction. We work in the category of based compactly generated spaces cg_* from Definition 8.5.13. The reduced n-fold suspension functor $\Sigma^n : cg_* \to cg_*$ is given by

$$\Sigma^n X = \mathbb{S}^n \wedge X.$$

Its right adjoint is the n-fold based loop functor, which sends X to $\underline{k\mathsf{Top}}_*(\mathbb{S}^n, X) = \Omega^n X$.

If Y is an n-fold based loop space, that is, $Y = \Omega^n X$ for some X in cg_*, then Y is a $C = \Omega^n \Sigma^n$-algebra whose structure map

$$\Omega^n \Sigma^n Y = \Omega^n \Sigma^n \Omega^n X \to \Omega^n X$$

is $\Omega^n(\varepsilon_X)$. The functor Σ^n is a right C-module, and thus, we can form the two-sided bar construction $B(\Sigma^n, \Omega^n \Sigma^n, \Omega^n X)$. This example features prominently in [**May72**]. Beck showed [**Be69**, Theorem (16)] that a space Y that is an algebra for the monad $\Omega^n \Sigma^n$ and is of the homotopy type of a CW complex can be identified with the n-fold based loop space of $B(\Sigma^n, \Omega^n \Sigma^n, Y)$. This bar construction hence provides an n-fold delooping of $Y = \Omega^n X$.

- An important example of a two-sided cobar construction is the cosimplicial object associated with two continuous maps of topological spaces:

$$
\begin{array}{ccc}
 & & E \\
 & & \downarrow{\scriptstyle p} \\
X & \xrightarrow{\;f\;} & B.
\end{array}
$$

Every topological space is a comonoid, thanks to the diagonal map of topological spaces. We set

$$\Omega(X, B, E)_n := X \times B^n \times E.$$

Then, X is a right B-comodule via the map

$$X \xrightarrow{\quad \Delta \quad} X \times X \xrightarrow{\quad 1_X \times f \quad} X \times B,$$

and dually, E is a left B-comodule via $(p \times 1_E) \circ \Delta : E \to E \times E \to B \times E$.

A special case of this situation is the path-loop fibration. Choose a basepoint $b_0 \in B$ and consider

where $PB \subset B^I$ is the space of continuous maps ω from the unit interval $I = [0, 1]$ to B with $\omega(0) = b_0$. Then, the fiber product $PB \times_B PB$ is equivalent to the based loop space of B, ΩB. More generally, for every fibration $p: E \to B$ and any continuous map $f: X \to B$, the cosimplicial object $[n] \mapsto X \times B^n \times E$ is the input for one way of constructing the Eilenberg–Moore spectral sequence, which tells you something about the (co)homology of the fiber product $X \times_B E$. You can find a general overview on the Eilenberg–Moore spectral sequence in [**McCl01**]. For approaches using the cobar construction and related cosimplicial techniques, see [**Sm70, Dw74, Bou87**] and [**Sh96**].

- Adams' cobar construction [**Ad56**] is of central importance. Despite its name, it is *similar* to the cobar construction in the sense of Remark 10.4.3, but it is slightly twisted. Consider a differential $\mathbb{N}_0$-graded coalgebra C_* over a commutative ground ring k with $C_0 = k$, $C_1 = 0$ and such that every C_n is a free k-module. Later, C_* will be the coalgebra of normalized simplicial chains $C_*(X)$ on a 1-reduced simplicial set X, that is, an X in sSets with $X_0 = X_1 = \{*\}$.

Consider the reduction $\bar{C}_* = \bigoplus_{n \geq 2} C_n$, which you can also write as the kernel of the counit map. Let $s^{-1}\bar{C}_*$ denote the desuspension of $\bar{C}_*$, with

$$(s^{-1}\bar{C}_*)_n = \bar{C}_{n+1}.$$

It is common to denote elements $x \in (s^{-1}\bar{C}_*)_n$ by $x = s^{-1}c$ for the unique element $c \in \bar{C}_n$ that represents x. Note that $s^{-1}\bar{C}_*$ inherits a differential $d_{s^{-1}\bar{C}_*}$ from C_* via $d_{s^{-1}\bar{C}_*}(s^{-1}c) := -s^{-1}d_{C_*}(c)$. Due to the shift of degrees, the comultiplication on C_* gives rise to a map of degree -1:

$$\Delta: (s^{-1}\bar{C}_*)_n = \bar{C}_{n+1} \to \bigoplus_{p+q=n+1} C_p \otimes_k C_q$$

$$= \bigoplus_{p+q=n+1} (s^{-1}C_*)_{p-1} \otimes_k (s^{-1}C_*)_{q-1}$$

and you reduce the outcome to end up in

$$\bigoplus_{\ell+m=n-1} (s^{-1}\bar{C}_*)_\ell \otimes_k (s^{-1}\bar{C}_*)_m = (s^{-1}\bar{C}_* \otimes_k s^{-1}\bar{C}_*)_{n-1}.$$

We set

$$\Omega(C_*)_n := (s^{-1}\bar{C}_*)^{\otimes_k n},$$

so for every $n \geq 0$, you obtain a graded k-module $(s^{-1}\bar{C}_*)^{\otimes n}$. This looks like a reduced and shifted version of the n-th cosimplicial degree of $\Omega(k, C, k)$, but for now, you consider the underlying graded k-module.

This family of graded k-modules $(\Omega(C_*)_n)_{n \geq 0}$ carries a multiplication via the concatenation of tensors; that is, there are morphisms

$$(\mu_{\Omega(C_*)})_{n,m} : \Omega(C_*)_n \otimes_k \Omega(C_*)_m \to \Omega(C_*)_{n+m}$$

that send generators $s^{-1}c_1 \otimes \cdots \otimes s^{-1}c_n \in \Omega(C_*)_n$ and $s^{-1}c_1' \otimes \cdots \otimes s^{-1}c_m' \in \Omega(C_*)_m$ to

$$s^{-1}c_1 \otimes \cdots \otimes s^{-1}c_n \otimes s^{-1}c_1' \otimes \cdots \otimes s^{-1}c_m'.$$

You take the internal differential of C_* and the diagonal of C_* and merge them into a total differential, $d_{\Omega(C_*)}$. As we use as multiplication the one on the tensor algebra

$$T(s^{-1}\bar{C}_*) = \bigoplus_{n \geq 0}(s^{-1}\bar{C}_*)^{\otimes_k n} = \bigoplus_{n \geq 0}\Omega(C_*)_n$$

and as the tensor algebra functor is left adjoint to the forgetful functor, the differential $d_{\Omega(C_*)}$ is uniquely determined by its value on $\Omega(C_*)_1 = s^{-1}\bar{C}_*$ and the requirement that it has to satisfy the Leibniz rule. One defines

$$d_{\Omega(C_*)}(s^{-1}c) := -s^{-1}d_{C_*}(c) + \sum_{i \in I} s^{-1}c_i \otimes s^{-1}c_i'$$

if $\Delta(c) = c \otimes 1 + 1 \otimes c + \sum_{i \in I} c_i \otimes c_i'$, so you discard the primitive part of the diagonal map and just keep the mixed terms. By slight abuse of notation, $(\Omega(C_*), d_{\Omega(C_*)})$ denotes the complex $\Omega(C_*) = T(s^{-1}\bar{C}_*)$, graded by internal degree with this differential.

Adams shows that the cobar construction of the simplicial singular chains of a 1-reduced space X is a model of the based loop space of X [**Ad56**, Theorem p. 410]:

$$H_*(\Omega(C_*(X)); d_{\Omega(C_*(X))}) \cong H_*(\Omega X).$$

One can upgrade this homology isomorphism to a statement at chain level. For instance, Hess and Tonks prove [**HT10**, Theorem p. 1861] that for every 1-reduced simplicial set X, there is a natural strong deformation retract of chain complexes

$$\Omega(C_*(X)) \underset{\psi}{\overset{\phi}{\rightleftarrows}} C_*(G(X)),$$

where $G(X)$ is the Kan loop group of X [**GJ09**, §V.5], which is a simplicial model for $\Omega|X|$. Here, $\psi \circ \phi$ is the identity, and there is a natural chain homotopy from $\phi \circ \psi$ to the identity map on $C_*(G(X))$.

In rational homotopy theory, Adams' cobar construction is crucial because it gives rise to one of the possible algebraic models of rational spaces [**Q69**].

10.5 Simplicial Homotopies

For topological spaces, the usual notion of a homotopy between two continuous maps $f, g \colon X \to Y$ requires a map $H \colon X \times [0, 1] \to Y$, restricting to f on level 0 and to g at level 1. The notion for a *homotopy between two maps of simplicial sets* $f, g \colon X \to Y$ is similar, requiring a morphism of simplicial sets

$$H \colon X \times \Delta_1 \to Y$$

restricting to f and g, respectively, under $i_0, i_1 \colon \Delta_0 \to \Delta_1$, where $i_j(0) = j$ induces a map $\Delta_0 \to \Delta_1$.

The combinatorial nature of the simplicial category lets us express the notion of homotopy in the following combinatorial manner.

Definition 10.5.1 Let $f, g \colon X \to Y$ be two maps of simplicial sets. A *simplicial homotopy from f to g* consists of a family of maps

$$h_i \colon X_n \to Y_{n+1}, \quad 0 \le i \le n,$$

satisfying $d_0 h_0 = f$, $d_{n+1} h_n = g$ and

$$d_i h_j = \begin{cases} h_{j-1} d_i & \text{for } i < j, \\ d_i h_{i-1} & \text{for } i = j, \\ h_j d_{i-1} & \text{for } i > j+1, \end{cases} \quad \text{and} \quad s_i h_j = \begin{cases} h_{j+1} s_i & \text{for } i \le j, \\ h_j s_{i-1} & \text{for } i > j. \end{cases}$$

Recall from Section 10.3 that elements in $\Delta_1([n]) = \Delta([n], [1])$ can be described as Dedekind cuts, and we label the cuts as $\{c_0, \dots, c_{n+1}\}$ with c_i

representing the cut under $i \in [n]$, that is, $c_i^{-1}(0) = \{0, \ldots, i-1\}$. The following result relates the two notions of a simplicial homotopy:

Proposition 10.5.2

- *Given a simplicial homotopy, as in Definition 10.5.1 with $h_0, \ldots,$ $h_n : X_n \to Y_{n+1}$ defines a homotopy $H : X \times \Delta_1 \to Y$ as*

$$H_n(x_n, c_j) = \begin{cases} g_n(x_n) & \text{for } j = 0, \\ f_n(x_n) & \text{for } j = n+1, \\ d_j h_{j-1}(x_n) & \text{for } 1 \leq j \leq n. \end{cases}$$

- *Conversely, given a homotopy $H : X \times \Delta_1 \to Y$ with $H \circ i_0 = f$ and $H \circ i_1 = g$, the family $h_i(x_n) = H_{n+1}(s_i(x_n), c_{i+1})$ for $0 \leq i \leq n$ defines a simplicial homotopy, as in Definition 10.5.1, from g to f.*

Mind the switch of f and g in the second claim.

Proof   The map i_0 maps the unique element in $\Delta_0([n])$ to c_{n+1}, and i_1 maps it to c_0. We record the following facts about face and degeneracy maps on Dedekind cuts:

$$d_i c_j = \begin{cases} c_{j-1} & \text{for } i \leq j-1, \\ c_j & \text{for } i \geq j, \end{cases} \quad \text{and} \quad s_i c_j = \begin{cases} c_{j+1} & \text{for } i \leq j-1, \\ c_j & \text{for } i \geq j. \end{cases} \quad (10.5.1)$$

In particular, $d_i c_0 = c_0$ and $d_i c_{n+1} = c_n$ for all i.

For the first claim, note that $H_n \circ i_0(x_n) = H_n(x_n, c_{n+1}) = f_n(x_n)$ by definition, and also, $H_n \circ i_1(x_n) = g_n(x_n)$. The remaining identities boil down to using the relations from (10.5.1) and the simplicial identities.

For the converse, given a homotopy $H : X \times \Delta_1 \to Y$ with $H \circ i_0 = f$ and $H \circ i_1 = g$, the assignment $h_i(x_n) = H_{n+1}(s_i(x_n), c_{i+1})$ satisfies

$$d_0 h_0(x_n) = d_0(H_{n+1}(s_0(x_n), c_1)) = H_n(x_n, d_0 c_1) = H_n(x_n, c_0)$$
$$= H \circ i_1(x_n) = g_n(x_n),$$

and similarly,

$$d_{n+1} h_n(x_n) = d_{n+1} H_{n+1}(s_n(x_n), c_{n+1}) = H_n(x_n, d_{n+1} c_{n+1}).$$

Here, c_{n+1} is a Dedekind cut in $\Delta([n+1], [1])$, so it sends $n+1$ to 1, but d_{n+1} applied to it sends all $i \in [n]$ to 0, and hence, $d_{n+1} h_n(x_n) = H_n(x_n, c_{n+1}) = f_n(x_n)$. $\square$

Exercise 10.5.3 Let C be a monad on the category of sets and let X be a C-algebra. (This can be extended to monads on cg.)

- Use the multiplication in the monad and the C-algebra structure on X to define a morphism of simplicial objects:

$$\varrho_X \colon B(C, C, X) \to c(X),$$

where $B(C, C, X)$ is the two-sided bar construction and where $c(X)$ is the constant simplicial object with value X.

- Use the unit of monad to define a morphism $\varphi_X \colon c(X) \to B(C, C, X)$, such that $\varrho_X \circ \varphi_X = 1_{c(X)}$ and such that there is a simplicial homotopy between $\varphi_X \circ \varrho_X$ and $1_{c(X)}$. (If you are desperate, refer to [**May72**, Proposition 9.8].)

Remark 10.5.4 If we express a monad C as a composite of adjoint functors $C = RL$, then

$$B(C, C, X)_n = (RL)^{\circ n+1}(X) = R(L(RL)^{\circ n}(X) = R(B(L, RL, X)_n).$$

In the case of $R = \Omega^k$ and $L = \Sigma^k$, this trick, together with Exercise 10.5.3, can be used to show that a $C = \Omega^k \Sigma^k$-algebra X is equivalent to $\Omega^k |B(\Sigma^k, \Omega^k \Sigma^k, X)|$ (see [**May72**, §13] or [**Be69**, Proof of Theorem (16)]).

10.6 Geometric Realization of a Simplicial Set

The geometric realization of a simplicial set was introduced by Milnor [**Mi57**].

Definition 10.6.1 Let X be a simplicial set. The *geometric realization of X*, $|X|$, is the topological space

$$|X| = \bigsqcup_{n \geq 0} X_n \times \Delta^n / \sim .$$

Here, we consider the sets X_n as discrete topological spaces, and Δ^n denotes the topological n-simplex

$$\Delta^n = \{(t_0, \ldots, t_n) \in \mathbb{R}^{n+1} | 0 \leq t_i \leq 1, \sum t_i = 1\}.$$

The spaces $\Delta^n, n \geq 0$ form a cosimplicial topological space with structure maps

$$\delta_i(t_0, \ldots, t_n) = (t_0, \ldots, t_{i-1}, 0, t_i, \ldots, t_n) \text{ for } 0 \leq i \leq n$$

and

$$\sigma_j(t_0, \ldots, t_n) = (t_0, \ldots, t_j + t_{j+1}, \ldots, t_n) \text{ for } 0 \leq i \leq n.$$

The quotient in the geometric realization is generated by the relations

$$(d_i(x), (t_0, \dots, t_n)) \sim (x, \delta_i(t_0, \dots, t_n)),$$
$$(s_j(x), (t_0, \dots, t_n)) \sim (x, \sigma_j(t_0, \dots, t_n)).$$

Remark 10.6.2 The geometric realization of a simplicial set X is nothing but the coend of the functor

$$H \colon \Delta^o \times \Delta \to \mathsf{Top},$$

with $H([n], [m]) = X_n \times \Delta^m$, using that $[n] \mapsto X_n$ is a contravariant functor from Δ to Sets and that $[m] \mapsto \Delta^m$ is a covariant functor from the category Δ to the category Top. Here, we use the embedding of Sets into Top.

If $f \colon X \to Y$ is a morphism of simplicial sets, that is, a natural transformation from X to Y, then f induces a continuous map of topological spaces

$$|f| \colon |X| \to |Y|,$$

where an equivalence class $[(x, t_0, \dots, t_n)] \in |X|$ is sent to the class $[(f(x), t_0, \dots, t_n)] \in |Y|$. This turns the geometric realization into a functor from the category of simplicial sets to the category of topological spaces.

Elements of the form $s_j(x)$ are identified with something of a lower degree in the geometric realization because of the relation

$$(s_j(x), (t_0, \dots, t_n)) \sim (x, \sigma_j(t_0, \dots, t_n)).$$

Hence, these elements do not contribute any geometric information to $|X|$. This might justify the name degenerate for such elements. Note that elements in X_0 are never degenerate.

An element $(y, (t_0, \dots, t_m)) \in X_m \times \Delta^m$ is called *nondegenerate* if y is not of the form $s_j(x)$ for any x and j and if $(t_0, \dots, t_m) \in \Delta^m$ is not a point on the boundary of the topological m-simplex.

Lemma 10.6.3 *Assume that X is not the empty simplicial set. Then, every element $(x, (t_0, \dots, t_n)) \in X_n \times \Delta^n$ is equivalent to a uniquely determined nondegenerate element in $|X|$.*

Proof Lemma 10.1.4 yields that we can write every $x \in X_m$ in a unique way in the form $x = s_{j_p} \circ \cdots \circ s_{j_1} y$ with a nondegenerate $y \in X_{m-p}$ and $0 \leq j_1 < \cdots < j_p < m$. We define a map ψ as

$$\psi[(s_{j_p} \circ \cdots \circ s_{j_1} y, t_0, \dots, t_m)] = [(y, \sigma_{i_1} \circ \cdots \circ \sigma_{i_p}(t_0, \dots, t_m))].$$

If $(u_0, \ldots, u_n)$ is a point in the boundary of $\triangle^n$ of the form

$$(u_0, \ldots, u_n) = \delta_{i_r} \circ \cdots \circ \delta_{i_1}(v_0, \ldots, v_{n-r}),$$

such that $(v_0, \ldots, v_{n-r})$ is a point in the interior of $\triangle^{n-r}$ and such that $0 \leq i_1 < \cdots < i_r \leq n$, then we define

$$\varrho[(x, (u_0, \ldots, u_n))] = [(d_{i_1} \circ \cdots \circ d_{i_r} x, (v_0, \ldots, v_{n-r}))].$$

The maps ϱ and ψ are self-maps of $|X|$, and the composite $\psi \circ \varrho$ sends a representative of a point in $|X|$ to a unique nondegenerate representative that is equivalent to the representative we started with. $\qquad \square$

With the help of the result, one can describe the structure of the topological space $|X|$ as a CW complex.

Proposition 10.6.4 ([**Mi57**, Theorem 1], [**May67**, Theorem 14.1]) *The geometric realization* $|X|$ *of a simplicial set* X *is a CW complex, such that every nondegenerate n-simplex corresponds to an n-cell.*

Remark 10.6.5 Every CW complex is a compactly generated Hausdorff space; in particular, $|X|$ is an object of cg (see Definition 8.5.13) for every simplicial set X.

Examples 10.6.6
(1) The topological 1-sphere is the quotient space $[0, 1]/0 \sim 1$. If we want to find a simplicial model for the 1-sphere, such that the geometric realization has the desired cell structure, then we should define a simplicial set $\mathbb{S}^1$ with one 0-simplex, 0, and one nondegenerate 1-simplex, 1. The simplicial identities force the existence of a 1-simplex $s_0(0)$, so we get two 1-simplices. For the cell structure, we do not need any further maps, so we just take these simplices and all the resulting elements that are given due to the simplicial structure maps. We then get $\mathbb{S}^1_n \cong [n]$ with face and degeneracy maps as follows:

$$[0] \rightleftarrows [1] \rightleftarrows [2] \cdots,$$

The map $s_i \colon [n] \to [n+1]$ is the unique monotone injection, whose image does not contain $i + 1$, while $d_i \colon [n] \to [n - 1]$ is given by $d_i(j) = j$ if $j < i, d_i(i) = i$ if $i < n$, and $d_n(n) = 0$ and $d_i(j) = j - 1$ if $j > i$.

The face maps glue the only nondegenerate 1-simplex 1 to the zero simplex $0 \in [0]$, and we obtain that the geometric realization, $|\mathbb{S}^1|$, is the topological 1-sphere.

(2) The geometric realization of the representable simplicial set Δ_n is $|\Delta_n| = \Delta^n$. This is a general fact about tensor products of functors and representable objects 15.1.5.

(3) Let X and Y be two simplicial sets. We already saw the product $X \times Y$, which is the simplicial set with $(X \times Y)_n = X_n \times Y_n$. The simplicial structure maps d_i and s_j are defined coordinatewise. Be careful, an n-simplex $(x, y) \in X_n \times Y_n$ of the form $(s_i x', s_j y')$ for $i \neq j$ might not be degenerate in $X_n \times Y_n$, despite the fact that both coordinates are degenerate.

You know that a product of two CW complexes does not have to carry a CW structure. How can we describe the geometric realization $|X \times Y|$?

Proposition 10.6.7 *Assume that X and Y are two simplicial sets, such that $|X| \times |Y|$ is a CW complex, with the CW structure induced by the one on $|X|$ and $|Y|$. Then,*

$$|X \times Y| \cong |X| \times |Y|.$$

Proof This proof is more or less the original one from [**Mi57**]. The projection maps $p_X \colon X \times Y \to X$ and $p_Y \colon X \times Y \to Y$ induce continuous maps

$$|p_X| \colon |X \times Y| \to |X|, \quad |p_Y| \colon |X \times Y| \to |Y|,$$

which combine to a continuous map $p \colon |X \times Y| \to |X| \times |Y|$. We construct a bijective map $|X| \times |Y| \to |X \times Y|$, which is continuous if $|X| \times |Y|$ is a CW complex.

Let (z, w) be a point in $|X| \times |Y|$ with a nondegenerate representative $(x, (t_0, \ldots, t_n))$ of z and $(y, (u_0, \ldots, u_m))$ of w. We consider the partial sums

$$t^i := \sum_{j=0}^{i} t_i, u^i := \sum_{j=0}^{i} u_i.$$

As some of the t_is and u_is might be zero, the sequence of partial sums is not strictly monotone. Some of the t^is might also coincide with some of the u^js. Let $r^0 < \cdots < r^q = 1$ be a strictly monotone ordering of the elements in the union

$$\{t^0, \ldots, t^n\} \cup \{u^0, \ldots, u^m\}.$$

We can interpret the r^is again as partial sums and can assign the element $(r^0, r^1 - r^0, \ldots, r^q - r^{q-1}) \in \Delta^q$ to it. This gives a description of $(t_0, \ldots, t_n)$ as $\sigma_{p_1} \circ \cdots \circ \sigma_{p_{n-a}}(r^0, r^1 - r^0, \ldots, r^q - r^{q-1})$, and similarly, we can express $(u_0, \ldots, u_m)$ as $\sigma_{\ell_1} \circ \cdots \circ \sigma_{\ell_{m-b}}(r^0, r^1 - r^0, \ldots, r^q - r^{q-1})$. Here, the sets $\{p_1, \ldots, p_{n-a}\}$ and $\{\ell_1, \ldots, \ell_{m-b}\}$ are disjoint and $p_1 < \cdots < p_{n-a}$, $\ell_1 < \cdots < \ell_{m-b}$. We define a map

$$\phi\colon |X| \times |Y| \to |X \times Y|$$

as

$$\phi(z, w) = \left[\left(\left(s_{p_{n-a}} \circ \cdots \circ s_{p_1} x, s_{\ell_{m-b}} \circ \cdots \circ s_{\ell_1} y\right), \left(r^0, r^1 - r^0, \ldots, r^q - r^{q-1}\right)\right)\right].$$

If $w' \in |X \times Y|$ has the nondegenerate representative $((x', y'), (t_0, \ldots, t_n))$, then $|p_X| w'$ has the nondegenerate representative $\psi(x', (t_0, \ldots, t_n))$ and $\psi(y', (t_0, \ldots, t_n))$ is a nondegenerate representative of $|p_Y| w'$. (Applying ψ is necessary because, for instance, $(x', y') = (s_i x, y')$ could occur.) In total, we get

$$\begin{aligned}
|p_X|(\phi(z, w)) &= |p_X| \left(\left[\left(\left(s_{p_{n-a}} \circ \cdots \circ s_{p_1} x, s_{\ell_{m-b}} \circ \cdots \circ s_{\ell_1} y\right),\right.\right.\right. \\
&\qquad\qquad \left.\left.\left.\left(r^0, r^1 - r^0, \ldots, r^q - r^{q-1}\right)\right)\right]\right) \\
&= \left[\psi\left(\left(s_{p_{n-a}} \circ s_{p_1} x, \left(r^0, r^1 - r^0, \ldots, r^q - r^{q-1}\right)\right)\right)\right] \\
&= [(x, (t_0, \ldots, t_n))] = z
\end{aligned}$$

and $|p_Y|(\phi(z, w)) = w$. Hence, $p \circ \phi = 1$. For a nondegenerate representative $((x', y'), (t_0, \ldots, t_n))$ of $w' \in |X \times Y|$, we get

$$\begin{aligned}
\phi \circ p \left[((x', y'), (t_0, \ldots, t_n))\right] &= \phi \left(\left[\psi\left((x', (t_0, \ldots, t_n))\right)\right],\right. \\
&\qquad\qquad \left.\left[\psi\left((y', (t_0, \ldots, t_n))\right)\right]\right) \\
&= \left[((x', y'), (t_0, \ldots, t_n))\right] = w'.
\end{aligned}$$

So, ϕ is an inverse of p.

If $|X| \times |Y|$ is a CW complex, then ϕ is continuous on every product of cells $[x \times \Delta^n] \times [y \times \Delta^m]$. In the interior, no coordinate in the simplex is zero and ϕ just reorganizes things. If one coordinate approaches zero, then two consecutive partial sums become equal, and in the sequence r^i, they will only be counted once. For the map ϕ, this has the effect of adding a map s_k. If several coordinates go to zero, then the simplicial identities ensure that the map is well-defined and therefore continuous. $\qquad\square$

As $|\Delta_1|$ is locally compact, we obtain the following compatibility result:

Corollary 10.6.8 *If $f, g\colon X \to Y$ are maps of simplicial sets that are homotopic, then $|f|$ is homotopic to $|g|$.*

Remark 10.6.9 One can always get a product description of $|X \times Y|$ as $|X| \times_k |Y|$. By the previous discussion, we know that $|\Delta_n \times \Delta_m| \cong |\Delta_n| \times |\Delta_m|$

and this is $|\Delta_n| \times_k |\Delta_m|$. There are also direct combinatorial proofs for that (see for example, [**GZ67**, III.3]). We will see that geometric realization is a left adjoint functor in Proposition 10.12.9, so $| - |$ preserves colimits. By (10.2.3), every simplicial set X is isomorphic to $\text{colim}_{\text{el}(X)} \Delta_n$. By Remark 8.5.17, the category cg is cartesian closed, and the product $(-) \times_k (-)$ preserves colimits in every argument. These properties, together with the interchange law for colimits from Proposition 3.5.1, give a chain of homeomorphisms

$$
\begin{aligned}
|X \times Y| &\cong |\text{colim}_{\text{el}(X)} \Delta_n \times \text{colim}_{\text{el}(Y)} \Delta_m| \\
&\cong |\text{colim}_{\text{el}(X) \times \text{el}(Y)} \Delta_n \times \Delta_m| \\
&\cong \text{colim}_{\text{el}(X) \times \text{el}(Y)} |\Delta_n \times \Delta_m| \\
&\cong \text{colim}_{\text{el}(X) \times \text{el}(Y)} |\Delta_n| \times_k |\Delta_m| \\
&\cong \text{colim}_{\text{el}(X)} |\Delta_n| \times_k \text{colim}_{\text{el}(Y)} |\Delta_m| \\
&\cong |X| \times_k |Y|.
\end{aligned}
$$

Remark 10.6.10 One can also apply geometric realization to simplicial spaces, that is, functors from Δ^o to Top. But beware that this might not have the properties you would expect. See Proposition 10.9.4 or [**Se74**, Appendix A] for more details.

10.7 Skeleta of Simplicial Sets

We consider the full subcategory $\Delta_{\leq n}$ of Δ with objects $[0], \ldots, [n]$. The inclusion functor

$$
\iota_n : \Delta_{\leq n} \to \Delta
$$

allows us to restrict simplicial sets X to $\Delta_{\leq n}$ by considering $X \circ \iota_n : \Delta_{\leq n}^o \to$ Sets.

Definition 10.7.1 The *n-skeleton of a simplicial set* X, $sk_n X$, is the left Kan extension of $X \circ \iota_n$ along ι_n.

Remark 10.7.2 This definition also allows us to express sk_n as a functor, left adjoint to the restriction functor. Precomposition with the inclusion functor

$\iota_n \colon \Delta_{\leq n} \to \Delta$ defines a functor

$$\iota_n^* \colon s\mathsf{Sets} \to s_{\leq n}\mathsf{Sets},$$

where $s_{\leq n}\mathsf{Sets}$ denotes the functor category $\mathsf{Fun}(\Delta_{\leq n}^o, \mathsf{Sets})$. The n-skeleton is then the left adjoint to ι_n^*.

So, first, you forget all information that X had in simplicial degrees greater than n, and then, you extend it, filling in degenerate elements in simplicial degrees greater than n. As degenerate simplices are identified to lower dimensional things, one gets the following immediately:

Proposition 10.7.3 *For every simplicial set X,*

$$|sk_n X| \cong sk_n |X| =: X^{(n)},$$

where $sk_n |X| = X^{(n)}$ denotes the n-skeleton of the CW complex $|X|$.

10.8 Geometric Realization of Bisimplicial Sets

Definition 10.8.1 A *bisimplicial set* is a functor $X \colon \Delta^o \times \Delta^o \to \mathsf{Sets}$. The morphisms in the category of bisimplicial sets are the natural transformations.

We abbreviate $X([p], [q])$ by $X_{p,q}$ and say that $X_{p,q}$ is in bisimplicial degree (p, q). By the exponential law,

$$\mathsf{Fun}(\Delta^o \times \Delta^o, \mathsf{Sets}) \cong \mathsf{Fun}(\Delta^o, \mathsf{Fun}(\Delta^o, \mathsf{Sets})),$$

and due to the symmetry of $\Delta^o \times \Delta^o$, we can also interpret X as a simplicial object in simplicial sets in two different ways.

Example 10.8.2 There are two easy ways to embed the category of simplicial sets into the category of bisimplicial sets. To any simplicial set Y, we can associate the bisimplicial set $X_Y([p], [q]) = Y_p$ that is constant in q-direction and $X^Y([p], [q]) = Y_q$ that is constant in p-direction.

Example 10.8.3 If Y and Z are two simplicial sets, then we can form their *external product*, $Y \boxtimes Z$, as a bisimplicial set, with

$$(Y \boxtimes Z)_{p,q} = Y_p \times Z_q \text{ and } (Y \boxtimes Z)(\varphi, \psi) = Y(\varphi) \times Z(\psi)$$

for all objects $[p]$ and $[q]$ in Δ and all $\varphi \in \Delta([p'], [p])$ and $\psi \in \Delta([q'], [q])$.

Exercise 10.8.4 Find a suitable bisimplicial set $Q(p, q)$, such that the set of morphisms of bisimplicial sets from $Q(p, q)$ to any bisimplicial set X are in natural bijection with the set $X_{p,q}$.

There are several things one could suggest as the geometric realization of a bisimplicial set X:

(1) We could define $|X|$ as the suitable coend.
(2) We could try to reduce the complexity of X by reducing X to a simplicial set and realizing the latter. The *diagonal simplicial set associated with the bisimplicial set X* is the functor $\text{diag}(X)\colon \Delta^o \to \text{Sets}$, given by

$$\text{diag}(X)_p = X([p], [p]), \text{ and } \text{diag}(X)(\varphi) = X(\varphi, \varphi)$$

for every object $[p]$ of Δ and every $\varphi \in \Delta([q], [p])$.
(3) We could do one step at a time: For a fixed $[p]$, the functor

$$X([p], -)\colon \Delta^o \to \text{Sets}$$

is a simplicial set and so is

$$X(-, [p])\colon \Delta^o \to \text{Sets}.$$

We could take the following two iterations of geometric realizations in sets and spaces: $|[q] \mapsto |[p] \mapsto X_{p,q}||$ or $|[p] \mapsto |[q] \mapsto X_{p,q}||$.

The good news is that it doesn't matter what you do.

Proposition 10.8.5 *All four topological spaces described earlier are homeomorphic to each other.*

Proof We first show with Quillen's method from [**Q73**, p. 10 aka p. 86] that the models from (2) and (3) are homeomorphic. To this end, we consider first special bisimplicial sets of the form $\Delta_p \boxtimes \Delta_q \times S$, where S is any set. Then, the geometric realization of the diagonal of $\Delta_p \boxtimes \Delta_q \times S$ is homeomorphic to $\Delta^p \times \Delta^q \times S$, because geometric realization commutes with products in this case (Proposition 10.6.7). If we consider the two-fold realization

$$|[n] \mapsto |[m] \mapsto \Delta_p([n]) \times \Delta_q([m]) \times S||, \qquad (10.8.1)$$

then in the inner realization, $\Delta_p([n])$ and S are constant, and we get that the above is homeomorphic to

$$|[n] \mapsto \Delta_p([n]) \times \Delta^q \times S|,$$

and this is also $\Delta^p \times \Delta^q \times S$. A symmetry argument gives the claim for the other two-fold realization.

You solved Exercise 10.8.4, and this implies that we can express every bisimplicial set X as the coequalizer:

$$\bigsqcup_{(\Delta\times\Delta)(([p],[q]),([p'],[q']))} \Delta_{p'} \boxtimes \Delta_{q'} \times X_{p,q} \rightrightarrows \bigsqcup_{([p],[q])\in\mathrm{Ob}(\Delta\times\Delta)} \Delta_p \boxtimes \Delta_q \times X_{p,q} \longrightarrow X.$$

Geometric realization is a left adjoint functor (see Proposition 10.12.9), and hence, it commutes with colimits, in particular with disjoint unions and coequalizers. This proves the claim.

Finally, it is not hard to see, for instance, that the two-fold realization, as in (10.8.1), has the same universal property as the coend in (1).   □

Note that the diagonal simplicial set associated with a bisimplicial set X^Y or X_Y, as in Example 10.8.2, that is constant in one simplicial direction is Y again.

Remark 10.8.6 We state the following fact without proof (see [**Ree∞**, p. 7] or [**BK72**, XII.4.2, p. 335]): Let $f: X \to Y$ be a morphism of bisimplicial sets, such that for every $p \geq 0$, the map $f_{p,*}: X_{p,*} \to Y_{p,*}$ is a weak equivalence. Then, the map $\mathrm{diag}(f): \mathrm{diag}(X) \to \mathrm{diag}(Y)$ is also a weak equivalence.

10.9 The Fat Realization of a (Semi)Simplicial Set or Space

Sometimes, you might want to use a variant of the geometric realization functor. An obvious reason is that there are sequences of objects X_0, X_1, ... that are only connected via face maps, but there are no degeneracy maps. Such functors are often called *semisimplicial objects*. In that situation, you cannot perform the geometric realization. The other situation that makes an alternative desirable is the situation where you want to perform the geometric realization of a simplicial space and this space has bad point set behavior.

Definition 10.9.1 ([**Se74**, Appendix A]) Let X be a simplicial set (or space), then the *fat realization of X, $||X||$,* is

$$||X|| = \bigsqcup_{n\geq 0} X_n \times \Delta^n / \sim,$$

where the quotient in the fat geometric realization is generated by the relation

$$(d_i(x), (t_0, \ldots, t_n)) \sim (x, \delta_i(t_0, \ldots, t_n)).$$

Remark 10.9.2 There are several alternative descriptions of $||X||$. One is to consider the *semisimplicial category, Δ,* whose objects are the objects of Δ, but we restrict to injective order-preserving maps. These are dual to the face maps used in the identifications in fat geometric realization. Thus, we can describe $||X||$ as the coend of the functor

$$H: \Delta^o \times \Delta \to \mathsf{Top},$$

with $H([p], [q]) = X_p \times \Delta^q$.

There is yet another description of the fat realization of a simplicial set or simplicial topological space (see, for instance, [**We05**, Proof of Proposition 1.3] or [**Se74**, p. 308]) as the ordinary geometric realization of a "fattened up" simplicial set.

Of course, $||X||$ also makes sense, if you start with a semisimplicial object, that is, a functor $X: \Delta^o \to \mathsf{Sets}$.

As we do not collapse degenerate simplices, the fat realization of a simplicial set is larger than the geometric realization.

Exercise 10.9.3 What is the fat realization of Δ_0?

Segal proves that the fat realization has some remarkable properties.

Proposition 10.9.4 ([**Se74**, Proposition A.1])

(1) If all the X_n are spaces of the homotopy type of a CW complex, then so is $||X||$.

(2) If $f: X \to Y$ is a morphism of simplicial topological spaces, such that all $f_n: X_n \to Y_n$ are homotopy equivalences, then $||f||$ is a homotopy equivalence.

(3) Fat realization commutes with finite products.

All these three properties don't hold for the geometric realization of simplicial topological spaces in general. Property (2) is fine if one works with bisimplicial sets instead of simplicial spaces (compare Remark 10.8.6).

Segal introduces the notion of a *good simplicial space* and shows that for such spaces X, the canonical map from the fat realization $||X||$ to the geometric realization $|X|$ is homotopy equivalence [**Se74**, A.1, A.4].

10.10 The Totalization of a Cosimplicial Space

A *cosimplicial space* is a cosimplicial object in simplicial sets or a simplicial object in cosimplicial sets, that is, a functor $Y: \Delta^o \times \Delta \to \mathsf{Sets}$. We denote the evaluation of Y on the object $([p], [q])$ of $\Delta^o \times \Delta$ as Y_p^q. Dual to the geometric realization of a bisimplicial set $X: \Delta^o \times \Delta^o \to \mathsf{Sets}$ aka a simplicial space, we can build the totalization of Y, $\mathrm{Tot}(Y)$ [**BK72**].

Example 10.10.1 Note that we can turn the set of function $\Delta([n], [m])$ into a cosimplicial space, because $\Delta(-, [m]) = \Delta_m$ is a simplicial set and $\Delta([n], -)$ is a cosimplicial set; thus, $\Delta(-, -) \colon \Delta^o \times \Delta \to$ Sets is a cosimplicial space.

Example 10.10.2 Let X and Z be two simplicial sets. Then, we can form the cosimplicial mapping space

$$([p], [q]) \mapsto \text{Sets}(X_q, Z_p).$$

For pointed simplicial sets, you could consider the variant that uses based maps.

Definition 10.10.3

(1) For two cosimplicial spaces Y_1 and Y_2, their *simplicial set of maps* from Y_1 to Y_2 is

$$\underline{hom}(Y_1, Y_2) \colon \Delta^o \to \text{Sets}, \quad \underline{hom}(Y_1, Y_2)_p = \text{Sets}^{\Delta^o \times \Delta}(Y_1 \times \Delta_p, Y_2).$$

(2) The *totalization of a cosimplicial space* is

$$\text{Tot}(Y) = \underline{hom}(\Delta, Y).$$

If we unravel the definition of $\text{Tot}(Y)$, then it is a simplicial set with

$$\text{Tot}(Y)_p = \text{Sets}^{\Delta^o \times \Delta}(\Delta(-, -) \times \Delta_p, Y).$$

Hence, in simplicial degree p, we consider the set of functions $f_n^q \colon \Delta([n], [q]) \times \Delta([n], [p]) \to Y_n^q$ that are natural in $[n]$ and $[q]$. For all $\varphi \in \Delta([q], [q'])$ and $\alpha \in \Delta([m], [n])$, the diagram

$$
\begin{array}{ccc}
\Delta([m], [q]) \times \Delta([m], [p]) & \xrightarrow{f_{m,q}} & Y_m^q \\
{\scriptstyle (\Delta(\alpha, 1_{[q]}), \Delta(\alpha, 1_{[p]}))} \big\uparrow & & \big\uparrow {\scriptstyle Y_\alpha^q} \\
\Delta([n], [q]) \times \Delta([n], [p]) & \xrightarrow{f_{n,q}} & Y_n^q \\
{\scriptstyle (\Delta(1_{[n]}, \varphi), 1_{\Delta([n],[p])})} \big\downarrow & & \big\downarrow {\scriptstyle Y_n^\varphi} \\
\Delta([n], [q']) \times \Delta([n], [p]) & \xrightarrow{f_{n,q'}} & Y_n^{q'}
\end{array}
$$

commutes. Hence, Tot is nothing but the equalizer of the diagram

$$\prod_{q \geq 0} \underline{s\text{Sets}}(\Delta(-, [q]), Y^q) \rightrightarrows \prod_{\varphi \in \Delta([q], [q'])} \underline{s\text{Sets}}(\Delta(-, [q]), Y^{q'}),$$

where $\underline{s\text{Sets}}(\Delta(-, [q]), Y^{q'})$ is the internal hom object in simplicial sets from Definition 10.2.10 and where the maps in the diagram send a family $(f_{n,q})_q$ to $(Y_n^\varphi \circ f_{n,q})_\varphi$ and $(f_{n,q'} \circ \Delta(1_{[n]}, \varphi))_\varphi$.

Remark 10.10.4 Bousfield and Kan develop a spectral sequence in [**BK72**, Chapter X], whose E^2 term consists of $\pi^s \pi_t Y$ for a (fibrant) cosimplicial space Y and which converges to the homotopy groups of $\text{Tot}(Y)$. This spectral sequence has many applications in the calculation of the homotopy groups of mapping spaces and obstruction theory, for instance, in [**GH04**, §§4,5].

10.11 Dold–Kan Correspondence

Let $A: \Delta^o \to \mathcal{A}$ be a simplicial object in an abelian category $\mathcal{A}$.

Definition 10.11.1
- The *normalized chain complex of* A is the chain complex $N_*(A)$ with $N_n(A) = \bigcap_{i=0}^{n-1} \ker(d_i)$ and boundary operator $d = (-1)^n d_n$.
- The *chain complex associated with* A is the chain complex $C_*(A)$ with $C_n(A) = A_n$ and boundary operator $d: A_n \to A_{n-1}$ being defined as $d = \sum_{i=0}^{n}(-1)^i d_i$.

The normalized chain complex is sometimes referred to as the Moore chain complex. Moore defined it in [**Mo54/55**, §4]. But beware, some people also refer to the unnormalized complex as the Moore complex.

There is a variant of the normalization functor that uses the auxiliary chain complex of degenerate elements $D_*(A)$ and defines $N_*(A)$ as $C_*(A)/D_*(A)$. For details, see [**GJ09**, III.2] or [**W94**, §§8.3, 8.4].

Dold and Kan discovered independently in the 1950s [**Do58**] that the normalization functor gives rise to an equivalence of categories.

Theorem 10.11.2 [**Do58**, Theorem 1.9] *The normalized chain complex is part of an equivalence of categories between the simplicial objects in $\mathcal{A}$ and the nonnegatively graded chain complexes over $\mathcal{A}$.*

The Yoneda lemma gives you a concise formula for what the inverse of N_*, Γ_N, has to be on objects. Assume that $\mathcal{A}$ is the category of k-modules for some commutative ring k. For every simplicial degree n, $\Gamma_N(A)_n$ corresponds to the maps of simplicial sets from Δ_n to the underlying simplicial set of $\Gamma_N(A)$ and hence to the maps of simplicial k-modules from $k\{\Delta_n\}$ to $\Gamma_N(A)$. As we claim that Γ_N is part of an equivalence, it is, in particular, an adjoint, so we get

$$\Gamma_N(A)_n = \mathsf{Ch}(k)_{\geq 0}(N_*(k\{\Delta_n\}), A).$$

You can also work out the simplicial structure maps of Γ_N by using this trick, but you could also look this up in [**Do58**, Definition 1.8].

Exercise 10.11.3 Assume that $f, g: A \rightarrow B$ are morphisms of simplicial k-modules and that there are maps $h_0, \ldots, h_n: A_n \rightarrow B_{n+1}$, satisfying the relations from Definition 10.5.1. Show that $C_*(f)$ is then chain homotopic to $C_*(g)$ with the chain homotopy $H = \sum_{i=0}^{n}(-1)^i h_i$.

10.12 Kan Condition

Definition 10.12.1 Let $X: \Delta^o \rightarrow$ Sets be a simplicial set. A functor $Y: \Delta^o \rightarrow$ Sets is a *simplicial subset of X* if $Y_n \subset X_n$ for every $n \geq 0$ and if for every morphism $f \in \Delta([n], [m])$, the induced morphism satisfies $Y(f) = X(f)$.

Remark 10.12.2 If there are elements $\{x_i | i \in I\}$ with $x_i \in X_{n_i}$, then we can form the simplicial subset of X generated by the x_is by considering all elements in X of the form $X(f)(x_i)$ for all morphisms f in Δ.

The representable functor Δ_n has several important subfunctors:

Definition 10.12.3
- The *ith face of Δ_n*, $\partial_i \Delta_n$, is the simplicial subset of Δ_n generated by δ_i: $[n-1] \rightarrow [n]$.
- The *boundary of Δ_n*, $\partial \Delta_n$, is the simplicial subset of Δ_n generated by $\{\delta_0, \ldots, \delta_n\} \subset \Delta([n-1], [n])$.
- The *simplicial k-horn of Δ_n*, Λ_n^k, for some k with $0 \leq k \leq n$, is the simplicial subset of Δ_n generated by $\{\delta_0, \ldots, \delta_{k-1}, \delta_{k+1}, \ldots, \delta_n\}$.

You can also think of the simplicial k-horn Λ_n^k as the simplicial subcomplex of Δ_n generated by all faces that contain the 0-simplex that maps 0 to k.

Exercise 10.12.4 Show that the geometric realization of the simplicial subsets defined earlier are what they should be: $|\partial_i \Delta_n| = \delta_i(\Delta^{n-1})$, $|\partial \Delta_n| = \partial \Delta^n$, and $|\Lambda_n^k| = \bigcup_{i=0, i \neq k}^{n} \delta_i(\Delta^{n-1})$.

Definition 10.12.5 A simplicial set X is a *Kan complex* if for all n and $0 \leq k \leq n$, all maps from Λ_n^k to X can be extended over Δ_n; that is, in all solid diagrams of the form

there is a morphism of simplicial sets $\xi: \Delta_n \rightarrow X$, such that $\xi \circ j = \alpha$.

Remark 10.12.6 Note that the condition of being a Kan complex has a very explicit description. We denote by $\check{x}$ the fact that x is omitted. Then, X is Kan if and only if for all n, all $0 \leq k \leq n$, and all n-tuples of $(n-1)$-simplices $(x_1, \ldots, x_{k-1}, \check{x}_k, x_{k+1}, \ldots, x_n)$, such that $d_i x_j = d_{j-1} x_i$ for $i < j$, $i, j \neq k$, there is an n-simplex x in X, such that $d_i(x) = x_i$. We call x the *horn filler*.

Lemma 10.12.7 *The simplicial sets* Δ_n *do not satisfy the Kan condition for* $n \geq 2$.

Proof We show that a specific map $\varphi \colon \Lambda_2^0 \to \Delta_2$ cannot be extended to a map $\tilde{\varphi} \colon \Delta_2 \to \Delta_2$. This argument extends to higher dimensions.

Define φ on the generators of Λ_2^0 as $\varphi(\delta_2) = \delta_2$, and let $\varphi(\delta_1)$ be the constant map with value zero.

Assume that an extension $\tilde{\varphi} \colon \Delta_2 \to \Delta_2$ exists. Then, $d_0 \tilde{\varphi}(0) = \tilde{\varphi} \delta_0(0) = \varphi \delta_2(1) = 1$ and $d_0 \tilde{\varphi}(1) = \tilde{\varphi} \delta_0(1) = \tilde{\varphi} \delta_1(1) = \varphi \delta_1(1) = 0$, and hence, the extension is not order preserving. $\square$

Definition 10.12.8 Let $Sing \colon \mathsf{Top} \to \mathsf{Sets}^{\Delta^o}$ be the functor that sends a topological space X to the simplicial set $Sing(X)$ with

$$Sing(X)_n = \mathsf{Top}(\Delta^n, X),$$

and where the simplicial structure is induced by the cosimplicial structure of the family of standard n-simplices.

Proposition 10.12.9
(1) The functor Sing is right adjoint to the geometric realization functor.
(2) For every topological space X, $Sing(X)$ is a Kan complex.

The proof here is similar to the one of Theorem 15.1.3, but note that in the geometric realization, we allow simplicial *sets* as input and view these sets as discrete topological spaces.

Proof

(1) The adjunction is a standard tensor-hom adjunction sending a continuous map

$$f \colon \bigsqcup_{n \geq 0} X_n \times \Delta^n / \sim \to Y$$

with compatible components $f_n \colon X_n \times \Delta^n \to Y$ to the map of simplicial sets

$$g \colon X \to Sing(Y),$$

whose degree n part $g_n \colon X_n \to \mathsf{Top}(\Delta^n, Y)$ is given as $f_n(x, -) \colon \Delta^n \to Y$.

(2) As the functor $Sing(-)$ is adjoint to geometric realization, the lifting property of $\Delta_0 \to Sing(X)$ with respect to $\Lambda_n^r \to \Delta_n$ is equivalent to the lifting property of the inclusion $\{*\} \to X$ with respect to $|\Lambda_n^r| \to |\Delta_n|$, but the latter is a strong deformation retract. $\qquad\square$

Proposition 10.12.10 [**Mo54/55**, Théorème 3] *Let G be a simplicial object in the category of groups. Then, the underlying simplicial set of G is a Kan complex.*

Proof Consider a compatible family of elements of G_{n-1} $(g_0, \ldots, g_{k-1}, g_{\ell-1}, \ldots, g_n)$ with $\ell \geq k + 2$, that is, $d_i g_j = d_{j-1} g_i$ for $i < j$. Assume that there is an n-simplex $g \in G_n$ with $d_i(g) = g_i$ for $i \leq k - 1$ and $i \geq \ell$. We consider $d_i d_{\ell-1}(g)$. If i smaller than k, then because of the simplicial identities and because of the compatibility of the g_is, we get

$$d_i d_{\ell-1}(g) = d_{\ell-2} d_i(g) = d_{\ell-2}(g_i) = d_i g_{\ell-1}. \tag{10.12.1}$$

Similarly, $d_i d_{\ell-1}(g) = d_i g_{\ell-1}$ if $i \geq \ell$.

We consider the element $s_{\ell-2}(g_{\ell-1} d_{\ell-1}(g^{-1}))g$. For $d_{\ell-1}$, we obtain

$$d_{\ell-1}\big(s_{\ell-2}(g_{\ell-1} d_{\ell-1}(g^{-1}))g\big)$$
$$= d_{\ell-1}(s_{\ell-2}(g_{\ell-1}))\, d_{\ell-1}\big(s_{\ell-2}\big(d_{\ell-1}\big(g^{-1}\big)\big)\big)\, d_{\ell-1}(g) = g_{\ell-1},$$

because $d_{\ell-1} \circ s_{\ell-2} = 1$ and because the structure maps are homomorphisms. Similarly, the relations in (10.12.1) imply

$$d_i \big(s_{\ell-2}\big(g_{\ell-1} d_{\ell-1}\big(g^{-1}\big)\big) g\big) = g_i$$

for $i \leq k - 1$ and $i \geq \ell$.

Thus, $s_{\ell-2}(g_{\ell-1} d_{\ell-1}(g^{-1}))g$ closes the gap one step further. $\qquad\square$

Remark 10.12.11 This result ensures that simplicial R-modules (R some associative ring), simplicial k-algebras (k some commutative ring with unit), simplicial Lie algebras, and many more simplicial objects that have an underlying simplicial group are Kan complexes.

Remark 10.12.12 For a Kan complex X, there is a combinatorially defined, well-defined notion of homotopy groups of X, such that $\pi_n(X, x) \cong \pi_n(|X|, x)$ for every zero simplex $x \in X_0$. See, for instance, [**May67**, I §3] or [**GJ09**, I §7] for more details. As simplicial k-modules A are always Kan complexes, $\pi_n(A, 0)$ can always be defined just by using the simplicial structure. A consequence [**GJ09**, Corollary III.2.7] of the Dold–Kan correspondence, Theorem 10.11.2, is that

$$\pi_n(A, 0) \cong H_n(N_*(A)) \cong H_n(C_*(A)).$$

For $n = 0$, the set $\pi_0(X)$ has a combinatorial description as the coequalizer of the diagram

$$X_1 \underset{d_1}{\overset{d_0}{\rightrightarrows}} X_0.$$

There is a canonical map $\varrho \colon |X| \to \operatorname{colim}_{\Delta^o} X$ for every simplicial set X, induced by the following commutative diagram:

$$
\begin{array}{ccccc}
\bigsqcup_{\alpha \in \Delta([m],[n])} \Delta^m \times X_n & \rightrightarrows & \bigsqcup_{n \geq 0} \Delta^n \times X_n & \longrightarrow & |X| \\
\downarrow{\scriptstyle pr_2} & & \downarrow{\scriptstyle pr_2} & & \downarrow{\scriptstyle \varrho} \\
\bigsqcup_{\alpha \in \Delta([m],[n])} X_n & \rightrightarrows & \bigsqcup_{n \geq 0} X_n & \longrightarrow & \operatorname{colim}_{\Delta^o} X.
\end{array}
$$

The map ϱ is nothing but the projection of $|X|$ to $\pi_0(X)$.

Exercise 10.12.13 Show that for a simplicial set X, the colimit of X, $\operatorname{colim}_{\Delta^o} X$, is isomorphic to $\pi_0(X)$.

Proposition 10.12.14 *The category Δ^o is sifted.*

Proof   We have to show that the colimit over Δ^o commutes with finite products of functors from Δ^o to Sets. By Exercise 10.12.13, we know that for every functor $X \colon \Delta^o \to$ Sets, the colimit is $\pi_0(X)$.

First, consider the terminal object, Δ_0. Here, the colimit is the terminal object in the category Sets because $|\pi_0(\Delta_0)| = 1$. Let $X, Y \colon \Delta^o \to$ Sets. Then, we claim that

$$\pi_0(X \times Y) \cong \pi_0(X) \times \pi_0(Y).$$

We establish this bijection by considering explicit maps. If (x, y) in $X_0 \times Y_0$ is equivalent to (x', y') in $X_0 \times Y_0$, then there is a zigzag of 1-simplices in $X \times Y$ connecting the two. But then, the first coordinate of the zigzag connects x to x' and the second connects y to y', so we get a well-defined map:

$$\varphi \colon \pi_0(X \times Y) \to \pi_0(X) \times \pi_0(Y); \quad [(x, y)] \mapsto ([x], [y]).$$

For the inverse,

$$\psi : \pi_0(X) \times \pi_0(Y) \to \pi_0(X \times Y),$$

consider a zigzag of 1-simplices in X connecting x and x' and another zigzag connecting y and y'. Then, their product does not necessarily give a zigzag of 1-simplices in $X \times Y$, but we can modify the two zigzags with the help of degenerate 1-simplices to morph them into the same shape. For instance, you can take first the zigzag for x and x' and combine it with degenerate 1-simplices $s_0 y$ in the second coordinate and then use $s_0 x'$ in the first coordinate and the zigzag for y and y' in the second coordinate. $\qquad \square$

Exercise 10.12.15 If R is a ring, M a left R-module, and N a right R-module, what is π_0 of the two-sided bar construction $B(N, R, M)$ from Definition 10.4.2?

10.13 Quasi-Categories and Joins of Simplicial Sets

Some lifting conditions of Kan complexes are different from others. Consider, for instance, the 2-simplex Δ_2 and its k-horns for $0 \le k \le 2$:

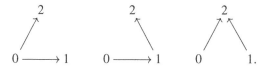

If we think of 1-simplices in a simplicial set X as oriented edges in X, then the lifting property for a 1-horn implies that two edges e_{01}, e_{12} in X that fit together can be composed up to homotopy. There is an edge e_{02} in X and a 2-simplex, that is, a triangle in X, such that e_{01}, e_{12}, and e_{02} are the edges of that triangle. The lifting properties for the 0- and 2-horns are much more drastic; they are not composition requirements but extension and lifting properties. For instance, if we have a 0-horn

then a horn filler gives you a left inverse for f up to homotopy. Dually, a horn filler for a 2-horn can give you a right inverse up to homotopy.

Definition 10.13.1 A *quasi-category* is a simplicial set X that has the lifting property with respect to all inner horns, that is, with respect to all Λ_n^k with $0 < k < n$.

This definition goes back to Boardman and Vogt [**BV73**, 4.8], who called quasi-categories *restricted Kan complexes* and considered them as "good substitutes for categories". Joyal advertised them [**Jo02, Jo-b∞**], and they are used as one model for infinity categories [**Lu09**].

Remark 10.13.2 If X is a quasi-category, then we think of the 0-simplices $x \in X_0$ as objects of X, and the 1-simplices are the morphisms of X. The inner horn-filler condition then shows that we might not have a composition of $f \in X_1$ and $g \in X_1$, even if the target of f is the source of g, but as we saw earlier, there is a 2-simplex in X that has f and g as the legs of a 1-horn, and we think of the third edge as a composition of f and g up to homotopy. Simplices of higher degrees then correspond to higher morphisms.

In order to stress the fact that quasi-categories behave like categories up to homotopy, they are often denoted by $\mathcal{C}$.

Exercise 10.13.3 Consider 1-simplices e_{01}, e_{12}, and e_{23} of a quasi-category X and call them f, g, and h. By the 1-horn-filler condition, we get a composition of g and f up to homotopy that we call e_{02} and a composition of h and g up to homotopy e_{13}. What does the horn-filler condition for $\Lambda_3^1 \subset \Delta_3$ and $\Lambda_3^2 \subset \Delta_3$ say?

We saw the join construction for categories in Definition 1.1.7. There is a related construction for simplicial sets, similar to the join construction in the category of topological spaces that you might have seen. There is a definition in terms of a Day convolution product [**Jo-c∞**, §3.2]. A hands-on definition is as follows:

Definition 10.13.4 Let X and Y be two simplicial sets. The *join of X and Y* is the simplicial set whose set of n-simplices is

$$(X * Y)_n = X_n \sqcup Y_n \sqcup \bigsqcup_{i+j=n-1} X_i \times Y_j.$$

The face maps $d_k \colon (X * Y)_n \to (X * Y)_{n-1}$ restricted to X_n and Y_n are the face maps of X and Y. On a summand $X_i \times Y_j$,

$$d_k(x, y) = \begin{cases} (d_k x, y), & \text{if } k \le i \text{ and } i \ne 0, \\ (x, d_{k-i-1} y), & \text{if } k > i \text{ and } j \ne 0. \end{cases}$$

If $i = 0$, then $d_0(x, y) = y$, and if $j = 0$, then $d_n(x, y) = x$.

For the degeneracy maps, you set

$$s_k(x, y) = \begin{cases} (s_k x, y), & \text{if } k \leq i, \\ (x, s_{k-i-1} y), & \text{if } k > i, \end{cases}$$

on the mixed terms, and you use the ordinary degeneracies on X_n and Y_n inside $(X * Y)_n$.

Remark 10.13.5 Joyal proves that the join of two quasi-categories is a quasi-category (see, for instance, [**Lu09**, 1.2.8.3]). Later, in Proposition 11.1.11, we will see that the nerve construction sends joins of categories to joins of simplicial sets.

Exercise 10.13.6 Check that the preceding definition of face and degeneracy maps actually defines a simplicial object.

Example 10.13.7 On representable functors, we obtain $\Delta_k * \Delta_\ell \cong \Delta_{k+\ell+1}$ for all $k, \ell \geq 0$. This can be seen by spelling out an explicit isomorphism. Note that $[k] \sqcup [\ell] \cong [k + \ell + 1]$ as sets. By definition, $(\Delta_k * \Delta_\ell)([n])$ is equal to

$$\Delta([n], [k]) \sqcup \Delta([n], [\ell]) \sqcup \bigsqcup_{i+j=n-1} \Delta([i], [k]) \times \Delta([j], [\ell]).$$

We define a map on each summand. Let

$$\varphi \colon \Delta([n], [k]) \to \Delta([n], [k] \sqcup [\ell]) = \Delta([n], [k + \ell + 1])$$

be the map that sends $\sigma \in \Delta([n], [k])$ to $i_{[k]}^{[k] \sqcup [\ell]} \circ \sigma$, where $i_{[k]}^{[k] \sqcup [\ell]}$ is the canonical inclusion. Similarly, let

$$\psi \colon \Delta([n], [\ell]) \to \Delta([n], [k] \sqcup [\ell])$$

be defined by $\psi(\tau) = i_{[\ell]}^{[k] \sqcup [\ell]} \circ \tau$, where $i_{[\ell]}^{[k] \sqcup [\ell]}(j) - j + k + 1$. On a mixed term $\Delta([i], [k]) \times \Delta([j], [\ell])$, we define

$$\zeta_{ij} \colon \Delta([i], [k]) \times \Delta([j], [\ell]) \to \Delta([n], [k] \sqcup [\ell])$$

as the map

$$\zeta_{ij}(\nu, \mu)(r) = \begin{cases} \nu(r), & \text{if } r \leq i \\ \mu(r - i - 1) + k + 1, & \text{for } r > i, \end{cases}$$

where $r \in [n]$.

It is easy to see that this gives a bijection of sets $(\Delta_k * \Delta_\ell)([n]) \cong \Delta_{k+\ell+1}([n])$, because for a given $f \in \Delta([n], [k + \ell + 1])$, you just cut f into pieces. If the image of f is contained in $[k]$, then you send it to the summand $\Delta([n], [k])$. If the image is contained in the complement of $[k]$ in $[k] \sqcup [l]$, then you send it to $\Delta([n], [\ell])$ by pushing the values down. If the image has nontrivial parts in $[k]$ and in its complement, then you cut f into two maps and send it to the corresponding summand $\Delta([i], [k]) \times \Delta([j], [\ell])$ if the value of f exceeds k at $i + 1$. This gives an inverse to the first map.

We leave it to the reader to check that this degreewise bijection is compatible with the simplicial structure maps.

Exercise 10.13.8 Draw a proof that $\Delta_1 * \Delta_1 \cong \Delta_3$. Show that $\Lambda_2^0 * \Delta_0 \cong \Delta_1 \times \Delta_1$. Again, draw the proof.

Definition 10.13.9 For a simplicial set X, $X * \Delta_0$ is called the *right cone on* X, and $\Delta_0 * X$ is the *left cone on* X.

10.14 Segal Sets

A concept that is closely related to quasi-categories is that of a Segal set. Consider the morphisms $i_j : [1] \to [n]$ in Δ for $1 \leq j \leq n$, sending 0 to $j - 1$ and 1 to j. Note that the i_js satisfy the condition that $i_j \circ \delta^0 = i_{j+1} \circ \delta^1$ for $j \leq n - 1$. We denote by

$$X[1] \times_{X[0]} \cdots \times_{X[0]} X[1]$$

the limit of the diagram

Definition 10.14.1 A simplicial set X is a *Segal set* if the map

$$(i_1, \ldots, i_n) : X_n \to X_1 \times_{X_0} \cdots \times_{X_0} X_1 \qquad (10.14.1)$$

is a bijection for all $n \geq 2$.

Exercise 10.14.2 Is every Segal set a quasi-category?

The Segal condition (10.14.1) means that X is determined by a subset of the 1-skeleton of X. For $n = 2$, this subset can be visualized as

and for $n = 3$, it looks like

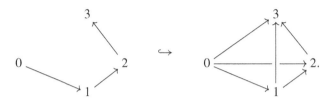

Proposition 10.14.3 *Let X be a Segal set with $X_0 = \{*\}$. Then, X_1 is an associative monoid with unit $s_0(*)$.*

Proof   As every map from X_1 to a point is the same, we have that $X_1 \times_{X_0} X_1 = X_1 \times X_1$. We define the multiplication $\mu \colon X_1 \times X_1 \to X_1$ as $d_1 \circ (i_1, i_2)^{-1}$. Note that in this low-dimensional case, we have $i_1 = d_2$ and $i_2 = d_0$.

For the associativity of μ, consider the diagram

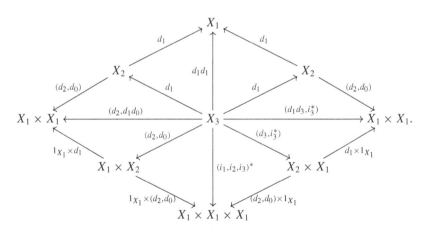

Going from the left via the middle to the right gives that

$$
\begin{aligned}
\mu \circ (1_{X_1} \times \mu) &= d_1 \circ (d_2, d_0)^{-1} \circ (1_{X_1} \times d_1) \circ \left(1_{X_1} \times (d_2, d_0)^{-1}\right) \\
&= d_1 d_1 \circ (i_1, i_2, i_3)^* \\
&= d_1 \circ (d_2, d_0)^{-1} \circ (d_1 \times 1_{X_1}) \circ \left((d_2, d_0)^{-1} \times 1_{X_1}\right) \\
&= \mu \circ (\mu \times 1_{X_1}).
\end{aligned}
$$

We show the left unit condition; the proof of the right unit condition follows from symmetry. The diagram

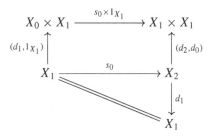

commutes because $d_1 s_0$ is the identity and

$$(d_2 s_0 x_1, d_0 s_0(x_1)) = (s_0 d_1 x_1, x_1) = (s_0*, x_1)$$

for all $x_1 \in X_1$, because X_0 is a one-point set. The evaluation $\mu(s_0*, x_1)$ is $d_1 \circ (d_2, d_0)^{-1}(s_0*, x_1)$, but we just found a preimage of (s_0*, x_1) under (d_2, d_0) in X_2, namely $s_0 x_1$; thus, $\mu(s_0*, x_1) = d_1(s_0 x_1) = x_1$ for all $x_1 \in X_1$. □

Remark 10.14.4 Proposition 10.14.3 has a variant where monoids are replaced by simplicial monoids. Julie Bergner compares reduced Segal spaces (see Definition 14.2.2) to simplicial monoids. She shows in [**B07**, Theorem 1.6] that there is a Quillen equivalence between the model category of simplicial monoids and the category of reduced simplicial spaces with an appropriate model structure, whose fibrant objects are reduced Segal spaces.

10.15 Symmetric Spectra

As an interlude, we present a symmetric monoidal model for the stable homotopy category that is build out of simplicial sets.

The stable homotopy category, that is, the homotopy category of spectra, has been around since the late 1950s and early 1960s. However, for a long time, there was no symmetric monoidal category of spectra modelling it. A particularly small and explicit model is the one of symmetric spectra, originally defined in [**HoSS00**]. Apart from its importance for modelling stable homotopy types, its definition and its applications have been extended to other contexts.

Let $(\mathcal{C}, \otimes, \mathbf{1})$ be a bicomplete closed symmetric monoidal category and let K be an object of $\mathcal{C}$. Hovey defines symmetric spectra in $\mathcal{C}$ with respect to an object K of $\mathcal{C}$ in [**Ho01**].

Let $\mathsf{Sym}(K)$ be the symmetric sequence in $\mathcal{C}$ whose nth level is $K^{\otimes n}$ and with Σ_n-action given by permutation of the $\otimes$-factors. This is actually a commutative monoid in $\mathcal{C}^\Sigma$ with the symmetric monoidal structure coming from the Day convolution product, as in Example 9.8.6.

Definition 10.15.1 The category of *symmetric spectra in $\mathcal{C}$ with respect to K*, $\mathsf{Sp}^\Sigma(\mathcal{C}, K)$, is the category of right $\mathsf{Sym}(K)$-modules in $\mathcal{C}^\Sigma$. Explicitly, a symmetric spectrum is a family of Σ_n-objects $X(\mathbf{n}) \in \mathcal{C}$, together with Σ_n-equivariant maps

$$X(\mathbf{n}) \otimes K \to X(\mathbf{n}+1)$$

for all $n \geq 0$, such that the composites

$$X(\mathbf{n}) \otimes K^{\otimes p} \to X(\mathbf{n}+1) \otimes K^{\otimes p-1} \to \cdots \to X(\mathbf{n}+\mathbf{p})$$

are $\Sigma_n \times \Sigma_p$-equivariant for all $n, p \geq 0$. Morphisms in $\mathsf{Sp}^\Sigma(\mathcal{C}, K)$ are morphisms of symmetric sequences that are compatible with the right $\mathsf{Sym}(K)$-module structure.

Symmetric spectra form a symmetric monoidal category $(\mathsf{Sp}^\Sigma(\mathcal{C}, K),$ $\wedge, \mathsf{Sym}(K))$, such that for $X, Y \in \mathsf{Sp}^\Sigma(\mathcal{C}, K)$,

$$X \wedge Y = X \square_{\mathsf{Sym}(K)} Y.$$

Here, we use the right action of $\mathsf{Sym}(K)$ on Y after applying the twist map in the symmetric monoidal structure on $\mathcal{C}^\Sigma$.

For the category of topological spaces or simplicial sets and K a model of the 1-sphere, this recovers the definition of symmetric spectra from [**HoSS00**]. For many important homotopy types, there are explicit models in symmetric spectra. We just mention two.

Examples 10.15.2

- Let S^1 be the standard simplicial model of the 1-sphere $\Delta_1/\partial\Delta_1$. The *symmetric sphere spectrum* $S = \mathsf{Sym}(S^1)$ has

$$S(\mathbf{n}) = S^n =: (S^1)^{\wedge n}.$$

- Let A be an abelian group. Then, the model for the *Eilenberg–Mac Lane spectrum of A, HA*, in Sp^Σ is

$$(HA)(\mathbf{n}) = A \otimes \bar{\mathbb{Z}}\{S^n\},$$

where $\bar{\mathbb{Z}}\{S^n\}_m$ is the free abelian group generated by the set S^n_m modulo the relation that all multiples of the basepoint are zero.

Other spectra such as many flavors of cobordism spectra and K-theory spectra have models in Sp^Σ. For more details and many more examples, we refer the reader to [**HoSS00**] and [**Schw∞**].

An important and amusing special case of general symmetric spectra is the one where you take the monoidal unit e as K. The following result goes back to Jeff Smith.

Proposition 10.15.3 [**RiS17**, Proposition 9.1] *Let $(\mathcal{C}, \otimes, e, \tau)$ be any symmetric monoidal category. Then, the category of symmetric spectra with respect to the unit object e, $\mathsf{Sp}^\Sigma(\mathcal{C}, e)$, is equivalent to the category of functors from $\mathcal{I}$ to $\mathcal{C}$, $\mathsf{Fun}(\mathcal{I}, \mathcal{C})$.*

Proof Let $X \in \mathsf{Sp}^\Sigma(\mathcal{C}, e)$. Then, there are Σ_n-equivariant maps $X(\mathbf{n}) \cong X(\mathbf{n}) \otimes e \to X(\mathbf{n}+\mathbf{1})$, such that the composite

$$\sigma_{n,p} \colon X(\mathbf{n}) \cong X(\mathbf{n}) \otimes e^{\otimes p} \to X(\mathbf{n}+\mathbf{p})$$

is $\Sigma_n \times \Sigma_p$-equivariant for all $n, p \geq 0$.

We send X to $\phi(X) \in \mathsf{Fun}(\mathcal{I}, \mathcal{C})$ with $\phi(X)(\mathbf{n}) = X(n)$. If $i = i_{p,n-p} \in \mathcal{I}(\mathbf{p}, \mathbf{n})$ is the standard inclusion, then we let $\phi(i) \colon \phi(X)(\mathbf{p}) \to \phi(X)(\mathbf{n})$ be $\sigma_{p,n-p}$. Every morphism $f \in \mathcal{I}(\mathbf{p}, \mathbf{n})$ can be written as $\xi \circ i$, where i is the standard inclusion and $\xi \in \Sigma_n$. For such ξ, the map $\phi(\xi)$ is given by the Σ_n-action on $X(n) = \phi(X)(\mathbf{n})$.

If $f = \xi' \circ i$ is another factorization of f into the standard inclusion, followed by a permutation, then ξ and ξ' differ by a permutation $\tau \in \Sigma_n$, which maps all j with $1 \leq j \leq p$ identically, that is, τ is of the form $\tau = 1_\mathbf{p} \oplus \tau'$ with $\tau' \in \Sigma_{n-p}$. As the structure maps $\sigma_{p,n-p}$ are $\Sigma_p \times \Sigma_{n-p}$-equivariant and as τ' acts trivially on $e^{\otimes(n-p)} \cong e$, the induced map $\phi(f) = \phi(\xi') \circ \phi(i)$ agrees with $\phi(\xi) \circ \phi(i)$.

The inverse of ϕ, ψ, sends an $\mathcal{I}$-diagram in $\mathcal{C}$, A, to the symmetric spectrum $\psi(A)$ whose nth level is $\psi(A)(\mathbf{n}) = A(\mathbf{n})$. The Σ_n-action on $\psi(A)(n)$ is given by the corresponding morphisms $\Sigma_n \subset \mathcal{I}(\mathbf{n}, \mathbf{n})$, and the structure maps of the spectrum are defined as

$$\psi(A)(n) \otimes e^{\otimes p} = A(\mathbf{n}) \otimes e^{\otimes p} \xrightarrow{\cong} A(\mathbf{n}) \xrightarrow{A(i_{n,p})} A(\mathbf{n}+\mathbf{p}) = \psi(A)(n+p)\cdot$$

The functors ϕ and ψ are well-defined and inverse to each other. □

11

The Nerve and the Classifying Space
of a Small Category

To any small category, you can associate a topological space that takes the data of the category (objects, morphisms, and composition of morphisms) and translates it into a CW complex. This is done in a two-stage process: First you construct a simplicial set out of your category, and then, you form its geometric realization.

11.1 The Nerve of a Small Category

The construction of a nerve of a small category $\mathcal{C}$ takes the data of the objects of $\mathcal{C}$ and all morphisms of $\mathcal{C}$, together with all their compositions, and creates a simplicial set out of it.

Definition 11.1.1

- For a small category $\mathcal{C}$, let $M_n(\mathcal{C})$ be the set

$$\left\{ C_0 \xrightarrow{f_1} C_1 \xrightarrow{f_2} \cdots \xrightarrow{f_n} C_n \,|\, C_i \text{ object of } \mathcal{C}, \, f_i \text{ morphism in } \mathcal{C} \right\}$$

of the n-tuples of composable morphisms in $\mathcal{C}$. We denote an element, as earlier, as $[f_n | \cdots | f_1]$.
- The *nerve of the category* $\mathcal{C}$ is the simplicial set $N\mathcal{C} \colon \Delta^{op} \to$ Sets, which sends $[n]$ to the set $M_n(\mathcal{C})$. The degeneracies insert identity morphisms

$$s_i[f_n | \cdots | f_1] = [f_n | \cdots | f_{i+1} | 1_{C_i} | f_i | \cdots | f_1], \quad 0 \le i \le n,$$

and the face maps compose morphisms

$$
d_i[f_n| \cdots |f_1] =
\begin{cases}
[f_n| \cdots |f_2], & i = 0, \\
[f_n| \cdots |f_{i+1} \circ f_i| \cdots |f_1], & 0 < i < n, \\
[f_{n-1}| \cdots |f_1], & i = n.
\end{cases}
$$

You can also interpret the ith face map as omitting the object C_i. If i is 0 or n, then the morphism next to C_i just dies with the object, whereas for $0 < i < n$, deleting the object causes the composition of the adjacent morphisms.

Proposition 11.1.2 *Nerves of small categories are quasi-categories.*

Proof The morphism $\varphi^i \colon [1] \to [n]$, $\varphi^i(0) = i$ and $\varphi^i(1) = i + 1$ is contained in Λ_n^k for $0 < k < n$ for all $0 \leq i < n$. We assume that we have a morphism $\psi \colon \Lambda_n^k \to N(\mathcal{C})$. Then, ψ applied to the $n + 1$ morphisms in $\Delta([0], [n])$ gives $n + 1$ objects $C_0, \ldots, C_n$ of $\mathcal{C}$, and ψ applied to φ^i gives a morphism f_{i+1} from C_i to C_{i+1} for all $0 \leq i < n$.

We define the extension $\tilde{\psi}$ of ψ to Δ_n by $\tilde{\psi}(1_{[n]}) = [f_n| \cdots |f_1]$.

In order to show that the restriction of $\tilde{\psi}$ to Λ_n^k agrees with ψ, it suffices to show that their restriction with respect to every map $\xi \colon [1] \to \{0, \ldots, j-1, j+1, \ldots, n\}$ agrees for every $j \neq k$, $0 \leq j \leq n$; in fact, we only need those maps that hit neighbors in the ordered set $\{0, \ldots, j-1, j+1, \ldots, n\}$. If these neighbors are of the form i and $i + 1$, then both restrictions agree by construction. Thus, we only need to consider the case that the image of ξ is $\{j-1, j+1\}$, and here, the cases $j = 0$ and $j = n$ lead to trivial cases. We can assume that n is at least 3, because for $n = 2$, the only nontrivial case is $j = k = 1$, but the ξ with image $\{0, 2\}$ is not contained in Λ_2^1.

For $\tilde{\psi}$, the restriction via ξ yields $f_{j+1} \circ f_j$, and by construction, the same is true for ψ, because for $n \geq 3$, the morphism $f \in \Delta([2], [n])$ with $f(0) = j - 1$, $f(1) = j$, and $f(2) = j + 1$, is in Λ_n^k for $0 < k < n$, and $\xi = d_1 f$ is sent to the composition of $d_2 f = f_j$ with $d_0 f = f_{j+1}$. $\qquad\square$

Remark 11.1.3 Note that the horn extensions in a nerve are actually unique. There is a partial converse of Proposition 11.1.2: If X is a simplicial set, such that for all n and for all $0 < k < n$, the diagram

$$
\begin{array}{ccc}
\Lambda_n^k & \xrightarrow{\alpha} & X \\
{\scriptstyle j}\downarrow & & \\
\Delta_n & &
\end{array}
$$

has a *unique* extension $\Delta_n \to X$, then there is a small category $\mathcal{C}$ and an isomorphism $X \cong N(\mathcal{C})$. A proof can be found in, for instance, [**Lu09**, 1.1.2.2].

Exercise 11.1.4 Show that a simplicial set is a Segal set if and only if it is isomorphic to the nerve of a small category.

Example 11.1.5 If we consider the poset $[n]$ as a category, then the nerve of $[n]$ is isomorphic to the representable functor Δ_n. We saw in Lemma 10.12.7 that Δ_n is *not* a Kan complex for $n \geq 2$. Thus, in general, we cannot hope that nerves are better than quasi-categories.

If our small category $\mathcal{C}$ has invertible morphisms, then we actually *can* guarantee the Kan condition. Heuristically, the horn fillers for Λ_n^0 and Λ_n^n require extensions or lifts, but if every morphism is invertible, this can be done.

Proposition 11.1.6 *Let $\mathcal{C}$ be a small groupoid. Then, $N\mathcal{C}$ is a Kan complex.*

A proof of this fact can be found in [**GJ09**, 3.5].

Lemma 11.1.7 *The nerve functor from the category of all small categories to the category of simplicial sets has a left adjoint.*

Proof Let X be a simplicial set. We have to construct a small category $L(X)$. The objects of $L(X)$ are the zero simplices of X, X_0. A $y \in X_1$ is considered a morphism from $d_1 y$ to $d_0 y$. The morphisms in $L(X)$ are freely generated by the 1-simplices in X, X_1, subject to the relations coming from 2-simplices. If there is an $x \in X_2$ with $d_0(x) = x_0$, $d_1(x) = x_1$, and $d_2(x) = x_2$, then $x_1 = x_0 \circ x_2$:

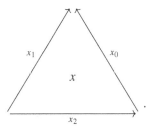

Degenerate 2-simplices encode the unit condition, and the associativity of the composition is ensured by the simplicial identities for the face maps. □

Remark 11.1.8 Note that the category associated with the nerve of a small category is isomorphic to the category, but the nerve of a category associated with a simplicial set X is "wildly wrong" [**LTW79**]; it associates with X a quasi-category with unique inner horn extensions, a Segal set.

Remark 11.1.9 Let $\mathcal{C}$ be a small category. Regarding $[n] = \{0, 1, \dots, n\}$ as a category associated with the poset $[n]$ with its natural ordering, we can define a variant of the nerve, $\mathcal{N}$, as being the simplicial category, such that $\mathcal{N}_n$ is the category of functors from the category $[n]$ to the category $\mathcal{C}$.

Later, we will need monoidal properties of the nerve functor. If $\mathcal{C} \times \mathcal{D}$ is a product of small categories, then objects and morphism in $\mathcal{C} \times \mathcal{D}$ are just pairs of morphisms. Hence, we obtain the following useful result:

Proposition 11.1.10 *The nerve functor N is strong symmetric monoidal.*

The nerve functor also respects the join construction.

Proposition 11.1.11 *Let $\mathcal{C}$ and $\mathcal{D}$ be two small categories and let $\mathcal{C} * \mathcal{D}$ be their join. Then, there is an isomorphism of simplicial sets*

$$N(\mathcal{C}) * N(\mathcal{D}) \cong N(\mathcal{C} * \mathcal{D}).$$

Proof   We follow [**Jo-c∞**, Proof of Corollary 3.3]. We define a map

$$\varphi \colon N(\mathcal{C}) * N(\mathcal{D}) \to N(\mathcal{C} * \mathcal{D})$$

using that

$$(N(\mathcal{C}) * N(\mathcal{D}))_n = N(\mathcal{C})_n \sqcup N(\mathcal{D})_n \sqcup \bigsqcup_{i+j=n-1} N(\mathcal{C})_i \times N(\mathcal{D})_j.$$

On the summands $N(\mathcal{C})_n$ and $N(\mathcal{D})_n$, we define φ by using the inclusions $\mathcal{C} \to \mathcal{C} * \mathcal{D}$ and $\mathcal{D} \to \mathcal{C} * \mathcal{D}$, which induce maps $N(\mathcal{C}) \to N(\mathcal{C} * \mathcal{D}) \leftarrow N(\mathcal{D})$. We express an element in $N(\mathcal{C})_i$ as a functor $x \colon [i] \to \mathcal{C}$ and similarly consider a $y \colon [j] \to \mathcal{D}$. Functoriality of the join then gives a map $x * y \colon [i] * [j] \to \mathcal{C} * \mathcal{D}$. We know from Exercise 1.4.12 that $[i] * [j] \cong [n]$. These maps combine φ, and again, it is straightforward to check that φ is an isomorphism of simplicial sets.   □

11.2 The Classifying Space and Some of Its Properties

Definition 11.2.1 The *classifying space of a small category* $\mathcal{C}$, $B\mathcal{C}$, is the geometric realization of the nerve of $\mathcal{C}$:

$$B\mathcal{C} = |N\mathcal{C}|.$$

The objects C of $\mathcal{C}$ give zero cells in $B\mathcal{C}$, and a nonidentity morphism from C to C' gives rise to an edge whose endpoints correspond to the objects C and C'. If $g \circ f$ is a composition of morphisms in $\mathcal{C}$, then in the classifying space, you will find a triangle, with edges corresponding to f, g, and $g \circ f$. Threefold compositions give rise to tetrahedra and so on.

The topological space BC is always a CW complex, and a functor $F: C \to D$ induces a continuous and cellular map of topological spaces:

$$BF: BC \to BD.$$

Hence, B is a functor from the category cat to the category Top of topological spaces.

Examples 11.2.2

- If X is a set and C is the corresponding discrete category, then the classifying space BC is X with the discrete topology.
- If G is a group and we consider the small category C_G associated with G, then the classifying space $B(C_G)$ is called the *classifying space of the group* G and is denoted by BG. If the group G is abelian, then the group composition is a group homomorphism, and it induces a functor $C_G \times C_G \to C_G$. We therefore obtain a map

$$BG \times BG \to B(C_G \times C_G) \to B(C_G) = BG,$$

and for abelian groups G, BG is a topological group. For instance, $B\mathbb{Z} \simeq \mathbb{S}^1$.

  If G is a topological group, then we can implement the topology into the construction of BG by endowing $G^n \times \triangle^n$ with the product topology.

  For instance, $B\mathbb{S}^1 \simeq \mathbb{C}P^\infty$, and this is an Eilenberg–Mac Lane space of type $(\mathbb{Z}, 2)$, $K(\mathbb{Z}, 2) = \mathbb{C}P^\infty \simeq B(B\mathbb{Z})$. In general, if A is a finitely generated abelian group, then the n-fold iterated classifying space construction is a model of the Eilenberg–Mac Lane space $K(A, n)$.

  If G is a discrete group, then the homology of the group G (with coefficients in $\mathbb{Z}$) is the singular homology $H_*(BG; \mathbb{Z})$.

- For a monoid, one can build $B(C_M)$. We will learn more about this space later (see, for instance, Theorem 13.4.6 and Proposition 13.4.4).
- Let us consider the category Σ. This has as objects the natural numbers (including zero), and the only morphisms are automorphisms with $\Sigma([n], [n]) = \Sigma_n$. Therefore, the classifying space of Σ has one component for every natural number, because the different objects are not connected by morphisms. Thus,

$$B\Sigma = \bigsqcup_{n \geq 0} B\Sigma_n.$$

- If we consider the poset $[n]$ as a category, then the nerve of $[n]$ is isomorphic to the representable functor $\triangle_n$ and $B[n] \cong \triangle^n$.

Proposition 11.2.3 *Viewed as a functor B: cat $\to$ cg, the functor B is strong symmetric monoidal.*

Proof   Let C and D be two small categories. As the nerve functor is strong symmetric monoidal (see Proposition 11.1.10) and as geometric realization commutes with products in the sense of Proposition 10.6.7 and Remark 10.6.9, we get a homeomorphism $\phi_{C,C'}\colon B(C \times C') \cong BC \times_k BC'$, so the functor B: cat $\to$ cg is strong symmetric comonoidal and strong symmetric monoidal. $\qquad\square$

Segal shows some properties of the classifying space of a (topological) category in **[Se74]**. For Quillen's work **[Q73]**, classifying spaces of categories and their properties are crucial.

Theorem 11.2.4

(1) For two functors $F, F'\colon C \to D$, a natural transformation $\tau\colon F \Rightarrow F'$ induces a homotopy between BF and BF'.

(2) If $C \underset{R}{\overset{L}{\rightleftarrows}} D$ is an adjoint pair of functors, then BC is homotopy equivalent to BD.

(3) In particular, an equivalence of categories gives rise to a homotopy equivalence of classifying spaces.

Proof

(1) We can view τ as a functor T from $C \times [1]$ to D: On objects, we define T as $T(C, 0) = F(C)$ and $T(C, 1) = F'(C)$. For a morphism $f \in C(C, C')$, we set $T(f, 1_0) = F(f)$ and $T(f, 1_1) = F'(f)$. For the only nontrivial morphism $0 < 1$ in $[1]$, we define $T(1_C, 0 < 1)$ as τ_C.

The composite

$$BC \times [0, 1] \cong B(C \times [1]) \xrightarrow{B(T)} BD$$

then gives the desired homotopy.

(2) For an adjoint pair of functors, we have the natural transformations $\eta\colon \mathrm{Id} \Rightarrow RL$ and $\varepsilon\colon LR \Rightarrow \mathrm{Id}$. We just showed that these give rise to a homotopy equivalence. $\qquad\square$

Corollary 11.2.5 *If a small category C has an initial or terminal object, then its classifying space is contractible.*

Proof   In both cases, there is an adjoint pair of functors between C and the category $[0]$, and $B[0]$ is a point. $\qquad\square$

Beware that an equivalence of categories does not give rise to homeomorphism in general. The following is an explicit example:

Example 11.2.6 Let G be a discrete group. We saw that in its translation category, $\mathcal{E}_G$, every object is terminal and initial, and thus, $B\mathcal{E}_G$ is contractible. In fact, $B\mathcal{E}_G$ is a model for the universal space for G-bundles, EG, and this is, in general, *not* homeomorphic to a point. For instance, $E\mathbb{Z}/2\mathbb{Z}$ is $\mathbb{S}^\infty = \operatorname{colim}_n \mathbb{S}^n$.

The simplicial structure on $N\mathcal{E}_G$ is as follows: An element in $(N\mathcal{E}_G)$ is a string

$$g_0 \xrightarrow{g_1 g_0^{-1}} g_1 \xrightarrow{g_2 g_1^{-1}} \cdots \xrightarrow{g_q g_{q-1}^{-1}} g_q,$$

but this can be simplified by just remembering the $(q+1)$-tuple of group elements $(g_0, \ldots, g_q)$. With this identification, the face maps omit a g_i, and the degeneracies double a g_i.

Proposition 11.2.7 *If $\mathcal{C}$ is a small category possessing binary products, then $B\mathcal{C}$ is contractible.*

Proof Let C be a fixed but arbitrary object of $\mathcal{C}$. We consider the functor

$$C \times (-): \mathcal{C} \to \mathcal{C}, \quad (C \times (-))(C') = C \times C'.$$

Projection to the first factor of the product gives rise to a natural transformation from $C \times (-)$ to the constant functor with value C, whereas projection to the second factor yields a natural transformation from $C \times (-)$ to the identity functor. □

Remark 11.2.8 We will see later in Proposition 11.3.10 that the classifying space of any small filtered category is contractible.

Exercise 11.2.9 Let $\mathcal{C}$ be a small category and let $\mathcal{C}^v$ be its dual category. Show that there is a homeomorphism between $B\mathcal{C}$ and $B\mathcal{C}^\circ$. Is this homeomorphism always induced by a functor from $\mathcal{C}$ to $\mathcal{C}^\circ$?

Remark 11.2.10 If $\mathcal{C}$ is a category with a proper class of objects but with a small skeleton $\mathcal{C}'$, then one might talk about $B\mathcal{C}$ but actually work with $B\mathcal{C}'$.

11.3 π_0 and π_1 of Small Categories

We can apply topological invariants to the space $B\mathcal{C}$, and we might get useful information about the category in this way. Some invariants that you have encountered in algebraic topology can be defined directly in terms of the objects and morphisms of a category. One example is the set of path components of a small category.

In Definition 1.4.16, we defined the notion of a small connected category, using the equivalence relation on the set of objects generated by the relation of being connected by a morphism.

Definition 11.3.1 Let $\mathcal{C}$ be a small category. We denote the set of equivalence classes of the morphism relation on the objects of $\mathcal{C}$ by $\pi_0\mathcal{C}$ and call this the *set of path components of the category* $\mathcal{C}$.

If the cardinality of the set $\pi_0\mathcal{C}$ is one, then we call the category $\mathcal{C}$ *path connected* or *connected*. In particular, categories with just one object are connected, for instance, the category associated with a monoid or group. By the very definition of $\pi_0\mathcal{C}$ and by the description of π_0 of a CW complex, we obtain the following:

Proposition 11.3.2 *The set of path components of the topological space* $B\mathcal{C}$, $\pi_0 B\mathcal{C}$, *is in bijection with the set of path components of* $\mathcal{C}$, $\pi_0\mathcal{C}$.

Example 11.3.3 Let X be a set and let G be a group, such that X carries a G-action. We can define the *translation category of the G-action on X*, $\mathcal{E}_G^X$, to be the category whose objects are the elements of X and

$$\mathcal{E}_G^X(x_1, x_2) = \{g \in G | gx_1 = x_2\}.$$

With this notation, the translation category of a group G is $\mathcal{E}_G = \mathcal{E}_G^G$.

Two elements x and x' of X are equivalent if and only if they lie in the same G-orbit, and thus,

$$\pi_0\mathcal{E}_G^X = X/G.$$

Exercise 11.3.4 ([W13, Example IV.3.3.2], but beware of the difference in notation!) Let $\mathcal{D}$ be a small category and let $F \colon \mathcal{D} \to \mathsf{Sets}$ be a functor. Show that

$$\pi_0(F\backslash\mathcal{D}) \cong \mathrm{colim}_{\mathcal{D}} F.$$

Our next aim is to describe the fundamental group of the space $B\mathcal{C}$ in terms of the morphisms of the category $\mathcal{C}$. To this end, we introduce the concept of maximal trees. In Proposition 11.5.6, we will see an alternative description.

Definition 11.3.5 Let $\mathcal{C}$ be a small category and let X be an arbitrary set of morphisms in $\mathcal{C}$.

- The *graph of X*, Γ_X, is the one-dimensional sub-CW complex of $B\mathcal{C}$ consisting of all edges in $B\mathcal{C}$ that correspond to morphisms in X.
- The set X is called a *tree* if Γ_X is a tree.
- If $\pi_0\mathcal{C}$ is trivial, then we call a tree X *maximal* if every object of $\mathcal{C}$ arises as the source or the target of a morphism in X.

We use maximal trees in order to describe $\pi_1(B\mathcal{C})$.

Proposition 11.3.6 *Let $\mathcal{C}$ be a small path-connected category and let X be a maximal tree in $\mathcal{C}$. The fundamental group of $B\mathcal{C}$ (with respect to any chosen object C of $\mathcal{C}$) has the following presentation: For every morphism f in $\mathcal{C}$, there is a generator $[f]$. The relations are given by the following list:*

- *A generator $[f]$ is trivial if $f \in X$.*
- *For every object C of $\mathcal{C}$, the generator $[1_C]$ is trivial.*
- *If $f: C_1 \to C_2$ and $g: C_2 \to C_3$ are morphisms in $\mathcal{C}$, then*

$$[g \circ f] = [g][f].$$

Proof The 1-skeleton of the CW complex $B\mathcal{C}$, $B\mathcal{C}^{(1)}$, corresponds to the graph Γ on the set of morphisms in $\mathcal{C}$, with the exception of the identity morphisms. The fundamental group of $B\mathcal{C}^{(1)}$ is the free group generated by all edges of Γ outside of the maximal tree X. Cellular approximation ensures that the fundamental group of $B\mathcal{C}$ is isomorphic to the fundamental group of the 2-skeleton $B\mathcal{C}^{(2)}$. Therefore, the only relations that occur are those coming from the composition of morphisms in $\mathcal{C}$. $\qquad\square$

Examples 11.3.7
- If G is a discrete group and $\mathcal{C}_G$ is its associated category, then every element $g \in G$ gives rise to a generator $[g]$ of $\pi_1(BG)$. Relations of the first kind do not occur, and the ones coming from composition identify $\pi_1(BG)$ as G.
- If M is a monoid and $\mathcal{C}_M$ is its associated category, then again, every element $m \in M$ gives rise to a generator $\pi_1(B\mathcal{C}_M)$, and the only nontrivial relations come from the composition of elements of m. We know that $\pi_1(B\mathcal{C}_M)$ is a group, so we have associated a group to the monoid M. We will encounter this group later again in Definition 13.2.6.

In an extension of the earlier result, we define the higher homotopy groups of a small category:

Definition 11.3.8 Let $\mathcal{C}$ be a small category, C be an object of $\mathcal{C}$, and n be greater than or equal to 0. The *nth homotopy group of $\mathcal{C}$ with respect to the basepoint C* is

$$\pi_n(\mathcal{C}; C) := \pi_n(B\mathcal{C}, [C]).$$

Here, $[C]$ denotes the 0-simplex associated with the object C.

Note that the category of small categories has filtered colimits. If $(F(D), D$ in $\mathcal{D})$ is a filtered system of categories, that is, a functor $F: \mathcal{D} \to$ cat from a small filtered category $\mathcal{D}$ to cat, then $\text{colim}_\mathcal{D} F$ is again a small category $\mathcal{C}$, with set of objects being $\text{Ob}(\mathcal{C}) = \text{colim}_\mathcal{D}\text{Ob}(F(D))$. Every element in the nerve $[f_n| \cdots |f_1] \in N_n\mathcal{C}$ can be represented by an element $[g_n| \cdots |g_1] \in N_n F(D)$ for some D, hence, we get that $\text{colim}_\mathcal{D} NF \cong N\mathcal{C}$.

Lemma 11.3.9 *Assume that $\mathcal{C}$ is a small category that is the filtered colimit of a functor $F: \mathcal{D} \to$ cat. For every D in $\mathcal{D}$, let C_D be an object in $F(D)$, such that for every $f \in \mathcal{D}(D, D')$, we have $F(f)(C_D) = C_{D'}$. Denote by C the equivalence class of the objects C_D in $\mathcal{C}$. Then, for all $n \geq 0$,*

$$\pi_n(B\text{colim}_\mathcal{D} F; [C]) \cong \text{colim}_\mathcal{D}(\pi_n(BF(D); [C_D])).$$

Proof   Geometric realization commutes with filtered colimits, and hence, $B\mathcal{C} \cong \text{colim}_\mathcal{D} BF$. As spheres are compact, the homotopy class of a map from a sphere to $B\mathcal{C}$ can be expressed as the equivalence class of a map from this sphere to some $BF(D)$.   □

Proposition 11.3.10 *The classifying space of any small filtered category is contractible.*

Proof   Let $\mathcal{D}$ be a small filtered category. Let $F: \mathcal{D} \to$ cat be the functor that sends an object D of $\mathcal{D}$ to $\mathcal{D} \downarrow D$, and a morphisms $f \in \mathcal{D}(D, D')$ is sent to $F(f)$ with

$$F(f)\big(\tilde{D}, g: \tilde{D} \to D\big) = \big(\tilde{D}, \tilde{D} \overset{g}{\longrightarrow} D \overset{f}{\longrightarrow} D'\big).$$

We claim that $\mathcal{D}$ is isomorphic to the filtered colimit of the small categories $\mathcal{D} \downarrow D$, $\text{colim}_\mathcal{D} F$.

There is a canonical functor θ from $\mathcal{D}$ to $\text{colim}_\mathcal{D} \mathcal{D} \downarrow D$ that sends an object D of $\mathcal{D}$ to the equivalence class of the object $(D, 1_D)$ of $\mathcal{D} \downarrow D$. A morphism $f \in \mathcal{D}(D, D')$ has $\theta(f) = [f]$. Note that $(D, 1_D)$ is equivalent to (D, f) in the colimit and f is a morphism in $\mathcal{D} \downarrow D'$ from (D, f) to $(D', 1_{D'})$.

The forgetful functor from each category $\mathcal{D} \downarrow D$ to $\mathcal{D}$ induces a functor

$$\pi: \text{colim}_\mathcal{D} F \to \mathcal{D}$$

that remembers the first coordinate. Note that π is well-defined because the equivalence relation does not affect the source of an object in $F(D')$ and also does not affect the morphisms.

It is easy to see that $\pi \circ \theta = \mathrm{Id}_{\mathcal{D}}$. For the converse, observe that $\theta \circ \pi$ applied to an equivalence class of $(D', g \colon D' \to D)$ is the equivalence class of $(D', 1_{D'})$, but $(D', 1_{D'})$ is equivalent to $(D', g \colon D' \to D) = F(g)(D', 1_{D'})$.

Each classifying space $B(\mathcal{D} \downarrow D)$ is contractible, because the object D is terminal in $\mathcal{D} \downarrow D$, and hence, $B\mathcal{D} \simeq B\mathrm{colim}_{\mathcal{D}} F$ is contractible as well. $\qquad\square$

11.4 The Bousfield Kan Homotopy Colimit

The idea of a homotopy colimit is to create a homotopy invariant notion of a colimit. See [**D∞**] for a rather gentle introduction. For functors to simplicial sets, there is an explicit model of a homotopy colimit introduced by Bousfield and Kan in [**BK72**].

Functors to the category of sets do not carry any actual homotopical information. In this case, the Bousfield Kan construction reduces to the nerve of a category.

Definition 11.4.1 Let $\mathcal{C}$ be a small category and $F \colon \mathcal{C} \to \mathrm{Sets}$ be a functor, and consider the category $F\backslash\mathcal{C}$ associated with F (see Definition 5.1.11). The *homotopy colimit of* F, hocolim$_F$, is the nerve of $F\backslash\mathcal{C}$, $N(F\backslash\mathcal{C})$.

Of course, if you prefer topological spaces, you can take $B(F\backslash\mathcal{C}) = |N(F\backslash\mathcal{C})|$.

Let us make explicit what $N(F\backslash\mathcal{C})$ is. In simplicial degree zero, we get the set of objects of $F\backslash\mathcal{C}$, (C, x) with $x \in F(C)$. We can identify this set with $\bigsqcup_C F(C)$. In degree one, we get the set of all morphisms in $F\backslash\mathcal{C}$, that is, all $f_1 \colon C_0 \to C_1$, such that $F(f_1)$ maps $x \in F(C_0)$ to $y \in F(C_1)$. The information in $F(C_1)$ is redundant, and we obtain

$$N(F\backslash\mathcal{C})_1 \cong \bigsqcup_{f_1 \colon C_0 \to C_1} F(C_0),$$

and in general,

$$N(F\backslash\mathcal{C})_n \cong \bigsqcup_{[f_n|\cdots|f_1]\in N(\mathcal{C})_n} F(C_0).$$

Here, the simplicial structure maps s_i only affect the nerve and so do the d_is for $i > 0$, but d_0 maps an $x \in F(C_0)$ in the component of $[f_n|\cdots|f_1] \in N(\mathcal{C})_n$ to $F(f_1)(x) \in F(C_1)$ in the component of $[f_n|\cdots|f_2] \in N(\mathcal{C})_{n-1}$. If, instead, F

takes values in the category of simplicial sets, then one can use the diagonal of a bisimplicial set and define the homotopy colimit as follows:

Definition 11.4.2 Let $\mathcal{C}$ be a small category.

(1) Let $F\colon \mathcal{C} \to s\mathrm{Sets}$ be a functor. Then, *the Bousfield–Kan homotopy colimit of F*, $\mathrm{hocolim}_{\mathcal{C}} F$, is the simplicial set with n-simplices

$$\coprod_{[f_n|\cdots|f_1]\in N(\mathcal{C})_n} F(C_0)_n.$$

(2) For a functor $F\colon \mathcal{C} \to \mathrm{Top}$, *the Bousfield–Kan homotopy colimit of F* is the simplicial topological space with n-simplices

$$\coprod_{[f_n|\cdots|f_1]\in N(\mathcal{C})_n} F(C_0).$$

As discussed earlier, the simplicial structure on $\mathrm{hocolim}_{\mathcal{C}} F$ is induced by the nerve of $\mathcal{C}$, the application of $F(f_1)$ to $F(C_0)$, and the simplicial structure of $F(C_0)$. For instance, for $x \in F(C_0)_n$, we get

$$d_0([f_n|\cdots|f_1], x) = ([f_n|\cdots|f_2], d_0(F(f_1)(x))).$$

One can still view $\mathrm{hocolim}_{\mathcal{C}} F$ as the nerve of $F\backslash\mathcal{C}$, bearing in mind that the $F(C)$s are themselves simplicial sets or topological spaces and that this structure has to be taken into account.

Bousfield and Kan actually give a different description of the homotopy colimit in [**BK72**, XII.5] first and then show in [**BK72**, XII.5.2] that it is equivalent to the one in Definition 11.4.2.

Remark 11.4.3

- Note that there is always a map

$$\coprod_{[f_n|\cdots|f_1]\in N(\mathcal{C})_n} F(C_0)_n \to \coprod_{[f_n|\cdots|f_1]\in N(\mathcal{C})_n} \Delta_0$$

induced by the morphism of simplicial sets from $F(C_0)$ to the terminal simplicial set Δ_0. Hence, there is always a continuous map $|\mathrm{hocolim}_{\mathcal{C}} F| \to B\mathcal{C}$, no matter what F is.
- If F is a simplicial set, that is, if $\mathcal{C} = \Delta^o$ and the target category is the category of sets, then the map from $N(F\backslash\Delta^o)$ to its set of path components is nothing but the map from the homotopy colimit of F to the colimit of F;

compare Exercise 10.12.13. For a functor $F: \mathcal{C} \to$ Top, we can always project the zeroth space of the homotopy colimit to the coequalizer of the diagram

$$\bigsqcup_{[f_1] \in N(\mathcal{C})_1} F(C_0) \mathrel{\substack{\longrightarrow \\ \longrightarrow}} \bigsqcup_{C \in N(\mathcal{C})_0} F(C),$$

and this gives a morphism between the simplicial topological space $\text{hocolim}_{\mathcal{C}} F$ and the constant simplicial space $\text{colim}_{\mathcal{C}} F$.

Examples 11.4.4

(1) Let $[0]$ be the category with one object and one morphism. Then, the category $F \backslash [0]$ for $F: [0] \to s\text{Sets}$ captures the simplicial set $F[0]$ as the only information. The nerve of $[0]$ has $N[0]_n = \{[1_0 | \cdots | 1_0]\}$, and hence, the homotopy colimit of F over $[0]$ is equivalent to the simplicial set $F[0]$. Similarly, if $\mathcal{C}$ is a small discrete category, then

$$\text{hocolim}_{\mathcal{C}} F \simeq \bigsqcup_{C \in \text{Ob}(\mathcal{C})} F(C).$$

(2) If $\mathcal{C}$ is the category of natural numbers viewed as a poset, then the homotopy colimit over $\mathcal{C}$ of any $F: \mathcal{C} \to s\text{Sets}$ gives the telescope of

$$F[0] \to F[1] \to F[2] \to \cdots.$$

(3) Let G be a group and let $F: \mathcal{C}_G \to$ Top be a functor. Then, we can identify F with the G-space $X = F(*)$. By definition, in simplicial degree n, we get

$$(\text{hocolim}_{\mathcal{C}_G} F)_n = \bigsqcup_{[g_n | \cdots | g_1] \in N_n(\mathcal{C}_G)} F(*) \cong G^n \times X.$$

The simplicial structure maps insert the neutral element in G in the G-coordinates or, in case of the face maps d_is, either use the multiplication in the group or the group action on X (for d_0). This simplicial space is isomorphic to $N\mathcal{E}_G \times_G X$ with $EG = |N\mathcal{E}_G|$, as in Example 11.2.6, where the equivalence relation on $N_q(\mathcal{E}_G) \times X$ for forming $N_q(\mathcal{E}_G) \times_G X$ is

$$((g_0, g_1, \ldots, g_q), x) \sim ((e, g_1 g_0^{-1}, \ldots, g_q g_0^{-1}), g_0 x),$$

and hence, the G-orbits $(N(\mathcal{E}_G) \times_G X)_q$ are represented by elements of the form $((e, g_1 g_0^{-1}, \ldots, g_q g_0^{-1}), (g_0 x))$. We can write $(e, g_1 g_0^{-1}, \ldots, g_q g_0^{-1})$ as $(e, h_1, h_1 h_2, \ldots, h_1 \cdot \ldots \cdot h_q)$ and can identify this with the element $(h_1, \ldots, h_q)$ in $N_q \mathcal{C}_G$, because the simplicial structure maps are now given by multiplication of group elements and insertion of the neutral element e.

Therefore,

$$\text{hocolim}_{\mathcal{C}_G} F \cong EG \times_G X,$$

and the homotopy colimit is the *Borel construction* of X with respect to G.

(4) For $X: \Delta^o \to s\text{Top}$, the fat realization from Definition 10.9.1 is a homotopy colimit under mild assumptions. Dugger shows, for instance, [**D**∞, Proposition 20.5] that the maps

$$\text{hocolim}_{\Delta^o} X \leftarrow \text{hocolim}_{\Delta^o_f} X \to ||X||$$

are weak equivalences if every X_n is cofibrant (that is, a retract of a cell complex). Here, Δ^o_f denotes the subcategory of Δ^o consisting of face maps.

For more examples, see [**BK72**, XII, §2] and [**T82**, §3].

Exercise 11.4.5 What is the geometric realization of the homotopy colimit of a functor $F: [1] \to \text{Top}$?

Remark 11.4.6 Dugger [**D**∞, Theorem 6.7, §§9, 10] provides a criterion that allows us to change the indexing category of a homotopy colimit without changing its homotopy type. Similar to Theorem 5.2.5, he shows that for a functor $\gamma: \mathcal{D} \to \mathcal{D}'$ between small categories, such that for all objects D of $\mathcal{D}$, the category $D' \downarrow \gamma$ is nonempty and has a contractible classifying space, the induced map

$$\text{hocolim}_{\mathcal{D}} \gamma^*(F) \to \text{hocolim}_{\mathcal{D}'} F$$

is a weak equivalence for all $F: \mathcal{D}' \to \text{Top}$.

Remark 11.4.7 Beatriz Rodríguez Gonzáles proves in [**RG14**, 4.6, p. 631] that the explicit Bousfield Kan homotopy colimit formula can be used in all model categories that satisfy simplicial descent [**RG12**, (2.4), p. 780]. In particular, the category of non-negatively graded chain complexes in an abelian category with quasi-isomorphisms as weak equivalences satisfies this property, and homotopy colimits can be computed as the total complex of the associated bicomplex of the simplicial chain complex

$$[n] \mapsto \bigoplus_{[f_n|\cdots|f_1] \in N(\mathcal{C})_n} F(C_0)_n. \tag{11.4.1}$$

This also holds for diagrams in unbounded chain complexes [**RiSa**∞, §2].

We will identify the E^1 term of one of the spectral sequences that computes the homology of this total complex with the help of the concept of the homology of small categories in Theorem 16.3.1.

For other approaches to homotopy colimits, see [**Rie14**, II.7], [**DwHKS04**], or [**Hi03**, §19].

11.4.1 Homotopy Limits

The *homotopy limit* of a functor $F : C \to s\mathrm{Sets}$ is defined as the totalization of the cosimplicial replacement of F. This is a cosimplicial space associated with F that takes products of $F(C)$ indexed over the nerve. See [**BK72**, XI.5] for details.

Remark 11.4.6 tells us that the following condition ensures that finite products commute with homotopy colimits:

Definition 11.4.8 Let $\mathcal{D}$ be a small category and let $\Delta : \mathcal{D} \to \mathcal{D} \times \mathcal{D}$ be the diagonal functor. We call $\mathcal{D}$ *homotopy sifted* if for all objects D_1 and D_2 of $\mathcal{D}$, the category $(D_1, D_2) \downarrow \Delta$ is nonempty and has a contractible classifying space.

Beware that the notion of homotopy sifted is *stronger* than the notion of sifted. The condition that the category $(D_1, D_2) \downarrow \Delta$ is nonempty and has a contractible classifying space is often called *homotopy terminal* and sometimes homotopy right cofinal [**Hi03**, Definition 19.6.1].

Rosický shows [**R07**, §4] that a category is homotopy sifted if and only if fibrant replacements of homotopy colimits commute with finite products.

Several of our examples of sifted categories are actually homotopy sifted.

Examples 11.4.9
- If $\mathcal{D}$ is a small category with finite coproducts, then the object $(D_1 \sqcup D_2, D_1 \sqcup D_2)$ is initial in the category $(D_1, D_2) \downarrow \Delta$ for all objects D_1, D_2 of $\mathcal{D}$. Hence, such a category is homotopy sifted.
- Filtered categories are homotopy sifted [**R07**, Proposition 3.8].
- The opposite of the simplicial category, Δ^o, is homotopy sifted [**R07**, Remark 4.5 (f)].

Remark 11.4.10 Interchange laws for (homotopy) limits and homotopy colimits are tricky. There are some concrete examples mentioned in [**V77**]. Rezk develops criteria for homotopy colimits commuting with homotopy pullbacks [**Rez∞**].

There is a statement in [**Hi∞**, Theorem 14.19] for certain types of diagrams (one category small and filtered and one "finite and acyclic" [**Hi∞**, Definition 14.15]). This result is close to the classical exchange result about colimits and limits from Theorem 3.5.6.

But if you have your general favorite small diagram categories $\mathcal{D}$ and $\mathcal{D}'$ and you want to know whether $\text{holim}_{\mathcal{D}}(\text{hocolim}_{\mathcal{D}'} X)$ is equivalent to $\text{hocolim}_{\mathcal{D}'}(\text{holim}_{\mathcal{D}} X)$, then life might be hard.

11.5 Coverings of Classifying Spaces

Coverings of classifying spaces of small categories have a very explicit description [**Q73**, §1]:

Proposition 11.5.1 *Let $\mathcal{C}$ be a small category and let $p: E \to B\mathcal{C}$ be a covering. Then, we can assign a functor $F: \mathcal{C} \to \text{Sets}$ to p, such that F sends every morphism in $\mathcal{C}$ to a bijection and such that the value of F on an object C of $\mathcal{C}$ is the fiber $p^{-1}[(C, 1)]$ with $[(C, 1)] \in B\mathcal{C}^{(0)}$.*

Proof   We have to define F on morphisms. For any $f \in \mathcal{C}(C_1, C_2)$, we consider the path $w_f: [0, 1] \to B\mathcal{C}$ that is given by $[f] \in N_1\mathcal{C}$ with $w_f(0) = [(C_1, 1)]$ and $w_f(1) = [(C_2, 1)]$. The path-lifting property of coverings gives a bijection $\sigma(w_f): p^{-1}[(C_1, 1)] \to p^{-1}[(C_2, 1)]$, such that a composition of morphisms results in the composition of bijections. The identity morphism 1_C gives the identity permutation and thus, F is a functor with the desired properties. □

Quillen [**Q73**, §1] states a converse to Proposition 11.5.1, and we provide a proof following the treatment in [**GZ67**, Appendix I].

Definition 11.5.2 A morphism $p: E \to B$ in sSets is a *simplicial covering* if $p_0: E_0 \to B_0$ is surjective and if for all $n \geq 0$ and for all commutative diagrams

$$
\begin{array}{ccc}
\Delta_0 & \xrightarrow{\alpha} & E \\
{\scriptstyle i}\downarrow & & \downarrow{\scriptstyle p} \\
\Delta_n & \xrightarrow{\beta} & B,
\end{array}
\tag{11.5.1}
$$

there is a unique $\xi: \Delta_n \to E$ with $p \circ \xi = \beta$ and $\xi \circ i = \alpha$. Here, i is the map induced by $\partial_n \circ \cdots \circ \partial_1: [0] \to [n]$, and thus, it sends the unique map $c_0^n \in \Delta_0([n])$ to the map $i(c_0^n)$ that maps all $j \in [n]$ to zero.

Thus, for every $e_0 \in E_0$ and $b_n \in B_n$ with $pe_0 = d_1 \circ \cdots \circ d_n b_n$, there is a unique $e_n \in E_n$ with $p_n e_n = b_n$ and $d_1 \circ \cdots \circ d_n e_n = e_0$.

Note that this definition implies that fibers of simplicial coverings are discrete. A fiber associated with a zero simplex $b_0 \in B_0$ is the fiber product $F = \Delta_0 \times_B E$, that is, the pullback of

$$
\begin{array}{ccc}
 & & E \\
 & & \downarrow p \\
\Delta_0 & \xrightarrow{\ b_0\ } & B.
\end{array}
$$

So, we can restrict attention to coverings of Δ_0. But there is only one morphism $\Delta_n \to \Delta_0$ for all $n \geq 0$, so any n-simplex in a covering space of Δ_0 is determined by one of its vertices, and thus, the fiber is discrete. A proof for the next result can be found in [**GZ67**, III.4, Appendix I.2, I.3].

Proposition 11.5.3 *The geometric realization of a simplicial covering is a covering of topological spaces.*

Remark 11.5.4 There is also a description of simplicial coverings as *locally trivial maps with typical discrete fiber* [**GZ67**, loc. cit.].

Theorem 11.5.5 *Let $\mathcal{C}$ be a small category, C be an object of $\mathcal{C}$, and $F: \mathcal{C} \to$ Sets be a functor that is* morphism-inverting, *that is, that sends every $f \in \mathcal{C}(C_1, C_2)$ to a bijection. Then, $p = B(\rho): B(F\backslash\mathcal{C}) \to B\mathcal{C}$ is a covering with typical fiber $F(C)$.*

Together with Proposition 11.5.1, this yields an equivalence of categories between the category of all coverings of $B\mathcal{C}$ and the category of all morphism-inverting functors $F: \mathcal{C} \to$ Sets.

Note that all the $F(C)$s are in bijection, because F is morphism-inverting.

Proof   We use Proposition 11.5.3 and show that $N(\rho): N(F\backslash\mathcal{C}) \to N\mathcal{C}$ is a simplicial covering. Let $C_0 \xrightarrow{u_1} C_1 \xrightarrow{u_2} \cdots \xrightarrow{u_n} C_n$ be an n-simplex of $N(\mathcal{C})$ and let (C_0, x_0) be in $N(F\backslash\mathcal{C})_0$. We define an n-simplex in $N(F\backslash\mathcal{C})$ out of these starting data as

$$
(C_0, x_0) \xrightarrow{u_1} (C_1, F(u_1)(x_0)) \xrightarrow{u_2} \cdots \xrightarrow{u_n} (C_n, F(u_n) \circ \cdots \circ F(u_1)(x_0)).
$$

Then, $N(\rho)_n$ maps this simplex to $C_0 \xrightarrow{u_1} C_1 \xrightarrow{u_2} \cdots \xrightarrow{u_n} C_n$ and $d_1 \circ \cdots \circ d_n$ applied to $((C_0, x_0) \xrightarrow{u_1} (C_1, F(u_1)(x_0)) \xrightarrow{u_2} \cdots \xrightarrow{u_n} (C_n, F(u_n) \circ \cdots \circ F(u_1)(x_0)))$ is (C_0, x_0).

This provides a unique lift ξ, as in (11.5.1). For $C \in N(\mathcal{C})_0$, the corresponding fiber $\Delta_0 \times_{N(\mathcal{C})} N(F \backslash \mathcal{C})$ is $\{x \in F(C)\} = F(C)$.   □

Associated with $\mathcal{C}$, there is a groupoid, $\mathcal{C}[\text{Mor}(\mathcal{C})^{-1}]$, obtained by formally inverting all morphisms in $\mathcal{C}$ [**GZ67**, I.1]. Morphism-inverting functors $F : \mathcal{C} \to$ Sets are nothing but functors from the category $\mathcal{C}[\text{Mor}(\mathcal{C})^{-1}]$ to Sets. For every object C of $\mathcal{C}$, let $G(C)$ be the group of automorphisms of C in the category $\mathcal{C}[\text{Mor}(\mathcal{C})^{-1}]$. We get an alternative description of $\pi_1(\mathcal{C})$.

Proposition 11.5.6 [**Q73**, p. 82 (aka p. 6 or p. 90)] *For all objects C of $\mathcal{C}$, there is an isomorphism of groups*

$$\pi_1(\mathcal{C}; C) \cong G(C).$$

Proof If $\mathcal{C}$ is connected, then the inclusion functor $\mathcal{C}_{G(C)} \to \mathcal{C}[\text{Mor}(\mathcal{C})^{-1}]$, which sends the unique object $*$ of $\mathcal{C}_{G(C)}$ to C, is part of an equivalence of categories. Coverings of $B\mathcal{C}$ are then in bijection with $G(C)$-sets; hence, $G(C) \cong \pi_1(B\mathcal{C}, [C]) = \pi_1(\mathcal{C}, C)$. If $\mathcal{C}$ is not connected, we restrict to the connected component of the object C.   □

11.6 Fibers and Homotopy Fibers

Let $F : \mathcal{C} \to \mathcal{D}$ be a functor between small categories. Quillen's Theorems A and B give us control over the fiber of F and the homotopy fiber of BF in good cases. We will first consider a categorical version of the actual fiber of F.

Definition 11.6.1 Let D be an object of $\mathcal{D}$. The *fiber of F over D*, $F^{-1}(D)$ is the category whose objects are the objects C of $\mathcal{C}$ with $F(C) = D$. A morphism $f \in F^{-1}(D)(C, C')$ is a morphism $f \in \mathcal{C}(C, C')$, such that $F(f) = 1_D$.

In many sources, the notation $\mathcal{C}_D$ for $F^{-1}(D)$ is used. Note that $F^{-1}(D)$ is the pullback in the diagram

$$
\begin{array}{ccc}
 & & \mathcal{C} \\
 & & \downarrow{\scriptstyle F} \\
[0] & \xrightarrow{\ D\ } & \mathcal{D}.
\end{array}
$$

The fiber of F over D describes the actual fiber on the level of classifying spaces.

Lemma 11.6.2 *Let* $F : \mathcal{C} \to \mathcal{D}$ *be a functor between small categories and let* D *be an object of* $\mathcal{D}$. *Then,*

$$B(F^{-1}(D)) = (BF)^{-1}([D]).$$

Proof If the fiber category is empty, then so is $(BF)^{-1}([D])$.

By the very definition of $F^{-1}(D)$, we get that $B(F^{-1}(D)) \subset (BF)^{-1}([D])$. Let $[([f_p \mid \cdots \mid f_1], (t_0, \cdots, t_p))]$ be an element of $(BF)^{-1}([D]) \subset B\mathcal{C}$; that is,

$$[([F(f_p) \mid \cdots \mid F(f_1))], (t_0, \ldots, t_p))]$$

is equivalent to $[D] \in B\mathcal{D}$. We saw in Lemma 10.6.3 that the element $[F(f_p) \mid \cdots \mid F(f_1)]$ can be uniquely written in the form

$$[F(f_p) \mid \cdots \mid F(f_1)] = s_{j_i} \circ \cdots \circ s_{j_1}[g_r \mid \cdots \mid g_1],$$

with $0 \le j_1 < \cdots < j_i \le p$ and such that none of the g_as is an identity morphism or such that $[F(f_p) \mid \cdots \mid F(f_1)]$ is of simplicial degree zero, that is, of the form $[D']$ for an object D' of $\mathcal{D}$. The first case cannot happen, as we assumed that the element is equivalent to $[D] = [([D], (1))]$, and hence, we can write $[([F(f_p) \mid \cdots \mid F(f_1)], (t_0, \ldots, t_p))]$ as $s_{j_p} \circ \cdots \circ s_{j_1}([D], (1)) = ([1_D \mid \ldots \mid 1_D], (r_0, \cdots, r_p))$ with one $r_b = 1$ and all others equal to 0 and $[([f_p \mid \ldots \mid f_1], (t_0, \ldots, t_p))]$ is in $B(F^{-1}(D))$. $\qquad\square$

One can construct a model for the homotopy fiber of the map $BF : B\mathcal{C} \to B\mathcal{D}$ in terms of comma categories. For an object D of $\mathcal{D}$, recall the definition of $F \downarrow D$ from Section 5.1: There is a functor $U : F \downarrow D \to \mathcal{C}$ sending a pair (C, f), to C; here, C is an object of $\mathcal{C}$ and $f \in \mathcal{D}(F(C), D)$. Dually, we can consider the comma category $D \downarrow F$ with objects (C, f) with C an object of $\mathcal{C}$ and $f \in \mathcal{C}(D, F(C))$. We will see in Theorem 11.7.7 under which assumptions $B(F \downarrow D)$ actually has the homotopy type of the homotopy fiber. There is always a map.

Proposition 11.6.3 *There is a natural transformation* $\tau : F \circ U \Rightarrow \Delta(D)$ *from* $F \circ U$ *to the constant functor with value* D. *Thus, the map* $BF \circ BU$ *is nullhomotopic, and there is a continuous map from* $B(F \downarrow D)$ *to the homotopy fiber of the map* $BF : B\mathcal{C} \to B\mathcal{D}$. *Similarly, there is a continuous map from* $B(D \downarrow F)$ *to the homotopy fiber of* BF.

Proof We define $\tau_{(C,f)} : FU(C, f) = F(C) \to D$ as f, and such a transformation gives rise to a homotopy

$$H : B(F \downarrow D) \times B[1] \to B\mathcal{D},$$

with $H(x, 0) = B(FU)(x)$ and $H(x, 1) = D$ for all $x \in B(F \downarrow D)$.

In such a situation, the map from $B(F \downarrow D)$ into the homotopy fiber of BF is standard. The homotopy fiber is the space

$$\mathrm{hfib}(BF) = \{(x, w) \in B\mathcal{C} \times B\mathcal{D}^{[0,1]} | w(0) = BF(x), w(1) = [D] \in B\mathcal{D}\},$$

and it comes with the map $\mathrm{hfib}(BF) \to B\mathcal{C}$, given by the projection onto the first coordinate. We define

$$\xi : B(F \downarrow D) \to \mathrm{hfib}(BF)$$

as $\xi(y) = (BU(y), H(y, -))$.

The proof for $D \downarrow F$ is similar. $\square$

Note that the fiber category $F^{-1}(D)$ is a full subcategory of $F \downarrow D$ (and of $D \downarrow F$), where an object C of $F^{-1}(D)$ corresponds to an object $(C, 1_D : F(C) \to D)$ (and $(C, 1_D : D \to F(C))$). Therefore, the inclusion functors induce continuous maps

$$BF^{-1}(D) \to B(F \downarrow D) \to \mathrm{hfib}(BF) \text{ and } BF^{-1}(D)$$
$$\to B(D \downarrow F) \to \mathrm{hfib}(BF).$$

We follow [**Q73, G-SGA1**] for the definition of a comparison property between the fiber $BF^{-1}(D)$ and $B(F \downarrow D)$ (respectively $B(D \downarrow F)$).

Definition 11.6.4 Let $F : \mathcal{C} \to \mathcal{D}$ be a functor between small categories.

- We call F *precofibered* if for every object D of $\mathcal{D}$, the inclusion functor $F^{-1}(D) \hookrightarrow F \downarrow D$ has a left adjoint L. If (C, f) is an object of $F \downarrow D$, then we denote the value of L on (C, f) by $L(C, f) = f_*(C)$.
- If L exists, then every $g : D \to D'$ in $\mathcal{D}$ induces a functor $F \downarrow D \to F \downarrow D'$, and hence, a *cobase-change functor* $g_* : F^{-1}(D) \to F^{-1}(D')$ by sending an object C of $F^{-1}(D)$ first to the object (C, g) in $F \downarrow D'$ and then to $L(C, g) = g_*(C)$.
- A precofibered functor F is *cofibered* if for every pair of composable morphisms $D \overset{f}{\longrightarrow} D' \overset{g}{\longrightarrow} D''$, we get

$$(g \circ f)_* = g_* \circ f_*.$$

- Dually, we call F *prefibered* if for every object D of $\mathcal{D}$, the inclusion functor $F^{-1}(D) \hookrightarrow D \downarrow F$ has a right adjoint R. If (C, f) is an object of $D \downarrow F$, then we denote the value of R on (C, f) by $R(C, f) = f^*(C)$.
- If R exists, then every $g : D \to D'$ in $\mathcal{D}$ induces a functor $D' \downarrow F \to D \downarrow F$, and hence, a *base-change functor* $g^* : F^{-1}(D') \to F^{-1}(D)$ by sending an object C of $F^{-1}(D')$ to $R(C, g) = g^*(C)$.

- A prefibered functor F is *fibered* if for every pair of composable morphisms

$$D \xrightarrow{\;f\;} D' \xrightarrow{\;g\;} D'',$$

we get $(g \circ f)^* = f^* \circ g^*$.

Remark 11.6.5 The existence of a left or right adjoint of the inclusion functor ensures that the actual fiber $F^{-1}(D)$ and $F \downarrow D$ or $D \downarrow F$ have homotopy equivalent classifying spaces.

We compare the notion of a pre-cofibered functor to a Grothendieck opfibration, which is heavily applied in [**Lu∞**, Chapter 2].

Definition 11.6.6 Let $F: \mathcal{C} \to \mathcal{D}$ be a functor and let $f \in \mathcal{C}(C_1, C_2)$ be a morphism. Then f is *F-cocartesian* if for all $h \in \mathcal{C}(C_1, C_3)$ and for all $v \in \mathcal{D}(D_2, D_3)$ with $F(C_2) = D_2$ and $F(C_3) = D_3$ with

$$F(h) = v \circ F(f),$$

there is a unique $g \in \mathcal{C}(C_2, C_3)$ with $v = F(g)$:

You can express the same fact by saying that the diagram

is a pullback square.

Exercise 11.6.7 Assume that $u \in \mathcal{D}(D_1, D_2)$ is a morphism and f is an F-cocartesian morphism $f \in \mathcal{C}(C_1, C_2)$ with $F(f) = u$. Prove that the object C_2 and the morphism f are unique up to unique isomorphism, so if $f' \in \mathcal{C}(C_1, C_2')$ is another F-cocartesian morphism with $F(f') = u$, then there is a unique isomorphism $g: C_2 \to C_2'$ with $g \circ f' = f$.

Definition 11.6.8 Let $F: \mathcal{C} \to \mathcal{D}$ be a functor. Then, F is a *Grothendieck opfibration* or a *cocartesian fibration* if for all objects C_1 of $\mathcal{C}$ and all morphisms $u \in \mathcal{D}(F(C_1), D_2)$, there is an F-cocartesian morphism $f \in \mathcal{C}(C_1, C_2)$ with $F(f) = u$.

Remark 11.6.9 Note that for any Grothendieck opfibration and any $u \in \mathcal{D}(D_1, D_2)$, there is a morphism $u_*: F^{-1}(D_1) \to F^{-1}(D_2)$, that is, a fiber transport map. Let C_1 be an object of $F^{-1}(D_1)$. Then, $u: D_1 = F(C_1) \to D_2$, so there is an F-cocartesian $f \in \mathcal{C}(C_1, C_2)$ with $F(f) = u$; in particular, C_2 is an object of $F^{-1}(D_2)$. So, we set $u_*(C_1) = C_2$.

If $h \in \mathcal{C}(C_1, C_1')$, with $F(h) = 1_{D_1}$ and if $f': C_1' \to C_2'$ is an F-cocartesian morphism of u, now viewed as a morphism $u: F(C_1') = D_1 \to D_2$, then $F(f' \circ h) = F(f') = u$, so there is a unique $g \in \mathcal{C}(C_2, C_2')$, with $F(g) = 1_{D_2}$, and so, one can set $u_*(h) = g$.

Example 11.6.10 The projection functor $U: \mathcal{C} \int F \to \mathcal{C}$ is a Grothendieck opfibration. For every object (C_1, X_1) of $\mathcal{C} \int F$ and every map $f \in \mathcal{C}(C_1, C_2) = \mathcal{C}(U(C_1, X_1), C_2)$, we can take the morphism $(f, 1_{F(f)(X_1)})$ as a U-cocartesian lift of f. First of all, $(f, 1_{F(f)(X_1)})$ is a morphism in $\mathcal{C} \int F$ from (C_1, X_1) to $(C_2, F(f)(X_1))$, and U maps it to f.

We have to show that $(f, 1_{F(f)(X_1)})$ is a U-cocartesian, so assume that (h_1, h_2) is a morphism in $\mathcal{C} \int F((C_1, X_1), (C_3, X_3))$ and that $v \in \mathcal{C}(C_2, C_3)$ with $v \circ f = h_1$. Then, we have no choice but to take

$$(g_1, g_2) = (v, h_2) \in (\mathcal{C} \int F)((C_2, F(f)(X_1)), (C_3, X_3))$$

as a lift of v.

Theorem 11.6.11 *Let $F: \mathcal{C} \to \mathcal{D}$ be a functor. Then, F is precofibered if and only if F is a Grothendieck opfibration.*

There is a dual statement comparing prefibered functors to Grothendieck fibrations. For a topologist, the name *cocartesian fibration* or *Grothendieck opfibration* might seem more natural than the name precofibered, because an opfibration has a fiber transport and a lifting property from the target category to the source category, and we might think of the latter as the total space of the functor. As Quillen uses *precofibered*, this term is widely spread in the algebraic K-theory community.

Proof   Assume that F is a Grothendieck opfibration, and let D be an arbitrary object of $\mathcal{D}$. We have to find a left adjoint to the inclusion functor $I: F^{-1}(D) \to \mathcal{C} \downarrow D$. We define a functor $L: F \downarrow D \to F^{-1}(D)$ via reflections (see Definition 2.4.13). For an object $(C, \alpha: F(C) \to D)$, let

$f: C \to C'$ be an F-cocartesian morphism with $F(f) = \alpha$, in particular with $F(C') = D$. We define $G_{(C, \alpha: F(C) \to D)}$ as C'. For the morphism $\eta_{(C, \alpha)}: (C, \alpha) \to I(C') = (C', 1_D: F(C') \to D)$, we can recycle f, because by the definition of f, the diagram

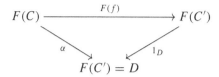

$$F(C) \xrightarrow{\quad F(f) \quad} F(C')$$

$$\alpha \searrow \qquad \swarrow 1_D$$

$$F(C') = D$$

commutes. We have to show that the pair (C', f) actually is a reflection, so let $\tilde{C}$ be an object of $F^{-1}(D)$ and let $h \in (C \downarrow D)((C, \alpha), (I(\tilde{C})))$ be an arbitrary morphism. Thus, $h: C \to \tilde{C}$ with $1_D \circ F(h) = F(h) = \alpha$. As f is cocartesian, there is a unique morphism $g \in C(C', \tilde{C})$ with $g \circ f = h$, and this is precisely the condition that $I(g) \circ \eta_{(C, \alpha)} = h$. Therefore, by Lemma 2.4.15, we obtain a functor $L: C \downarrow D \to F^{-1}(D)$, which, by Proposition 2.4.16, is left adjoint to I.

Now assume that F is precofibered and that L is a left adjoint to the inclusion functor $I: F^{-1}(D) \to C \downarrow D$. Let $u: F(C_1) = D_1 \to D_2$ be a morphism in D. The set $C_2 = L(C_1, u)$. As L is a left adjoint to I, the identity morphism

$$1_{C_2} \in F^{-1}(D_2)(C_2, C_2) = F^{-1}(D_2)(L(C_1, u), C_2)$$

corresponds to a morphism

$$f \in (C \downarrow D_2)((C_1, u), (C_2, 1_{D_2}: F(C_2) = D_2 \to D_2)),$$

and this $f: C_1 \to C_2$ satisfies

$$u = 1_{D_2} \circ F(f) = F(f),$$

and hence, f is F-cocartesian. $\qquad \square$

11.7 Theorems A and B

For the proof of Theorem A, we use a variant of the twisted arrow category from Definition 4.5.1, where we twist the morphisms, but here, we involve a functor as well.

Definition 11.7.1 Let $F: C \to D$ be a functor between small categories. The *category of twisted morphisms under F, F ⋊ D*, is the category whose objects are triples (C, D, f), where C is an object of C, D is an object of D, and

$f \in \mathcal{D}(F(C), D)$. A morphism in $F \rtimes \mathcal{D}$ from (C, D, f) to (C', D', f') is a pair of morphisms $(\alpha \in \mathcal{C}(C', C), \beta \in \mathcal{D}(D, D'))$, such that the diagram

$$
\begin{array}{ccc}
F(C) & \xleftarrow{F(\alpha)} & F(C') \\
{\scriptstyle f}\downarrow & & \downarrow{\scriptstyle f'} \\
D & \xrightarrow{\beta} & D'
\end{array}
$$

commutes.

For $F = \mathrm{Id}_{\mathcal{D}}$, this is precisely the definiton of a twisted arrow category, so

$$\mathrm{Id}_{\mathcal{D}} \rtimes \mathcal{D} = \mathcal{D}^{\tau}. \tag{11.7.1}$$

Theorem 11.7.2 ([Q73, Theorem A]) *Let $F: \mathcal{C} \to \mathcal{D}$ be a functor between small categories. If $B(F \downarrow D)$ is contractible for every object D of $\mathcal{D}$, then $BF: B\mathcal{C} \to B\mathcal{D}$ is a homotopy equivalence.*

Proof We first expand the nerve of the category $F \rtimes \mathcal{D}$ into a bisimplicial set. Let $X(F)_{p,q}$ be the set consisting of pairs

$$\left(C_p \xrightarrow{\alpha_p} \cdots \xrightarrow{\alpha_1} C_0; \; F(C_0) \xrightarrow{f_0} D_0 \xrightarrow{\beta_1} \cdots \xrightarrow{\beta_q} D_q\right)$$

of composable morphisms in $\mathcal{C}$ and $\mathcal{D}$.

- The diagonal simplicial set $\mathrm{diag}(X)$ has p-simplices:

$$\left(C_p \xrightarrow{\alpha_p} \cdots \xrightarrow{\alpha_1} C_0; \; F(C_0) \xrightarrow{f_0} D_0 \xrightarrow{\beta_1} \cdots \xrightarrow{\beta_p} D_p\right).$$

Applying F to the first component and filling in the diagram with compositions give

$$
\begin{array}{ccccccc}
F(C_0) & \xleftarrow{F(\alpha_1)} & F(C_1) & \xleftarrow{F(\alpha_2)} & \cdots & \xleftarrow{F(\alpha_p)} & F(C_p) \\
{\scriptstyle f_0}\downarrow & & \downarrow & & & & \downarrow \\
D_0 & \xrightarrow{\beta_1} & D_1 & \xrightarrow{\beta_2} & \cdots & \xrightarrow{\beta_p} & D_p.
\end{array}
$$

Thus, $\mathrm{diag}(X)_p \cong N_p(F \rtimes \mathcal{D})$.

- The projection to the first component defines a morphism of simplicial sets

$$\mathrm{pr}_2^q: X(F)_{p,q} \to N_p \mathcal{C}^o$$

for every fixed q.

- For a fixed p, we can take the geometric realization of the simplicial set $[q] \mapsto X(F)_{p,q}$ and obtain

$$\coprod_{C_p \xrightarrow{\alpha_p} \cdots \xrightarrow{\alpha_1} C_0} B(F(C_0) \downarrow \mathcal{D}).$$

But $B(F(C_0) \downarrow \mathcal{D}) \simeq *$, because $(F(C_0), 1_{F(C_0)})$ is an initial object in $F(C_0) \downarrow \mathcal{D}$. Therefore,

$$|\mathrm{pr}_2| \colon \left(\coprod_{C_p \xrightarrow{\alpha_p} \cdots \xrightarrow{\alpha_1} C_0} B(F(C_0) \downarrow \mathcal{D}) \right)$$

$$\to \left(\coprod_{C_p \xrightarrow{\alpha_p} \cdots \xrightarrow{\alpha_1} C_0} * \right) = N_p \mathcal{C}^o$$

is a weak equivalence for every p. Using the results from Remark 10.8.6 and Proposition 10.8.5 gives that there is a weak equivalence

$$p_2 \colon |\mathrm{diag}(X)| = B(F \rtimes \mathcal{D}) \to B\mathcal{C}^o.$$

- If we fix q and take geometric realization in p-direction, we obtain that the projection onto the first factor, pr_1, induces a morphism

$$\left(\coprod_{D_0 \xrightarrow{\beta_1} \cdots \xrightarrow{\beta_q} D_q} B(F \downarrow D_0) \right) \to \left(\coprod_{D_0 \xrightarrow{\beta_1} \cdots \xrightarrow{\beta_q} D_q} * \right),$$

and by assumption, the spaces $B(F \downarrow D_0)$ are contractible for all objects D_0 of $\mathcal{D}$, and hence, there is a weak equivalence

$$p_1 \colon |\mathrm{diag}(X)| = B(F \rtimes \mathcal{D}) \to N\mathcal{D}.$$

- Note that the category $\mathrm{Id}_{\mathcal{D}} \downarrow D$ has a contractible classifying space for all objects D of $\mathcal{D}$, because $\mathrm{Id}_{\mathcal{D}} \downarrow D$ is isomorphic to $\mathcal{D} \downarrow D$, and here, D is terminal. Hence, we obtain analogous weak equivalences, with F replaced by the identity functor on $\mathcal{D}$.

- We consider the commutative diagram

$$
\begin{array}{ccccc}
\mathcal{C}^o & \xleftarrow{\text{pr}_1} & F \rtimes \mathcal{D} & \xrightarrow{\text{pr}_2} & \mathcal{D} \\
{\scriptstyle F^o}\downarrow & & {\scriptstyle F_*}\downarrow & & \Vert \\
\mathcal{D}^o & \xleftarrow{\text{pr}_1} & \text{Id} \rtimes \mathcal{D} & \xrightarrow{\text{pr}_2} & \mathcal{D},
\end{array}
$$

where $F_*(C, D, f) = (F(C), D, f)$. The horizontal functors give rise to weak equivalences, and hence, BF^o is a weak equivalence and so is BF.

$\square$

Corollary 11.7.3 *If $F: \mathcal{C} \to \mathcal{D}$ is precofibered or prefibered and if the actual fiber $BF^{-1}(D)$ is contractible for all objects D of $\mathcal{D}$, then the induced map $BF: B\mathcal{C} \to B\mathcal{D}$ is a homotopy equivalence.*

Example 11.7.4 A twisted arrow category, $\mathcal{D}^\tau$ (see Definition 4.5.1), is cofibered over $\mathcal{D}^o$ via the source functor and over $\mathcal{D}$ via the target functor. Recall that $\mathcal{D}^\tau$ has as objects all $f \in \mathcal{D}(D, D')$, and a morphism from f to $g \in \mathcal{D}(D_1, D_1')$ is a pair of morphisms $h: D_1 \to D$ and $k: D' \to D_1'$ with $g = k \circ f \circ h$. (Thus, $\mathcal{D}^\tau$ is $\text{Id}_\mathcal{D} \rtimes \mathcal{D}$ in the notation of Definition 11.7.1.)

We show the claim for the target functor, $T: \mathcal{D}^\tau \to \mathcal{D}$, that sends f to D' and a pair (h, k) to k.

We define the left adjoint L to the inclusion $T^{-1}(D') \hookrightarrow \mathcal{D}^\tau \downarrow D'$ as the functor that takes an object $(\sigma: D \to D'', h: D'' \to D')$ of $\mathcal{D}^\tau \downarrow D'$ to the object $h \circ \sigma$ of the fiber $T^{-1}(D')$. A morphism (α, α'), as in the commuting diagram

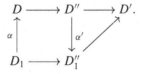

is sent to the morphism $(\alpha, 1_{D'})$ in $T^{-1}(D')$. By construction, L is left adjoint to the inclusion functor.

Exercise 11.7.5 Show that $F \rtimes \mathcal{D}$ is cofibered over $\mathcal{C}^o$ and $\mathcal{D}$.

We consider the inclusion functor ι from the category $\mathcal{C}_{\mathbb{N}_0}$ to the category $\mathcal{C}_{\mathbb{Z}}$.

Proposition 11.7.6 $B\iota: B\mathbb{N}_0 \to B\mathbb{Z}$ *is a homotopy equivalence.*

Proof   What is $\iota \downarrow *$? As $\iota(*) = *$, the objects of $\iota \downarrow *$ can be identified with the integers, and a morphism from $x \in \mathbb{Z}$ to $y \in \mathbb{Z}$ is a natural number n with

$y + n = x$. Hence, $\iota \downarrow *$ is the category of the ordered set of the integers. This is a filtered small category, and hence, by 11.3.10, $B(\iota \downarrow *)$ is contractible. $\square$

We state Theorem B without a proof and refer the reader to the excellent original source [**Q73**], to [**Sr96**, §6] for an extended account of Quillen's proof, and to [**GJ09**, IV.5] for a proof, using model structures and bisimplicial techniques.

Theorem 11.7.7 [**Q73**, Theorem B] *Let $F: \mathcal{C} \rightarrow \mathcal{D}$ be a functor between small categories, such that for every morphism $f \in \mathcal{D}(D, D')$, the induced functor $F \downarrow D \rightarrow F \downarrow D'$ is a homotopy equivalence. Then, for any object D, of $\mathcal{D}$, the diagram*

$$
\begin{array}{ccc}
B(F \downarrow D) & \xrightarrow{\ BU\ } & B\mathcal{C} \\
{\scriptstyle B\varepsilon}\downarrow & & \downarrow{\scriptstyle BF} \\
B(\mathcal{D} \downarrow D) & \xrightarrow{\ B\pi\ } & B\mathcal{D}
\end{array}
$$

is homotopy cartesian. Here, U sends a pair (C, f) to C, ε evaluates (C, f) to $(F(C), f)$, and π projects to the first component.

Here, *homotopy cartesian* means that the map from $B(F \downarrow D)$ to the homotopy fiber product of

$$
\begin{array}{ccc}
 & & B\mathcal{C} \\
 & & \downarrow{\scriptstyle BF} \\
B(\mathcal{D} \downarrow D) & \xrightarrow{\ B\pi\ } & B\mathcal{D}
\end{array}
$$

is a homotopy equivalence. Note that $B(\mathcal{D} \downarrow D)$ is contractible. In such a situation, the diagram is homotopy cartesian if and only if the map from $B(F \downarrow D)$ to the homotopy fiber of BF is a homotopy equivalence. Thus, under the assumptions of Theorem B, $B(F \downarrow D)$ is a valid model for the homotopy fiber, and we get a long exact sequence of homotopy groups

$$
\cdots \xrightarrow{\ \delta\ } \pi_n(B(F \downarrow D); [(C, f)]) \xrightarrow{\ \pi_n(BU)\ } \pi_n(B\mathcal{C}; [C]) \xrightarrow{\ \pi_n(BF)\ } \pi_n(B\mathcal{D}; [F(C)]) \xrightarrow{\ \delta\ } \cdots
$$

for all choices of objects (C, f) of $B(F \downarrow D)$.

Dually, we have an analogous statement for $D \downarrow F$. In the precofibered and prefibered cases, we can, of course, replace $B(F \downarrow D)$ with $B(F^{-1}(D))$.

Recall (for instance, from [**Ha02**, Section 4.K]) that a *quasi-fibration* is a continuous map $p: E \rightarrow B$ with B path-connected, such that the inclusion of each fiber $p^{-1}(b)$ into the homotopy fiber of p is a weak homotopy equivalence. Here, we are dealing with CW complexes, so we can upgrade such a weak homotopy equivalence to an actual homotopy equivalence.

Corollary 11.7.8 [**Q73**, Corollary to Theorem B] *Suppose that* $F: \mathcal{C} \to \mathcal{D}$ *is prefibered (respectively precofibered) and that for every morphism* $f \in \mathcal{D}(D, D')$, *the induced base-change functor* $f^*: F^{-1}(D') \to F^{-1}(D)$ *(respectively cobase-change functor) is a homotopy equivalence. Then,* $BF: B\mathcal{C} \to B\mathcal{D}$ *is a quasi-fibration.*

The following example is mentioned in [**BDR04**]:

Example 11.7.9 Let M be a monoid, let G be a group, and assume that $f: M \to G$ is a morphism of monoids. We obtain an associated functor $F: \mathcal{C}_M \to \mathcal{C}_G$ with $F(*) = *$ and $F(m) = f(m)$. We can identify the category $F \downarrow *$ with the category whose objects are morphisms $g \in \mathcal{C}_G(*, *)$ and whose morphisms are commutative diagrams

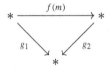

that is, $g_2 f(m) = g_1$. Every $g \in G$ induces an automorphism of categories on $F \downarrow *$, and hence, Quillen's Theorem B applies, and $B(F \downarrow *)$ is equivalent to the homotopy fiber of $BM \to BG$.

Exercise 11.7.10 Let M be an abelian monoid and $G(M)$ be its Grothendieck group, as in Definition 13.2.1. Generalizing Proposition 11.7.6, what are possible assumptions on M that guarantee that the canonical map $j: M \to G(M)$ induces a homotopy equivalence $Bj: BM \to BG(M)$?

Exercise 11.7.11 What happens in Example 11.7.9 if f is the inclusion of a subgroup $H < G$? What about a quotient map $f: G \to G/N$ of groups? That should look familiar.

11.8 Monoidal and Symmetric Monoidal Categories, Revisited

11.8.1 Monoidal Categories via Grothendieck Opfibrations

For defining a monoidal category $(\mathcal{C}, \otimes, 1)$, you can write down the explicit axioms that the monoidal product has to satisfy and state the necessary coherence conditions. This is the approach that we presented in Definition 8.1.4. A different approach is to package the data of a monoidal category into the requirement that a certain functor is a Grothendieck opfibration. We first describe how every monoidal category gives rise to such a functor with the help of an auxiliary category.

Definition 11.8.1 Let $(\mathcal{C}, \otimes, e)$ be a monoidal category. The category $\mathcal{C}^{\otimes}$ has as objects finite sequences $(C_1, \ldots, C_n)$ of objects C_i of $\mathcal{C}$ for all $n \in \mathbb{N}_0$, where we take the empty sequence for $n = 0$. A morphism $(\varphi; (f_i)_{i=1}^m)$ from $(C_1, \cdots, C_n)$ to $(C_1', \cdots, C_m')$ consists of a $\varphi \in \Delta([m], [n])$ and $f_i \in \mathcal{C}(C_{\varphi(i-1)+1} \otimes \ldots \otimes C_{\varphi(i)}, C_i')$, where we set $C_{\varphi(i-1)+1} \otimes \ldots \otimes C_{\varphi(i)} = e$ if $\varphi(i-1) = \varphi(i)$.

Note that we do not assume that $\mathcal{C}$ is strict monoidal, so a priori, we have to fix a way of placing parentheses in $C_{\varphi(i-1)+1} \otimes \cdots \otimes C_{\varphi(i)}$. But Mac Lane's coherence result (see Remark 8.1.8) implies that such choices don't matter, because $C_{\varphi(i-1)+1} \otimes \cdots \otimes C_{\varphi(i)}$ is defined up to canonical isomorphism.

Example 11.8.2 Let us spell out the effect of $\varphi = \delta_j \colon [m] \to [m + 1]$ and $\varphi = \sigma_j \colon [m] \to [m - 1]$. For δ_j, we have $\delta_j(i - 1) + 1 = \delta(i)$ for all i but $i = j$. So, possible morphisms are $C_i \to C_i'$ for $i \neq j$ and $C_j \otimes C_{j+1} \to C_j'$. For σ_j, we also have $\sigma_j(i - 1) + 1 = \sigma(i)$ for all $i - 1 \neq j$ and $\sigma_j(j) + 1 = j = \sigma_j(j + 1)$, so possible morphisms are $C_i \to C_i'$ for $i - 1 \neq j$ and $e \to C_{j+1}'$.

Exercise 11.8.3 Make the identity morphisms and the composition in $\mathcal{C}^{\otimes}$ explicit.

Let $\varphi \in \Delta([4], [4])$ be the morphism that has $\varphi^{-1}(0) = \{0, 1\}$, $\varphi^{-1}(1) = \{2\}$ and $\varphi^{-1}(4) = \{3, 4\}$. What are possible morphisms in $\mathcal{C}^{\otimes}$ with first coordinate φ?

There is a canonical projection functor $p \colon \mathcal{C}^{\otimes} \to \Delta^o$ that sends an object $(C_1, \ldots, C_n)$ to $[n]$ and $(\varphi; (f_i)_{i=1}^m)$ to φ.

Proposition 11.8.4 *If $\mathcal{C}$ is a monoidal category, then p is a Grothendieck opfibration.*

Proof Let $(C_1, \ldots, C_n)$ be an arbitrary object in $\mathcal{C}^{\otimes}$ and let $\varphi \in \Delta([m], [n])$. We set $C_i' = C_{\varphi(i-1)+1} \otimes \cdots \otimes C_{\varphi(i)}$ and choose the morphism

$$(\varphi; 1_{C_{\varphi(0)+1} \otimes \cdots \otimes C_{\varphi(1)}}, \ldots, 1_{C_{\varphi(m-1)+1} \otimes \ldots \otimes C_{\varphi(m)}}) =: (\varphi; 1, \ldots, 1)$$

as a lift of φ in $\mathcal{C}^{\otimes}$. We have to show that this morphism is p-cocartesian. So, let $(\beta; (f_i)_{i=1}^{\ell}) \in \mathcal{C}^{\otimes}((C_1, \ldots, C_n), (C_1'', \ldots, C_\ell''))$ with $\beta \colon [\ell] \to [n]$ in Δ. Assume that $\gamma \in \Delta([\ell], [m])$, such that $\beta = \varphi \circ \gamma = \gamma \circ^o \varphi$. We have to show that there is a unique $(\gamma; (g_i)_{i=1}^{\ell})$ in $\mathcal{C}^{\otimes}$ lifting γ. But as the C_i's are just tensor words in the C_is and as $\beta = \varphi \circ \gamma$, we know where these tensor words have to be sent in order to end up in the C_i''s. $\square$

In the preceding proof, one could have taken any isomorphism $C'_i \cong C_{\varphi(i-1)+1} \otimes \cdots \otimes C_{\varphi(i)}$; taking the identity was not necessary, just lazy.

Remark 11.8.5 We know that Grothendieck opfibrations have a fiber transport. Note that for $p: C^\otimes \to \Delta^o$, the fiber at [1], $C_{[1]} := p^{-1}([1])$, is isomorphic to C. There is the additional effect that we can recover all of the fibers $C^\otimes_{[n]}$ from C itself. Let $\iota_{i-1,i}: [1] \to [n]$ for $1 \le i \ne n$ denote the unique map in Δ with image $\{i - 1, i\}$. Then, the $\iota_{i-1,i}$s induce an equivalence of categories

$$C^\otimes_{[n]} \cong \prod_{i=1}^n C^\otimes_{[1]} \cong \prod_{i=1}^n C.$$

Note that $C^\otimes_{[0]}$ is the category with the empty sequence and an identity morphism, so $C^\otimes_{[0]}$ is isomorphic to the category [0]. The important thing is that we can reverse the process.

Theorem 11.8.6 *If $p: \mathcal{D} \to \Delta^o$ is a Grothendieck opfibration, such that for all $n \ge 2$, the morphism $(\iota_{0,1}, \ldots, \iota_{n-1,n})$ induces an equivalence of categories*

$$\mathcal{D}_{[n]} \to \prod_{i=1}^n \mathcal{D}_{[1]},$$

and $\mathcal{D}_{[0]} \cong [0] =: \mathcal{D}^0_{[1]}$. Then, $\mathcal{D}_{[1]} =: C$ is a monoidal category.

Proof   We consider the morphism $\delta_1: [1] \to [2]$ in Δ and an object (C_1, C_2) in $C \times C$. We use the equivalence $C \times C \cong \mathcal{D}_{[2]}$ and choose an inverse to $(\iota_{0,1}, \iota_{1,2})$, together with δ_1, to induce a map

$$\otimes: C \times C \cong \mathcal{D}_{[2]} \to \mathcal{D}_{[1]} = C,$$

which we denote by $\otimes$, and we define $C_1 \otimes C_2$ in this way.

As p is a Grothendieck opfibration, for a fixed choice of (C_1, C_2) and for δ_1, the lift is unique up to unique isomorphism by Exercise 11.6.7. On the level of morphisms, we define $f \otimes g$ for $f \in C(C_1, C'_1)$ and $g \in C(C_2, C'_2)$ by precomposition in $C \times C$ by (f, g). This constructs $\otimes$ as a functor. The identity $\delta_1 \circ \delta_1 = \delta_2 \circ \delta_1$ ensures that $\otimes$ is associative up to isomorphism, because $\delta_1 \circ \delta_1$ corresponds to $(C_1, C_2, C_3) \mapsto (C_1 \otimes C_2) \otimes C_3$, whereas $\delta_2 \circ \delta_1$ sends (C_1, C_2, C_3) to $C_1 \otimes (C_2 \otimes C_3)$. We call the resulting natural associativity isomorphism α. The pentagon axiom (8.1.2) for α, as in Definition 8.1.4, then follows from the equality in Δ for expressing the morphism from [1] to [4] that sends 0 to 0 and 1 to 4:

$$\delta_1 \circ \delta_2 \circ \delta_1 = \delta_1 \circ \delta_1 \circ \delta_1 = \delta_2 \circ \delta_1 \circ \delta_1 = \delta_2 \circ \delta_2 \circ \delta_1 = \delta_3 \circ \delta_2 \circ \delta_1.$$

The morphism $\sigma_0 \in \Delta([1], [0])$ induces a functor $I : [0] \to \mathcal{C}$ (unique up to isomorphism), and we define e as the object $I(0)$ of $\mathcal{C}$. The equality $\sigma_0 \circ \delta_1 = 1_{[1]} = \sigma_1 \circ \delta_1$ translates into natural isomorphisms $e \otimes C \cong C \cong C \otimes e$. The triangle axioms for these isomorphisms arise from the equality

$$\sigma_0 \circ \delta_1 \circ \delta_1 = \delta_1 = \sigma_1 \circ \delta_1 \circ \delta_1.$$

$\square$

11.8.2 Symmetric Monoidal Categories

For symmetric monoidal categories, Segal's category of finite pointed sets, Γ, is used for encoding the symmetric monoidal structure. We follow [**Lu∞**] and [**Gro20**] and use the same name as in Definition 11.8.1 for the resulting auxiliary category.

Definition 11.8.7 Let $(\mathcal{C}, \otimes, e, \tau)$ be a symmetric monoidal category. The category $\mathcal{C}^\otimes$ has as objects n-tuples of objects of $\mathcal{C}$, $(C_1, \ldots, C_n)$ for $n \geq 0$, and, again, for $n = 0$ we take the empty sequence. A morphism $(f; (g_i)_{i=1}^m)$ from $(C_1, \ldots, C_n)$ to $(C_1', \ldots, C_m')$ consists of an $f \in \Gamma([n], [m])$ and morphisms

$$g_i : \bigotimes_{j \in f^{-1}(i)} C_j \to C_i'$$

in $\mathcal{C}$. If $f^{-1}(i) = \varnothing$, then $g_i : e \to C_i'$.

Here, the tensor product is uniquely determined up to isomorphism, and the coherence result for symmetric monoidal categories ensures that the maps g_i are defined up to canonical isomorphism. We fix a choice that respects the natural ordering on the fibers $f^{-1}(i)$. Note that we only consider the fibers away from the basepoint 0. As in the monoidal case, there is a projection functor $p : \mathcal{C}^\otimes \to \Gamma$ that sends an object $(C_1, \ldots, C_n)$ to $[n]$ and takes a morphism $(f; (g_i)_{i=1}^m)$ to f.

For $1 \leq i \leq n$, let $\varrho_i \in \Gamma([n], [1])$ be the morphism that sends i to 1 and all other elements of $[n]$ to 0. The symmetric monoidal structure on $\mathcal{C}$ then ensures that the projection functor p is a Grothendieck opfibration and that the fiber $\mathcal{C}_{[n]}$ is equivalent to $\mathcal{C}_{[1]}^n$ via the morphism $(\varrho_1, \ldots, \varrho_n)$. We also obtain the analog of Theorem 11.8.6.

Theorem 11.8.8 *Let $p : \mathcal{D} \to \Gamma$ be a Grothendieck opfibration and let $\mathcal{C}$ be $\mathcal{D}_{[1]}$. If for all $n \geq 0$, the fiber $\mathcal{D}_{[n]}$ is equivalent to $\mathcal{C}^n$, such that the equivalence is induced by the $\varrho_i s$, then $\mathcal{C}$ is a symmetric monoidal category.*

We refer to [**Lu∞**, Chapter 2] for a proof.

11.8.3 Monoidal Structures on Quasi-Categories

It goes well beyond the scope of this book to give an introduction to (symmetric) monoidal quasi-categories, because the definitions need background from [**Lu09**, Chapters 1, 2]. Therefore, we only give a brief overview of how to generalize (symmetric) monoidal structures to the context of ∞-categories.

Joyal and Lurie generalize the concept of cocartesian morphisms and Grothendieck opfibrations to the setting of quasi-categories in [**Jo-c∞**, p. 173] and [**Lu09**, Chapter 2]. Crucial for this is the transfer of the concept of slice categories in the sense of Joyal 5.1.14. Joyal develops this generalization in [**Jo-c∞**, 3.3]. Recall the join of simplicial sets from Definition 10.13.4.

The following is a simplicial analog of Proposition 5.1.13.

Proposition 11.8.9 *For a fixed simplicial set Y, the functor $(-) * Y \colon s\mathsf{Sets} \to Y \downarrow s\mathsf{Sets}$ has a right adjoint, and dually, the functor $X * (-) \colon s\mathsf{Sets} \to X \downarrow s\mathsf{Sets}$ has a right adjoint for every simplicial set X.*

Proof   We prove the first claim: If such a right adjoint, R exists, then for every object $f \colon Y \to Z$ in $Y \downarrow s\mathsf{Sets}$, we get in simplicial degree n that

$$R(f \colon Y \to Z)_n = s\mathsf{Sets}(\Delta_n, R(f \colon Y \to Z))$$
$$\cong (Y \downarrow s\mathsf{Sets})(\Delta_n * Y, Z),$$

so we have no choice but to set $R(f \colon Y \to Z)_n := (Y \downarrow s\mathsf{Sets})(\Delta_n * Y, Z)$. $\square$

Example 11.8.10   The most important examples in this context are simplices in Z; thus, $f \colon \Delta_k \to Z$ for some $k \geq 0$, for instance, points in Z ($k = 0$) or morphism in a quasi-category, that is, $f \colon \Delta_1 \to Z$.

For $f \colon \Delta_k \to Z$, we obtain for $R(f \colon Y \to Z)_k$ maps of simplicial sets from $\Delta_n * \Delta_k$ to Z that extend f. But $\Delta_n * \Delta_k \cong \Delta_{n+k+1}$ (see Example 10.13.7), so we just obtain maps of simplicial sets from Δ_{n+k+1} to Z restricting to f on $\Delta_k \hookrightarrow \Delta_{n+k+1}$; these are elements in Z_{n+k+1} restricting to the simplex that corresponds to f.

For $k = 0$, we obtain morphisms from the right cone on Δ_n, $\Delta_n * \Delta_0 \cong \Delta_{n+1}$ to Z, so $(n+1)$-simplices of Z, whose restriction to the cone point is the point corresponding to f.

Dually, for the right adjoint of $X * (-)$, we get for an $f \colon \Delta_k \to Z$ an $(n+k+1)$-simplex in Z restricting to f on $\Delta_k \hookrightarrow \Delta_k * \Delta_n \cong \Delta_{n+k+1}$.

Definition 11.8.11 [**Jo-c∞**, §3.3] [**Lu09**, 1.2.9.2]   Let X, Y, and Z be simplicial sets and let $f \colon X \to Z$, and $g \colon Y \to Z$ be morphisms in $s\mathsf{Sets}$.

- We denote the value of the right adjoint of $(-)*Y$ on $g: Y \to Z$, $R(g: Y \to Z)$ by $Z_{/g}$.
- The value of the right adjoint of $X * (-)$ on $f: X \to Z$, $R(f: X \to Z)$ is denoted by $Z_{f/}$.
- If Z is a quasi-category, then $Z_{/g}$ is the ∞-*category of objects over* g, and $Z_{f/}$ is the ∞-*category of objects under* f.

Recall from Proposition 11.1.2 that nerves of small categories are automatically quasi-categories. There is a compatibility result relating the slices of small categories to the simplicial sets $Z_{/g}$ and $Z_{f/}$ via the nerve functor.

Proposition 11.8.12 **[Jo-c∞**, Proposition 3.13] *Let $\mathcal{C}$ and $\mathcal{D}$ be small categories and let $F: \mathcal{C} \to \mathcal{D}$ be a functor. There are canonical isomorphisms $N(\mathcal{D}_{F/}) \cong N(\mathcal{D})_{N(F)/}$ and $N(\mathcal{D}_{/F}) \cong N(\mathcal{D})_{/N(F)}$.*

Proof We prove the first claim. An n-simplex in $N(\mathcal{D}_{F/})$ is a map $\Delta_n \to N(\mathcal{D}_{F/})$. As $\Delta_n \cong N[n]$, this corresponds to a functor from the category $[n]$ to $\mathcal{D}_{F/}$. By adjunction (Proposition 5.1.13), such functors are in bijection with functors from $[n] * \mathcal{C}$ to $\mathcal{D}$ that extend F. As the nerve functor preserves joins (see Proposition 11.1.11), this corresponds to maps of simplicial sets $\Delta_n * N(\mathcal{C}) \to N(\mathcal{D})$ that extend $N(F)$. By adjunction, this is nothing but an n-simplex in $N(\mathcal{D})_{N(F)/}$. $\square$

The crucial point is the generalization of cocartesian morphisms and Grothendieck opfibrations to the context of quasi-categories. We just record the definition and refer the interested reader to [**Jo-c∞**], [**Lu09**, §2.4.2], and [**Gro20**, §3.2] for more details.

Definition 11.8.13 [**Gro20**, Definitions 3.6, 3.7] Let X and Y be quasi-categories and let $p: X \to Y$ be a map of simplicial sets.

(1) An $f: x_1 \to x_2 \in X$ (that is, an $f \in X_1$ with $d_1(f) = x_1$ and $d_0(f) = x_2$) is *p-cocartesian* if the map

$$X_{f/} \to X_{x_1/} \times_{Y_{p(x_1)/}} Y_{p(f)/}$$

is an acyclic fibration of simplicial sets.
(2) The map p is a *cocartesian fibration* if p is an inner fibration and if for every $x_1 \in X_0$ and every $\alpha: p(x_1) = y_1 \to y_2$ in Y, there is a p-cocartesian morphism $f: x_1 \to x_2$ in X with $p(f) = \alpha$.

Here, an inner fibration is a map of simplicial sets that has the right lifting property with respect to inner horn inclusions.

The definition of (symmetric) monoidal ∞-categories is now a direct transfer of Theorems 11.8.6 and 11.8.8.

Definition 11.8.14

- A *monoidal ∞-category* is a quasi-category X, together with a cocartesian fibration $p\colon X \to N(\Delta^o)$, such that for all $n \geq 0$, the inclusions $\iota_{i-1,i}$ induce a categorical equivalence of ∞-categories $X_{[n]} \to X_{[1]}^n$. Often, $X_{[1]}$ is then referred to as the *underlying ∞-category of X*.

- A *symmetric monoidal ∞-category* is a quasi-category X, together with a cocartesian fibration $p\colon X \to N(\Gamma)$, such that for all $n \geq 0$, the projections ϱ_i induce a categorical equivalence of ∞-categories $X_{[n]} \to X_{[1]}^n$. Then, $X_{[1]}$ is the *underlying ∞-category of X*.

For the notion of categorical equivalence, see [**Gro20**, Definition 1.24] or [**Lu09**, Definition 1.1.5.14]. It is common to denote a symmetric monoidal ∞-category by $\mathcal{C}^{\otimes}$ and to denote the underlying ∞-category by $\mathcal{C} = \mathcal{C}_{[1]}^{\otimes}$, in order to stress the similarity to the construction from Definition 11.8.7.

Remark 11.8.15 Segal's construction from Section 14.4 gives rise to a functor from Γ to the category of simplicial sets for every symmetric monoidal category. For a symmetric monoidal ∞-category X, the fiber transport induces functors of ∞-categories $f_!\colon X_{[n]} \to X_{[m]}$ for every $f \in \Gamma([n],[m])$, and we know that each $X_{[n]}$ is equivalent to $X_{[1]}^n$. This is an ∞-categorical analog of Segal's construction from Definition 14.4.1, where the choices of isomorphisms are implemented in the objects.

If $\mathcal{C}$ is a small (symmetric) monoidal category, then $N(\mathcal{C})$ is a (symmetric) monoidal ∞-category ([**Lu∞**, Example 2.1.1.21]). For more interesting examples of symmetric monoidal ∞-categories, see [**Lu∞**]. Thomas Nikolaus and Steffen Sagave show in [**NS17**] that all (presentably) symmetric monoidal ∞-categories are represented by symmetric monoidal model categories.

12

A Brief Introduction to Operads

Operads were introduced in the 1970s in topology when Michael Boardman, Rainer Vogt, Peter May, and others [**BV68, BV73, May72**] worked on a systematic understanding of iterated loop spaces. These are spaces of the form $\Omega^n X$, where X is a pointed topological space and $\Omega^n X$ is the topological space of pointed maps from an n-sphere, $\mathbb{S}^n$, to X. Operads had a renaissance in the 1990s, and have never been out of fashion since then. We will only give a very superficial introduction to operads. For more comprehensive overviews, see, for instance, [**Lo94, MSS02, LoVa12**] and [**Fr09**].

12.1 Definition and Examples

Let $(\mathcal{C}, \otimes, e, \tau)$ be a symmetric monoidal category. We will ignore associativity isomorphisms in the following. This is justified by the coherence theorem, which ensures that no matter which convention you choose for setting parentheses, it will amount to the same definition. If you feel uncomfortable with this, then just think of $\mathcal{C}$ as being a permutative category, but most applications use operads in symmetric monoidal categories that are *not* strict monoidal, such as the category of k-modules for some commutative ring k or the category of chain complexes over k. The concept of an operad is easy, but it can be confusing at first sight.

If $\mathcal{C}$ is, in addition, a concrete category, then it is legitimate to visualize the operad axioms by thinking about an element $w_m \in O(m)$ as a machine that allows for m slots of input and that has one output.

We can stack n such machines with k_i inputs and one output on top of one machine with n inputs, in order to create a combined machine with $k_1 + \cdots + k_n$ inputs and one output.

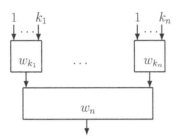

This is what the composition morphisms in an operad do.

The axioms of an operad are then the natural ones that arise from iterated composition and by a relabelling of the input slots.

Definition 12.1.1 Let $(\mathcal{C}, \otimes, e, \tau)$ be a symmetric monoidal category. An *operad in* $\mathcal{C}$ is a collection of objects $O(n)$ of $\mathcal{C}$ for $n \in \mathbb{N}_0$, together with a unit morphism $\eta \colon e \to O(1)$, a right Σ_n-action on $O(n)$ for all n, and composition morphisms

$$\gamma \colon O(n) \otimes O(k_1) \otimes \cdots \otimes O(k_n) \to O\left(\sum_{i=1}^{n} k_i\right)$$

for $n \geq 1$ and $k_i \geq 0$. The composition maps are associative, unital, and equivariant in the following sense:

- Let k be $\sum_{i=1}^{n} k_i$ and let m_i be the sum $k_1 + \cdots + k_i$. Starting from

$$O(n) \otimes \left(\bigotimes_{i=1}^{n} O(k_i)\right) \otimes \left(\bigotimes_{j=1}^{k} O(\ell_j)\right),$$

one can first use γ on $O(n) \otimes (\bigotimes_{i=1}^{n} O(k_i))$ to get to $O(k) \otimes (\bigotimes_{j=1}^{k} O(\ell_j))$ and then use another instance of γ to map to $O(\sum_{j=1}^{k} \ell_j)$. We require that this morphism is equal to the one where we first just permute the tensor factors

$$\left(\bigotimes_{i=1}^{n} O(k_i)\right) \otimes \left(\bigotimes_{j=1}^{k} O(\ell_j)\right),$$

such that every $O(k_i)$ comes next to $O(\ell_{m_{i-1}+1}) \otimes \cdots \otimes O(\ell_{m_i})$. We can evaluate γ on these terms to get to $O(n) \otimes (\bigotimes_{i=1}^{n} O(\ell_{m_{i-1}+1} + \cdots + \ell_{m_i}))$ and then apply γ again to end up in $O\left(\sum_{j=1}^{k} \ell_j\right)$:

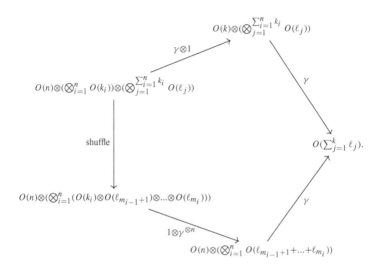

Here, the shuffle map is induced by the symmetry τ of $\mathcal{C}$.

- The unit map $\eta \colon e \to O(1)$ fits into the following commutative diagrams:

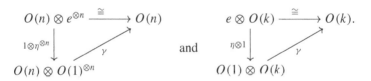

and

- We require the following two equivariance conditions:

  (1) If $\sigma \in \Sigma_n$, then we denote by $\sigma(k_1, \ldots, k_n)$ the permutation in Σ_k that permutes the blocks $k_{i-1} + 1, \ldots, k_i$ for $1 \le i \le n$ as σ permutes the numbers $1, \ldots, n$. Then, the following diagram must commute:

$$
\begin{array}{ccc}
O(n) \otimes O(k_1) \otimes \cdots \otimes O(k_n) & \xrightarrow{\sigma \otimes \sigma^{-1}} & O(n) \otimes O(k_{\sigma(1)}) \otimes \cdots \otimes O(k_{\sigma(n)}) \\
\gamma \downarrow & & \downarrow \gamma \\
O(k) & \xrightarrow{\sigma(k_{\sigma(1)}, \ldots, k_{\sigma(n)})} & O(k).
\end{array}
$$

(2) We also need the permutation $\tau_1 \oplus \cdots \oplus \tau_n \in \Sigma_{k_1 + \cdots + k_n}$ for $\tau_i \in \Sigma_{k_i}$ for $1 \le i \le n$, which is the concatenation of the τ_is. Then, the diagram

$$
\begin{array}{ccc}
O(n) \otimes O(k_1) \otimes \cdots \otimes O(k_n) & \xrightarrow{\;1_{O(n)} \otimes \tau_1 \otimes \cdots \otimes \tau_n\;} & O(n) \otimes O(k_1) \otimes \cdots \otimes O(k_n) \\
\Big\downarrow{\scriptstyle \gamma} & & \Big\downarrow{\scriptstyle \gamma} \\
O(k) & \xrightarrow{\;\tau_1 \oplus \cdots \oplus \tau_n\;} & O(k)
\end{array}
$$

is commutative.

The object $O(n)$ is often called the *n-ary part of the operad*. In the first equivariant condition, σ^{-1} permutes the tensor factors $O(k_i)$ by using the symmetry τ of the symmetric monoidal structure.

Examples 12.1.2

- Our first example is *the* prototypical operad. In fact, we will see later that any operad action can be described by it. Assume that $\mathcal{C}$ is closed symmetric monoidal, and let C be any object of $\mathcal{C}$. Then, the *endomorphism operad of* C, $\mathrm{End}_{\mathcal{C}}(C)$, has as n-ary part

$$
\mathrm{End}_{\mathcal{C}}(C)(n) := \underline{\mathcal{C}}(C^{\otimes n}, C),
$$

where $\underline{\mathcal{C}}(C^{\otimes n}, C)$ is the internal morphism object in $\mathcal{C}$. The map γ is just given by the composition of internal morphism objects. The right Σ_n-action on $\mathrm{End}_{\mathcal{C}}(C)(n)$ is given by the left Σ_n-action on $C^{\otimes n}$.

For instance, if $\mathcal{C}$ is the category of k-modules for some commutative ring k with unit and M is a k-module, then γ sends an $f \in k\text{-mod}(M^{\otimes n}, M)$ and $g_i \in k\text{-mod}(M^{\otimes k_i}, M)$ $(1 \le i \le n)$ to the composition

$$
\gamma(f \otimes g_1 \otimes \cdots \otimes g_n) = f \circ (g_1 \otimes \cdots \otimes g_n) \in k\text{-mod}(M^{\otimes(k_1 + \cdots + k_n)}, M),
$$

$$
M^{\otimes \sum_{i=1}^{n} k_i} = M^{\otimes k_1} \otimes \cdots \otimes M^{\otimes k_n} \xrightarrow{\;g_1 \otimes \cdots \otimes g_n\;} M \otimes \cdots \otimes M = M^{\otimes n}
$$
$$
\Big\downarrow{\scriptstyle f}
$$
$$
M.
$$

- Let $\mathcal{C}$ be the category Sets of sets. The collection of the symmetric groups $(\Sigma_n)_{n \ge 0}$ forms an operad called the *associative operad*, As, in Sets with $\mathrm{As}(n) = \Sigma_n$. Here, we follow the convention that Σ_0 and Σ_1 are both the trivial group. The composition morphisms

$$
\gamma \colon \Sigma_n \times \Sigma_{k_1} \times \cdots \times \Sigma_{k_n} \to \Sigma_{k_1 + \cdots + k_n}
$$

are determined by the equivariance condition of Definition 12.1.1. Explicitly, a $(\sigma, \tau_1, \ldots, \tau_n) \in \Sigma_n \times \Sigma_{k_1} \times \cdots \times \Sigma_{k_n}$ is sent to

$$\gamma(\sigma, \tau_1, \ldots, \tau_n) = (\tau_{\sigma^{-1}(1)} \oplus \ldots \oplus \tau_{\sigma^{-1}(n)}) \circ \sigma(k_1, \ldots, k_n).$$

Of course, we also get an associative operad in Top by considering the Σ_ns as discrete spaces. We can also transfer this operad to the category of k-modules for some commutative ring with unit k by taking the free k-module generated by Σ_n, $k\{\Sigma_n\}$ as the n-ary part of the operad. We denote this operad by As, no matter in which of these (and many more) categories we consider it.

- The operad in Sets that has the one-point set $*$ with trivial Σ_n-action in arity n is the operad Com. Composition is given by the obvious identification of products of $*$ with $*$. Again, we can transfer this operad to the category of k-modules by taking the free k-module generated by $*$ as $\mathrm{Com}(n)$ for every $n \geq 0$.

Remark 12.1.3 The zeroth part of an operad, $O(0)$, contracts inputs. For instance, the composition morphism

$$O(n) \otimes O(0) \cong O(n) \otimes O(0) \otimes e^{\otimes n-1} \xrightarrow{\;1_{O(n)} \otimes 1_{O(0)} \otimes \eta^{\otimes n-1}\;} O(n) \otimes O(0) \otimes O(1)^{\otimes n-1}$$
$$\downarrow \gamma$$
$$O(n-1)$$

reduces something with n inputs to something with $n - 1$ inputs.

Definition 12.1.4 An operad O in $(\mathcal{C}, \otimes, e, \tau)$ is *unital* if $O(0) \cong e$. For any unital operad O, we call the map

$$\varepsilon(n) \colon O(n) \cong O(n) \otimes e^{\otimes n} \cong O(n) \otimes O(0)^{\otimes n} \to O(0)$$

that is induced by the isomorphism $O(0) \cong e$ and the composition map γ, the *augmentation of the operad O*.

Definition 12.1.5 Let O and P be two operads in $\mathcal{C}$. A *morphism of operads* $f \colon O \to P$ is a family of Σ_n-equivariant morphisms $f(n) \colon O(n) \to P(n)$ that respect the unit morphisms η_O of O and η_P of P and the composition morphisms γ_O and γ_P; that is, we require that $f(1) \circ \eta_O = \eta_P$ and the commutativity of

$$O(n) \otimes O(k_1) \otimes \cdots \otimes O(k_n) \xrightarrow{f(n) \otimes f(k_1) \otimes \cdots \otimes f(k_n)} P(n) \otimes P(k_1) \otimes \cdots \otimes P(k_n)$$

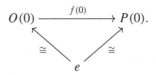

for all $k_1 + \cdots + k_n = k$. The operads in $\mathcal{C}$ form a category.

If O and P are unital operads, then we require that $f(0)$ is a compatible isomorphism

$$O(0) \xrightarrow{f(0)} P(0).$$

Example 12.1.6 Note that $*$ is terminal in Sets. If $\mathcal{O}$ is any operad in Sets, then the unique map $\mathcal{O}(n) \to *$ defines a morphism of operads $\mathcal{O} \to \mathsf{Com}$.

12.2 Algebras Over Operads

Definition 12.2.1 An object A of $(\mathcal{C}, \otimes, e, \tau)$ is an *algebra over an operad O* (an O-algebra) in $\mathcal{C}$ if there are morphisms $\theta_n \in \mathcal{C}(O(n) \otimes A^{\otimes n}, A)$ for all n that are associative, unital, and equivariant in the following sense:

(1) The action maps are associative. For all $k = \sum_{i=1}^{n} k_i$, the diagram

$$O(n) \otimes O(k_1) \otimes \cdots \otimes O(k_n) \otimes A^{\otimes k} \xrightarrow{\gamma \otimes 1} O(k) \otimes A^{\otimes k}$$

commutes.

(2) The element e acts as an identity via η and θ_1:

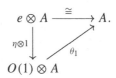

(3) The symmetric group action on the operad and on n-fold tensor powers of A is compatible for all n:

$$O(n) \otimes A^{\otimes n} \xrightarrow{\ \sigma \otimes \sigma^{-1}\ } O(n) \otimes A^{\otimes n} \qquad (12.2.1)$$

with θ_n and θ_n to A

commutes for all $\sigma \in \Sigma_n$.

A *morphism of O-algebras* $f : A \to B$ is a morphism $f \in \mathcal{C}(A, B)$, such that

$$
\begin{array}{ccc}
O(n) \otimes A^{\otimes n} & \xrightarrow{\ \theta_n\ } & A \\
\downarrow{\scriptstyle 1 \otimes f^{\otimes n}} & & \downarrow{\scriptstyle f} \\
O(n) \otimes B^{\otimes n} & \xrightarrow{\ \theta_n\ } & B
\end{array}
$$

commutes for all n.

We denote the category of O-algebras by O-alg.

In the preceding definition, we follow the convention that 0-fold tensor powers are the unit of the monoidal structure: $A^{\otimes 0} = e$; thus, we always have a morphism $\theta_0 : O(0) \to A$. If $O(0)$ is an initial object in $\mathcal{C}$, then this is no extra datum, but if O is unital, then $\theta_0 : e \to A$.

Definition 12.2.2 If O is an unital operad, then an O-algebra A is a *unital O-algebra*.

Examples 12.2.3 Assume that $\mathcal{C}$ is closed symmetric monoidal and is cocomplete.

- Every object C of $\mathcal{C}$ is an algebra over an operad, namely over its endomorphism operad. In this case, the action map

$$\theta_n : \mathrm{End}_{\mathcal{C}}(C^{\otimes n}, C) \otimes C^{\otimes n} \to C$$

is just the evaluation map.
- As $\mathcal{C}$ is closed, tensoring with $A^{\otimes n}$ commutes with colimits; in particular, it commutes with direct sums. Associative monoids in $\mathcal{C}$ are then algebras over the operad As with $\mathrm{As}(n) = \bigsqcup_{\sigma \in \Sigma_n} e$. For $e = e_\sigma$ in the component of $\sigma \in \Sigma_n$, the map θ_n restricted to e_σ is a multiplication $A^{\otimes n} \to A$ that first acts by permutation with σ on $A^{\otimes n}$. If $\mathcal{C}$ is the category of k-modules, then

$$\theta_n(e_\sigma; a_1 \otimes \cdots \otimes a_n) = a_{\sigma^{-1}(1)} \cdots \cdots a_{\sigma^{-1}(n)}.$$

- Commutative monoids in C are algebras over the operad Com with $\mathsf{Com}(n) = e$. The equivariance condition from (12.2.1) then ensures that the product $\theta_n : A^{\otimes n} \cong e \otimes A^{\otimes n} \to A$ satisfies $\theta_n \circ \sigma = \theta_n$ for all permutations $\sigma \in \Sigma_n$. In the category of k-modules, this gives the condition that

$$a_{\sigma^{-1}(1)} \cdots a_{\sigma^{-1}(n)} = a_1 \cdots a_n$$

for all $\sigma \in \Sigma_n$ and for all $a_i \in A$.

Exercise 12.2.4 Assume that C is a closed symmetric monoidal category. Show that giving an O-algebra structure to an object A of C is equivalent to defining a morphism of operads from the operad O to the endomorphism operad of A.

Exercise 12.2.5 Show that for every unital operad in a closed symmetric monoidal cocomplete category, the augmentation map $\varepsilon(n) : O(n) \to O(0) \cong e$ from Definition 12.1.4 is a morphism of operads $\varepsilon : O \to \mathsf{Com}$.

For example, if C is the category of k-modules, then the morphism of operads $\varepsilon : \mathsf{As} \to \mathsf{Com}$ that is induced by sending every $\sigma \in \Sigma_n$ to $*$ arises in this way.

Lemma 12.2.6 *Let C and D be symmetric monoidal categories and let $F : C \to D$ be a lax symmetric monoidal functor. Then, for every operad O in C, $F(O)$ is an operad in D. If A is an O-algebra in C, then $F(A)$ is an $F(O)$-algebra in D.*

Proof We define the structure maps of $F(O)$ as

$$F(O(n)) \otimes F(O(k_1)) \otimes \cdots \otimes F(O(k_n)) \xrightarrow{\varphi} F(O(n) \otimes O(k_1) \otimes \cdots \otimes O(k_n))$$
$$\Big\downarrow{\scriptstyle F(\gamma)}$$
$$F(O(\textstyle\sum_{i=1}^{n})).$$

Here, φ is the structure map of the lax monoidal functor F. The morphism $\eta_F : e_D \to F(e_C)$ induces the unit map of the operad via $F(\eta_O) \circ \eta_F : e_D \to F(O(1))$. As F is lax symmetric monoidal, these structure maps turn $F(O)$ into an operad. For the action of $F(O)$ on $F(A)$, we take

$$F(O(n)) \otimes F(A)^{\otimes n} \xrightarrow{\varphi} F(O(n) \otimes A^{\otimes n}) \xrightarrow{F(\theta_n)} F(A). \qquad \square$$

Note that if O is unital, then $F(O)$ doesn't have to be unital, unless $e_D \cong F(e_C)$.

12.2.1 Monads Associated with an Operad

Again, we assume that our symmetric monoidal category $(\mathcal{C}, \otimes, e, \tau)$ is closed and cocomplete. In particular, $(-) \otimes C$ commutes with colimits for all objects C of $\mathcal{C}$.

Lemma 12.2.7 *If O is an operad in $\mathcal{C}$ and X is an object of $\mathcal{C}$, then*

$$F_O(X) := \bigsqcup_{n \geq 0} O(n) \otimes_{\Sigma_n} X^{\otimes n}$$

defines a functor from $\mathcal{C}$ to the category of O-algebras in $\mathcal{C}$.

Sketch of Proof The structure map θ of the O-action on $F_O(X)$ is induced by the operad product γ and by the concatenation of tensor powers of X. $\square$

Exercise 12.2.8 Check the details in the preceding sketch of a proof.

Proposition 12.2.9 *The functor $F_O(-) \colon \mathcal{C} \to O\text{-alg}$ is left adjoint to the forgetful functor $U \colon O\text{-alg} \to \mathcal{C}$.*

Proof A morphism $f \in \mathcal{C}(X, U(A))$ gives rise to a morphism $\tilde{f} \in O\text{-alg}(F_O(X), A)$. We define the restriction of $\tilde{f}$ to $O(n) \otimes_{\Sigma_n} X^{\otimes n}$ to be the composite

$$O(n) \otimes_{\Sigma_n} X^{\otimes n} \xrightarrow{\; 1_{O(n)} \otimes f^{\otimes n} \;} O(n) \otimes_{\Sigma_n} A^{\otimes n} \xrightarrow{\; \theta_n \;} A.$$

This is indeed a map of O-algebras. For the converse, a morphism of O-algebras $g \colon F_O(X) \to A$ restricts to a morphism

$$X \cong e \otimes X \xrightarrow{\; \eta \otimes 1_X \;} O(1) \otimes X \longrightarrow \bigsqcup_{n \geq 0} O(n) \otimes_{\Sigma_n} X^{\otimes n} \xrightarrow{\; g \;} A$$

in $\mathcal{C}$. This yields

$$O\text{-alg}(F_O(X), A) \cong \mathcal{C}(X, U(A)),$$

and this isomorphism is natural in X and A. $\square$

Definition 12.2.10 The monad associated with an operad O in a closed symmetric monoidal cocomplete category $\mathcal{C}$ is the composite $\mathsf{O} = U \circ F_O(-) \colon \mathcal{C} \to \mathcal{C}$.

Remark 12.2.11 If A is an O-algebra, then we can use the monad of O to build a two-sided bar construction, as in Definition 10.4.2 $B(\mathsf{O}, \mathsf{O}, UA)$. This is a simplicial resolution of UA in the sense of Exercise 10.5.3.

Exercise 12.2.12 Assume that O is a nonunital operad. Show that an object A of $\mathcal{C}$ is an O-algebra if and only if it is an algebra for the monad O.

12.3 Examples

12.3.1 The Barratt–Eccles Operad

The definition of the Barratt–Eccles operad is based on the definition of the associative operad, As, in Sets from Example 12.1.2. We will first present its definition in the symmetric monoidal category of small categories, cat.

Recall that we defined for $(\sigma, \tau_1, \ldots, \tau_n) \in \Sigma_n \times \Sigma_{k_1} \times \cdots \times \Sigma_{k_n}$ the composition in As as

$$\gamma(\sigma, \tau_1, \ldots, \tau_n) = (\tau_{\sigma^{-1}(1)} \oplus \cdots \oplus \tau_{\sigma^{-1}(n)}) \circ \sigma(k_1, \ldots, k_n). \qquad (12.3.1)$$

We consider the collection of translation categories $(\mathcal{E}_{\Sigma_n})_{n \geq 0}$. Note that $\mathcal{E}_{\Sigma_0} \cong [0] \cong \mathcal{E}_{\Sigma_1}$. The composition, as in (12.3.1), then gives a morphism from the sets of objects of the category $\mathcal{E}_{\Sigma_n} \times \mathcal{E}_{\Sigma_{k_1}} \times \cdots \times \mathcal{E}_{\Sigma_{k_n}}$ to the objects of the category $\mathcal{E}_{\Sigma_{k_1 + \cdots + k_n}}$. The composition γ from (12.3.1) can be prolonged to a compatible composition on the level of morphisms, and this yields the following:

Lemma 12.3.1 *The composition morphisms γ, as in (12.3.1), give $(\mathcal{E}_{\Sigma_n})_{n \geq 0}$ the structure of a unital operad in* cat.

We call $E_\Sigma = (\mathcal{E}_{\Sigma_n})_{n \geq 0}$ the *Barratt–Eccles operad* in cat.

The application of Lemma 12.2.6 allows us to transfer the Barratt–Eccles operad to several different settings.

Proposition 12.3.2

(1) The nerve of the Barratt–Eccles operad, $(N\mathcal{E}_{\Sigma_n})_n$, is a unital operad in the category of simplicial sets with monoidal structure $\times$.

(2) The classifying space of the Barratt–Eccles operad, $(|N\mathcal{E}_{\Sigma_n}|)_n$, is a unital operad in the category of topological spaces cg.

(3) For any commutative ring with unit k, the free simplicial k-module generated by the nerve of the Barratt–Eccles operad, $(k\{N\mathcal{E}_{\Sigma_n}\})_n$, is a unital operad in the category of simplicial k-modules with the simplicial tensor product as monoidal structure.

(4) The chain complexes associated with $k\{N\mathcal{E}_{\Sigma_n}\}$, $(C_*(k\{N\mathcal{E}_{\Sigma_n}\}))_n$ and the normalized chain complexes, $(N(k\{N\mathcal{E}_{\Sigma_n}\}))_n$, form operads in the category of non-negatively graded chain complexes over k with the tensor product of chain complexes as the monoidal structure. Here, $(N(k\{N\mathcal{E}_{\Sigma_n}\}))_n$ is a unital operad.

Exercise 12.3.3 Can one also take the cellular chain complex of the $|N\mathcal{E}_{\Sigma_n}|$s instead of $(N(k\{N\mathcal{E}_{\Sigma_n}\}))_n$ to end up in the category of chain complexes over k?

12.3.2 Little n-Cubes Operad

The little n-cubes operad $\mathscr{C}_n$ was introduced by Boardman and Vogt in [**BV68, BV73**] as a tool for understanding n-fold loop spaces. They observed that for any topological space Y, the n-fold loop space $\Omega^n Y$ has an action of $\mathscr{C}_n$. May showed in [**May72**, Theorem 13.1] that if X is a connected space with an action of $\mathscr{C}_n$, then there exists a space Y and a zigzag of weak equivalences of $\mathscr{C}_n$-algebras connecting X and $\Omega^n Y$. Explicitly,

$$X \leftarrow B(C_n, C_n, X) \rightarrow B(\Omega^n \Sigma^n, C_n, X) \rightarrow \Omega^n B(\Sigma^n, C_n, X)$$

is a chain of weak homotopy equivalences of $\mathscr{C}_n$-algebras. Here, C_n denotes the monad associated with $\mathscr{C}_n$. Thus, $Y = B(\Sigma^n, C_n, X)$ is an n-fold delooping. There is an equivalence $X \rightarrow B(C_n, C_n, X)$, but this is *not* a map of $\mathscr{C}_n$-algebras.

For any $n \geq 1$, let I^n be the n-dimensional unit cube.

Definition 12.3.4

- An *open little n-cube* is a linear embedding $f : \overset{\circ}{I^n} \rightarrow \overset{\circ}{I^n}$ with parallel axes, that is, $f = (f_1, \ldots, f_n)$ and each f_i is a function $f_i : \overset{\circ}{I} \rightarrow \overset{\circ}{I}$ of the form $f_i(t) = (v_i - u_i)t + u_i$ with $u_i, v_i \in I$ and $v_i > u_i$.
- Let $1 \leq n < \infty$. The *operad of little n-cubes*, $\mathscr{C}_n$, has as $\mathscr{C}_n(r)$ (for $r \geq 1$) the space of r-tuples $(c_1, \ldots, c_r)$ of little n-cubes, such that the images of the c_is are pairwise disjoint. The topology on $\mathscr{C}_n(r)$ is given as a subspace of $\mathrm{Top}(\bigsqcup_{i=1}^{r} \overset{\circ}{I^n}, \overset{\circ}{I})$. Let $\mathscr{C}_n(0)$ be the point corresponding to the unique map from the empty set into $\overset{\circ}{I^n}$.
- The composition of the operad $(\mathscr{C}_n(r))_{r \geq 0}$ is given by the composition of functions. For $c \in \mathscr{C}_n(m)$ and $(c_i \in \mathscr{C}_n(r_i))_{i_1}^m$, one defines $\gamma(c; c_1, \ldots, c_m) := c \circ (c_1 \sqcup \cdots \sqcup c_m)$:

$$\bigsqcup_{j=1}^{r_1} \overset{\circ}{I^n} \sqcup \cdots \sqcup \bigsqcup_{j=1}^{r_m} \overset{\circ}{I^n} \xrightarrow{c_1 \sqcup \cdots \sqcup c_m} \bigsqcup_{j=1}^{m} \overset{\circ}{I^n} \xrightarrow{c} \overset{\circ}{I^n}.$$

The identity map $1_{\overset{\circ}{I^n}} \in \mathscr{C}_n(1)$ is the unit of the operad, and the symmetric group acts by permuting the factors in a tuple.

Note that the Σ_r-action on $\mathscr{C}_n(r)$ is free.

A picture for an element of the little square operad with $r = 3$ might look as follows. We will draw the cubes with their boundary boxes.

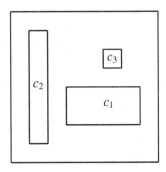

The action of $\mathscr{C}_n$ on an n-fold based loop space is easy to describe. Let $X = \Omega^n Y$ for $n \geq 1$ and let $\alpha_1, \ldots, \alpha_r$ be r based n-loops in Y. We can view the α_i as maps from the n-dimensional unit cube to Y, which send the boundary to the basepoint of Y. If ω is an element of $\mathscr{C}_n(r)$, then we define the action of ω on $(\alpha_1, \ldots, \alpha_r)$ to be the based n-loop, which evaluates α_1 on the first n-cube up to α_r on the rth n-cube and which sends the complement of the cubes to the basepoint.

In our example, one can depict that as follows:

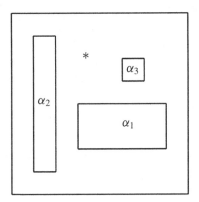

In the operad of little n-cubes $\mathscr{C}_n(r)$, $r \geq 0$, we can shrink the inner cubes to their centers. This gives a homotopy equivalence between $\mathscr{C}_n(r)$ and the ordered configuration space of r points in $\overset{\circ}{I}{}^n$.

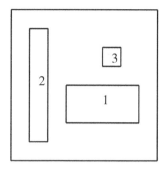 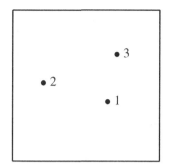

This equivalence is Σ_r-equivariant.

Definition 12.3.5 Let $\sigma(r)\colon \mathscr{C}_n(r) \to \mathscr{C}_{n+1}(r)$ be the map that sends an r-tuple of little n-cubes $(c_1, \ldots, c_r)$ to $(c_1 \times 1_{\overset{\circ}{I}}, \ldots, c_r \times 1_{\overset{\circ}{I}})$. We define

$$\mathscr{C}_\infty(r) = \operatorname{colim}_n \mathscr{C}_n(r).$$

The maps $\sigma(r)$ are morphisms of operads, and hence, $\mathscr{C}_\infty(r)$ is an operad as well.

Definition 12.3.6 An E_∞-operad in the category of topological spaces is a unital operad $\mathcal{O}$ in spaces, such that for each $\mathcal{O}(n)$, the augmentation map is homotopy equivalence and every $\mathcal{O}(n)$ has a free Σ_n-action.

Note that for every E_∞-operad $\mathcal{O}$ in spaces, the space $\mathcal{O}(n)$ is contractible and $\mathcal{O}(n)/\Sigma_n$ is a classifying space for Σ_n.

Example 12.3.7 The operad $\mathscr{C}_\infty$ is an E_∞-operad (see, for instance, [**May72**, Theorem 4.8]).

Example 12.3.8 The Barratt–Eccles operad $(|N\mathcal{E}_{\Sigma_n}|)_n$ is an E_∞-operad. Each space $|N\mathcal{E}_{\Sigma_n}|$ is a contractible space with a free Σ_n-action. Therefore, each space $|N\mathcal{E}_{\Sigma_n}|$ is an $E\Sigma_n$, and $|N\mathcal{E}_{\Sigma_n}|/\Sigma_n$ is a classifying space $B\Sigma_n$.

Remark 12.3.9 You saw that there are different possibilities to encode E_∞-structures (via the Barratt–Eccles operad or the operad $\mathscr{C}_\infty$). There are many more. Also, for k-fold loop spaces for $1 < k < \infty$, the operad $\mathscr{C}_k$ isn't the only choice. In [**Smi89**], Smith, for instance, constructs a series of simplicial sets

$$FX = \Gamma^{(1)}X \subset \Gamma^{(2)}X \subset \cdots \subset \Gamma X,$$

where FX is Milnor's model of $\Omega\Sigma|X|$ [**Mi55**] and ΓX is the Barratt–Eccles model of $\Omega^\infty\Sigma^\infty|X|$ [**BE74**]. Smith shows [**Smi89**, Theorem 1.1] that the geometric realization of $\Gamma^{(k)}X$ is a model for $\Omega^k\Sigma^k|X|$ for all $1 \le k$. Here, $\Gamma^{(k)}X$ is constructed like ΓX but with respect to a combinatorially defined suboperad $C_k\Sigma_n$ of the Barratt–Eccles operad. Smith conjectured that the geometric realization of $C_k\Sigma_n$ is equivalent to the ordered configuration space of n points in $\mathbb{R}^k$ [**Smi89**, p. 334], and Kashiwabara proved this in [**Kash93**]. Hence, the operad $(C_k\Sigma_n)_n$ is equivalent to $(\mathscr{C}_k(n))_n$. There are many more models. For a beautiful overview over some of them, see [**Be97**].

12.3.3 Homology of Iterated Loop Spaces

If $X = \Omega^n Y$ and $n \ge 2$, then the homology of X carries a very rich structure.

Kudo and Araki study the Pontrjagin rings $H_*(\Omega^N\mathbb{S}^m; \mathbb{F}_2)$, $0 < N < m$, [**KA56b**] and define H_n-spaces. For X an H_n-space and for $0 \le i \le n$, they establish [**KA56a**] the existence of homology operations

$$Q_i\colon H_q(X; \mathbb{F}_2) \to H_{2q+i}(X; \mathbb{F}_2).$$

Browder describes [**Br60**] $H_*(\Omega^n\Sigma^n Z; \mathbb{F}_2)$ as an algebra in terms of $H_*(Z; \mathbb{F}_2)$. For odd primes, Dyer and Lashof extend the definition of the operations Q_is to the homology, with coefficents in $\mathbb{F}_p$ for p odd, and they prove partial results about $H_*(\Omega^n\Sigma^n Z; \mathbb{F}_p)$ [**DL62**]. Milgram [**M66**] gives a description of $H_*(\Omega^n\Sigma^n Z; \mathbb{F}_p)$ as an algebra, depending only on the homology of Z and n. A complete description of the homology operations on iterated loop spaces and of $H_*(\Omega^n\Sigma^n Z; k)$ for $k = \mathbb{Q}$ and $k = \mathbb{F}_p$ is obtained by Cohen [**CLM76**, Chapter III].

The homology $H_*(\mathscr{C}_{n+1}, \mathbb{Q})$ forms an operad in the category of graded $\mathbb{Q}$-vector spaces, the operad that codifies n-Gerstenhaber algebras.

Definition 12.3.10 An n-Gerstenhaber algebra over $\mathbb{Q}$ is a (non-negatively) graded $\mathbb{Q}$-vector space G_* with a map $[-, -]\colon G_* \otimes G_* \to G_*$ that raises degree by n and a graded commutative multiplication of degree zero on G_*, such that $[-, -]$ satisfies a graded version of the Jacobi relation and graded antisymmetry (that is, $[x, y] = -(-1)^{qr}[y, x]$ for $x \in G_{q-n}$ and $y \in G_{r-n}$). In addition, there is a Poisson relation

$$[x, yz] = [x, y]z + (-1)^{q(r-n)}y[x, z].$$

Cohen showed that the rational homology of any space $X = \Omega^{n+1}Y$ is an n-Gerstenhaber algebra and that

$$H_*(\mathscr{C}_{n+1}Z; \mathbb{Q}) \cong nG(\bar{H}_*(Z; \mathbb{Q}))$$

for any space Z. Here, $nG(-)$ denotes the free n-Gerstenhaber algebra functor.

For prime fields $\mathbb{F}_p$, the situation is way more involved. Note that $\mathscr{C}_2(2) \simeq \mathbb{S}^1$ as a Σ_2-space and consider

$$\Sigma_2 \to \mathbb{S}^1 \to \mathbb{S}^1 = \mathbb{R}P^1.$$

We get two operations. The fundamental class of $\mathbb{S}^1$ corresponds to a Lie bracket of degree one, λ, on $H_*(\Omega^2X; \mathbb{F}_2)$, and the class of $\mathbb{R}P^1 \sim \mathbb{S}^1$ gives rise to an operation

$$\xi : H_m(\Omega^2X; \mathbb{F}_2) \to H_{2m+1}(\Omega^2X; \mathbb{F}_2).$$

For this note that $x \otimes x$ is invariant under the Σ_2-action, and thus, we have

$$x \mapsto \kappa([\mathbb{R}P^1] \otimes (x \otimes x)).$$

Think of this as being "half the circle" giving rise to "half the Lie bracket $[x, x]$", aka the restriction on x.

The complete description of $H_*(\mathscr{C}_{n+1}Z; \mathbb{F}_p)$ and $H_*(\Omega^{n+1}\Sigma^{n+1}Z; \mathbb{F}_p)$ as free objects built out of the reduced homology of Z involves a restricted n-Gerstenhaber structure, allowable modules, and algebras over the Dyer–Lashof algebra and a compatible coalgebra structure. See [**CLM76**, Chapter III] for the full picture.

12.3.4 Stasheff's Associahedra

The associahedra form a *non-symmetric operad*. This is a sequence of objects that satisfy the axioms of Definition 12.1.1, but one doesn't require an action of the symmetric groups, and consequently, there is no equivariance condition that has to be satisfied. The associahedra are a family of polytopes that describe objects that are not necessarily strictly associative but satisfy associativity up to coherent homotopies – and that's what is called an A_∞-*algebra*. An early approach to such structures is due to Sugawara [**Su57**], who formulated everything in terms of the appropriate properties of higher homotopies.

Stasheff introduced a nonpolytopal version of the associahedra in [**Sta63**, part I, §2, §6]. Later, other descriptions of the associahedra were introduced; see, for instance, [**Lo04**], [**GKZ94**, §7.3], and [**Lu∞**, §4.1.6]. We will give a description in terms of planar binary rooted trees.

Consider a set with a product $X \times X \to X$ that we just denote by concatenation. Then, for three elements $x, y, z \in X$, the terms $(xy)z$ and $x(yz)$ might differ if the product is not associative. A typical example is the set of based loops in a topological space W, that is, loops that start and end at a fixed basepoint $w_0 \in W$. Then, the concatenation of loops x and y

$$(x * y)(t) = \begin{cases} x(2t), & 0 \le t \le 1/2, \\ y(2t - 1), & 1/2 \le t \le 1, \end{cases}$$

is associative up to homotopy but not strictly associative. Planar binary rooted trees encode this situation. We will not draw an edge for the root. In the following picture, the left tree has three leaves, so it can digest three inputs and you compose the two right inputs first before you compose the result with the left input, so this tree stands for $x(yz)$, whereas the tree on the right-hand side encodes $(xy)z$. The interval indicates that there is a homotopy between these two ways of forming the product. The tree in the middle of the interval stands for this 1-cell that one might denote by xyz in order to express that up to homotopy, one does not have to set parentheses.

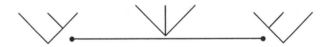

If one now forms a product of four elements, there are already five different ways of setting pairs of parentheses in order to form a meaningful product. If we want to form the product of u, v, x, and y in this fixed order, then we could form $u(v(xy))$, $(uv)(xy)$, $((uv)x)y$, $(u(vx))y$, and $u((vx)y)$. These correspond to the vertices in the following pentagon that are labelled with planar binary rooted trees. There are the homotopies between adjacent terms that correspond to 1-cells in the polytope. These are labelled by trees with one internal vertex, whereas the 2-cell in the middle is labelled by the corolla with four leaves; this tree has no internal vertex but just a root.

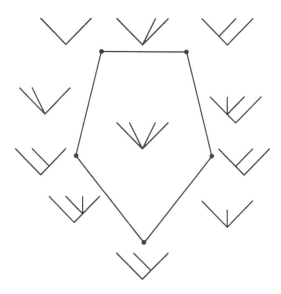

These pictures should look familiar if you compare them to the pentagon axiom (8.1.2) in the definiton of a monoidal category from Definition 8.1.4.

Definition 12.3.11 [**Lo04**, §2.4 and §2.5] For $n \geq 2$, the *nth Stasheff polytope*, K_n, is the $(n-2)$-dimensional polytope, whose vertices correspond to planar rooted binary trees with n leaves and whose unique n-cell corresponds to the corolla tree with n leaves. Its k-dimensional cells can be labelled by planar rooted trees with $n - k + 1$ internal vertices ($1 \leq k \leq n$). We set $K_1 = \{*\}$.

Stasheff's original definition can be found in [**Sta63**, part I, §2, §6]. A picture of K_5 is as follows:

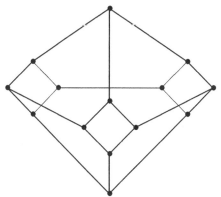

Exercise 12.3.12 Fill in the trees in the preceding picture. (If that seems too much work, then fill in trees for one of the squares.)

The number of ways in which $n + 1$ factors can be completely parenthesized is the *nth Catalan number*, C_n. Here, Catalan stands for Eugène Charles Catalan, a Belgian mathematician who lived from 1814 to 1894. For more background on the combinatorial aspects of Catalan numbers, see, for instance, [**St99**, p. 212]. You saw that C_4 is 14. A closed formula for C_n is

$$C_n = \frac{1}{n+1}\binom{2n}{n}.$$

Exercise 12.3.13 What is C_5?

Every point in K_n can be uniquely represented by a planar rooted trees with n leaves if one assigns a length in $[0, 1]$ to each internal edge. The fact that the polytopes $(K_n)_{n \geq 1}$ form a nonsymmetric operad can be seen by grafting trees. One can use the polytopes K_n for describing spaces with a product that is associative up to coherent homotopy.

Definition 12.3.14 A topological space is an A_∞-space if it is an algebra over the operad K_n.

For a beautiful explicit description of the structure of the K_n, see [**Lo04**] (but beware that his K^n is our K_{n+2}). Similar to the relationship between the K_ns and monoidal categories, one can describe permutative categories in terms of *permutahedra*, $(P_n)_{n \geq 1}$. Here, P_n is the convex hull of the points in $\mathbb{R}^n$ with coordinates $(\sigma^{-1}(1), \ldots, \sigma^{-1}(n))$ for $\sigma \in \Sigma_n$ (see, for instance, [**Lo04, Ka93**]). If one drops the strictness assumption and considers general symmetric monoidal categories, then one arrives at the family of *permutoassociahedra*, KP_n. The vertices of KP_n correspond to permuted words with a complete set of parentheses. Explicit CW models of the KP_ns are constructed in [**Ka93**], where Kapranov also shows that KP_n is an $(n-1)$-dimensional ball, thus, in particular, contractible.

12.4 E_∞-monoidal Functors

Some functors are monoidal but not symmetric monoidal, so they will not map commutative monoids in the source category to commutative monoids in the target category. An important example is the inverse Γ_N of the normalization functor $N: s(k\text{-mod}) \to \mathrm{Ch}(k)_{\geq 0}$. In this case, Γ_N maps differential graded commutative algebras to simplicial algebras that are commutative up to coherent homotopy.

Definition 12.4.1

- A unital operad O in the category of chain complexes is an E_∞-*operad* if the $O(n)$s are degreewise projective k-modules and if the augmentation maps

$$\varepsilon(n)\colon O(n) \to O(0) \cong S^0(k)$$

are quasi-isomorphisms; that is, the map of operads $\varepsilon = (\varepsilon(n))_{n\geq 0}\colon O \to$ Com is a quasi-isomorphism in every arity. Here, Com is the operad with $\mathsf{Com}(n) = S^0(k)$ for all $n \geq 0$.

- Similarly, a unital operad O in the category of simplicial k-modules is an E_∞-*operad* if the $O(n)$s are degreewise projective and if the augmentation map ε to the operad Com is a weak equivalence in every arity. Here, $\mathsf{Com}(n) = c(k)$, where $c(k)$ is the constant simplicial k-module with value k.

One can always replace such an operad O by a weakly equivalent one P, such that every $P(n)$ satisfies a Σ_n-freeness condition [**Ri03**, p. 96]. This is why we don't require this condition here. If you want to define E_∞-operads in other categories with weak equivalences, then you have to be careful about this issue.

Definition 12.4.2 [**Ri03**, Definition 2.3] A functor $F\colon \mathsf{Ch}(k)_{\geq 0} \to sk\text{-mod}$ is E_∞-*monoidal* if there is an E_∞-operad O in simplicial k-modules, such that there are natural maps

$$\theta_{n, M_1, \ldots, M_n}\colon O(n)\hat{\otimes}F(M_1)\hat{\otimes}\cdots\hat{\otimes}F(M_n) \to F(M_1 \otimes \cdots \otimes M_n)$$

for all $M_1, \ldots, M_n \in \mathsf{Ch}(k)_{\geq 0}$ and such that the natural unitality, associativity, and equivariance conditions are satisfied.

Remark 12.4.3 [**Ri03**, Proposition 2.4] The important feature of E_∞-monoidal functors is that they map commutative monoids to algebras over an E_∞-operad. If C_* is a differential graded commutative algebra, then $F(C_*)$ is an O-algebra with the structure maps

$$O(n)\hat{\otimes}F(C_*)^{\hat{\otimes}n} \xrightarrow{\;\theta_{n,C_*,\ldots,C_*}\;} F(C_*^{\otimes n}) \xrightarrow{\;F(\mu_{C_*})\;} F(C_*),$$

where μ_{C_*} denotes the multiplication in C_*.

In [**Ri03**, Theorem 4.1], we show that Γ_N is an E_∞-monoidal functor.

Remark 12.4.4 One can generalize the concept of E_∞-monoidal functors to $\mathcal{P}$-monoidal functors, where $\mathcal{P}$ is an operad. For details on this, see [**AM10**, §4.3.3].

13

Classifying Spaces of Symmetric Monoidal Categories

We relate monoidal structures on categories to multiplicative properties of the corresponding classifying spaces. In this chapter, all spaces are objects of cg, and we abbreviate the k-ification of the product of two spaces $X \times_k Y$ with $X \times Y$.

13.1 Commutative H-Space Structure on BC for C Symmetric Monoidal

Definition 13.1.1 An *H-space* is a topological space X, with a chosen point $x_0 \in X$ and a continuous map $\mu \colon X \times X \to X$, such that the two maps $\mu(x_0, -)$ and $\mu(-, x_0)$ are homotopic to the identity on X via homotopies, preserving the basepoint. An H-space is *associative* if μ is associative up to homotopy, and it is *commutative* if μ is homotopy commutative. An H-space is called *group-like* if there is a continuous map $\chi \colon X \to X$, such that $\mu \circ (1_X \times \chi) \circ \Delta$ is homotopic to the identity. Here, Δ denotes the diagonal map on X. Group-like associative H-spaces are also called *H-groups*.

Remark 13.1.2 In order to avoid pathological behavior, one should assume that (X, x_0) is well-pointed, that is, the inclusion of $\{x_0\}$ into X is a cofibration.

If X is a CW complex and x_0 is a 0-cell of X, then the requirement that $\mu(x_0, -)$ and $\mu(-, x_0)$ are homotopic to the identity on X via pointed homotopies can be relaxed to the property that $\mu(x_0, -)$ and $\mu(-, x_0)$ are just homotopic to the identity on X. You can also demand that x_0 is a strict two-sided unit. See, for instance, [**Ha02**, §3.C] for the equivalence of these notions.

304

Exercise 13.1.3 Prove that $\pi_1(X, x_0)$ is abelian if X is an H-space.

Every based loop space is a group-like H-space. Let Y be a topological space with basepoint y_0 and let

$$\Omega Y := k\underline{\mathsf{Top}}_*((\mathbb{S}^1, 1), (Y, y_0))$$

be the based loop space on Y (with the compact open topology). Then, the concatenation of loops is a homotopy-associative multiplication, and time reversal of loops gives an inverse up to homotopy. Higher based loop spaces

$$\Omega^n Y := k\underline{\mathsf{Top}}_*((\mathbb{S}^n, 1), (Y, y_0)), n \geq 2,$$

are also group-like H-spaces, and for $n \geq 2$, loop concatenation is homotopy commutative. The map μ is induced by the pinch map $\mathbb{S}^n \longrightarrow \mathbb{S}^n \vee \mathbb{S}^n$, which collapses the equator of the sphere to a point. You should read Adams' book [**Ad78**]. It is an excellent book that tells you the story of loop spaces, and it is fun to read.

Symmetric monoidal structures on a (small) category directly translate into H-space structures on the corresponding classifying space.

Theorem 13.1.4 *Let $\mathcal{C}$ be a small symmetric monoidal category. Then, $B\mathcal{C}$ is an associative and commutative H-space.*

Proof We define the multiplication μ as

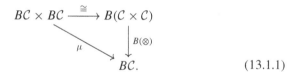

$$(13.1.1)$$

We choose $x_0 = [e] \in B\mathcal{C}^{(0)} \subset B\mathcal{C}$ as a basepoint of $B\mathcal{C}$. The two natural isomorphisms λ and ρ give natural transformations between the functors

$$\mathcal{C} \cong \{e\} \times \mathcal{C} \hookrightarrow \mathcal{C} \times \mathcal{C} \xrightarrow{\mu} \mathcal{C} \text{ and } \mathcal{C} \cong \mathcal{C} \times \{e\} \hookrightarrow \mathcal{C} \times \mathcal{C} \xrightarrow{\mu} \mathcal{C}$$

and the identity functor. These natural transformations yield the necessary homotopies in order to identify x_0 as a unit up to homotopy. The natural isomorphism α for the associativity in $\mathcal{C}$ is a natural transformation between $\otimes \circ (\mathrm{Id} \times \otimes)$ and $\otimes \circ (\otimes \times \mathrm{Id})$, and hence, $B\mathcal{C}$ is a homotopy associative H-space. Homotopy commutativity follows from the properties of the symmetry isomorphism τ, which gives rise to a homotopy between $\otimes$ and $\otimes \circ (1, 2)$, where $(1, 2) \colon \mathcal{C} \times \mathcal{C} \to \mathcal{C} \times \mathcal{C}$ denotes the permutation of the two factors. $\square$

We can upgrade this structure to an action of the Barratt–Eccles operad.

Theorem 13.1.5 [**May74**, Theorem 4.9] *Let C be a small permutative category. Then, the classifying space BC is a unital algebra in* cg *over the Barratt–Eccles operad* $(B\mathcal{E}_{\Sigma_n})_n$.

Sketch of Proof   By Lemma 8.3.6 from [**May74**, §4], we have Σ_n-equivariant functors

$$\tau_n \colon \mathcal{E}_{\Sigma_n} \times C^n \to C$$

for every n. Applying the classifying space functor B and using the continuous map $B\mathcal{E}_{\Sigma_n} \times (BC)^n \to B(\mathcal{E}_{\Sigma_n} \times C^n)$ give

$$\theta_n \colon B\mathcal{E}_{\Sigma_n} \times (BC)^n \to BC$$

for every n. As these maps only use the property of B to being lax symmetric monoidal (see Proposition 11.2.3) and the action of Σ_n on C^n, these θ_ns are compatible and define an operad action. You can find a detailed proof of the necessary coherence properties in [**May74**, Lemmata 4.4 and 4.5]. The unit of the permutative structure on C gives a distinguished basepoint in BC, and hence, BC is unital. $\square$

Remark 13.1.6   By Proposition 8.3.4, we can rigidify every symmetric monoidal category to a permutative one, and the corresponding classifying spaces are homeomorphic. This implies that the classifying space of every symmetric monoidal category carries an E_∞-structure.

Remark 13.1.7   If C is a strict monoidal category (without any symmetries), then the canonical map from (13.1.1) gives BC the structure of an associative topological monoid.

Exercise 13.1.8   Let M be a monoid. Consider the category C_M with one object and M as morphism set. When is C_M a monoidal category? What does this say about general classifying spaces of groups?

Remark 13.1.9   For any symmetric monoidal category C, the set of path components of C carries the structure of an abelian monoid with the product

$$[C] + [D] := [C \otimes D] \tag{13.1.2}$$

for objects C, D of C. The neutral elements is $[e]$.

Examples 13.1.10

• We saw earlier that the classifying space of the category Σ splits as

$$B\Sigma \cong \bigsqcup_{n \geq 0} B\Sigma_n$$

and $\pi_0(\Sigma) = \mathbb{N}_0$. This category is symmetric monoidal with

$$n \otimes m := n + m,$$

where $\tau_{n,m} \in \Sigma_{n+m}$ is the shuffle map sending the first n elements of $\{1, \ldots, n + m\}$ to the last n elements. The H-space structure is induced by the inclusion

$$\Sigma_m \times \Sigma_n \hookrightarrow \Sigma_{m+n}.$$

- Similarly, let R be an associative ring with unit and let $F(R)$ be the category whose objects are the natural numbers $n \in \mathbb{N}_0$ and whose morphisms are given by

$$F(R)(n, m) = \begin{cases} GL_n(R), & m = n, \\ \varnothing, & m \neq n. \end{cases}$$

Again, we get that $\pi_0(F(R)) = \mathbb{N}_0$ and

$$BF(R) \cong \bigsqcup_{n \geq 0} BGL_n R.$$

The H-space structure is induced by the block sum of matrices

$$GL_m(R) \times GL_n(R) \hookrightarrow GL_{m+n}(R), \quad (A, B) \mapsto \begin{pmatrix} A & 0 \\ 0 & B \end{pmatrix},$$

where 0 stands for zero matrices of a suitable size.
- Assume that $\mathcal{C}$ is a closed symmetric monoidal category, such that the Picard groupoid of $\mathcal{C}$, Picard($\mathcal{C}$), from Definition 8.4.6 is a small category. Let PIC($\mathcal{C}$) be the classifying space of the symmetric monoidal category Picard($\mathcal{C}$). The group of path components $\pi_0(\text{PIC}(\mathcal{C}))$ is the Picard group of $\mathcal{C}$, as in Remark 8.4.8.

In all of these examples, the classifying space carries an E_∞-structure.

13.2 Group Completion of Discrete Monoids

Definition 13.2.1 Let M be an abelian monoid. The *Grothendieck group on* M, $G(M)$ is an abelian group, together with a morphism of monoids $j \colon M \to G(M)$, which satisfies the following universal property. For every group G and for every morphism of monoids $f \colon M \to G$, there is a unique homomorphism of groups $f' \colon G(M) \to G$ with $f' \circ j = f$.

Note that the image of f consists of commuting elements in G because M is abelian, and similarly, the image of f' consists of commuting elements in G. Thus, the universal mapping property actually takes place in an abelian setting. The group $G(M)$ is also called the *group completion of the abelian monoid M*. There are explicit constructions of Grothendieck groups. Thanks to abstract nonsense, we know that any two such constructions are isomorphic.

Let us denote the product on M as addition. Take the product $M \times M$ and define the following addition on $M \times M$:

$$(m_1, n_1) + (m_2, n_2) := (m_1 + m_2, n_1 + n_2)$$

for $m_1, n_1, m_2, n_2 \in M$. We define the following equivalence relation on $M \times M$: A pair (m_1, n_1) is equivalent to a pair (m_2, n_2) if there is an $\ell \in M$ with

$$m_1 + n_2 + \ell = m_2 + n_1 + \ell.$$

The addition on $M \times M$ is well-defined on equivalence classes, and every class of the form $[(m, m)]$ is neutral with respect to addition. Hence, the additive inverse of the class of a pair (m, n) is the class of the pair (n, m). It might help to think of (m, n) as $m - n$.

Exercise 13.2.2 Prove that $(M \times M)/\sim$ has the universal property of $G(M)$.

Example 13.2.3 The Grothendieck group of the natural numbers $(\mathbb{N}_0, +, 0)$ is the group of the integers. An explicit isomorphism was given earlier.

In this case, the morphism $j \colon \mathbb{N}_0 \to G(\mathbb{N}_0)$ is injective. This is not always the case, but it *does* hold if the monoid M has the cancellation property, that is, if $m + p = n + p$ for $m, n, p \in M$, it implies that $m = n$.

The assignment $M \mapsto G(M)$ with any reasonable concrete model of $G(M)$ defines a functor from the category of abelian monoids to the category of abelian groups. The universal property means that this functor is left adjoint to the forgetful functor.

An alternative construction of $G(M)$ is as a quotient of the free abelian group generated by M:

$$G(M) = \mathbb{Z}\{M\}/\langle(m + n) - (m) - (n); n, m \in M\rangle.$$

We write $[m]$ for the equivalence class of $m \in M$ in $G(M)$ in this model. An element in $G(M)$ is the class of a sum $\sum_{i=1}^{\ell} \alpha_i[m_i]$ with integers α_i and $m_i \in M$. If you sort the elements in the sum in those with nonnegative α_i and those with negative α_i, then it is easy to see that elements of the form $[m] - [n]$ with $m, n \in M$, generate the group $G(M)$.

Exercise 13.2.4 Find an example of a nontrivial abelian monoid M with trivial $G(M)$.

Example 13.2.5 Let X be a compact Hausdorff space. Let $\mathsf{Vect}_{\mathbb{R}}(X)$ denote the set of isomorphism classes of real vector bundles over X of finite rank. Then, the Whitney sum of vector bundles induces an addition on $\mathsf{Vect}_{\mathbb{R}}(X)$. For two isomorphism classes $[\xi]$ and $[\chi]$ in $\mathsf{Vect}_{\mathbb{R}}(X)$, we define

$$[\xi] + [\chi] := [\xi \oplus \chi].$$

The class of the zero vector bundle serves as a neutral element with respect to addition, and there is a natural isomorphism of bundles $\xi \oplus \chi \cong \chi \oplus \xi$, and therefore, $(\mathsf{Vect}_{\mathbb{R}}(X), +, [0])$ is an abelian monoid. The Grothendieck group of this monoid is the *zeroth topological real K-theory group of X*:

$$KO^0(X) = G(\mathsf{Vect}_{\mathbb{R}}(X), +, [0]).$$

For instance, the representative of the real vector bundle over $\mathbb{S}^1$ associated with the Möbius strip gives rise to a nontrivial element.

If you consider complex vector bundles of finite rank instead of real ones, then you get the *zeroth topological complex K-theory group of X*:

$$KU^0(X) = G(\mathsf{Vect}_{\mathbb{C}}(X), +, [0]).$$

Sometimes, we have to handle nonabelian monoids. There is an analogous process for defining a universal group.

Definition 13.2.6 [BP72, 2.2] Let M be a monoid. The *universal group, $U(M)$, generated by M* is the quotient $F(M)/N$, where $F(M)$ is the free group generated by M, and N is the normal subgroup generated by all products xyz^{-1} for $x, y, z \in M$ with $xy = z$ in M.

Similar to the Grothendieck group, $U(M)$ has a universal property. If $f : M \to G$ is a morphism of monoids and G is a group, then there is a unique group homomorphism $\tilde{f} : U(M) \to G$. In particular, if M is an abelian monoid, then the universal properties of $G(M)$ and $U(M)$ guarantee that there is an isomorphism $U(M) \cong G(M)$.

13.3 Grayson–Quillen Construction

Let C be a small symmetric monoidal category. Our aim is to construct a category $C^{-1}C$, together with a functor $C \to C^{-1}C$, such that $\pi_0(C^{-1}C)$ is the group completion of the abelian monoid $\pi_0(C)$.

Quillen defined a group completion process for categories [**Gr76**].

Definition 13.3.1 Let $(C, \otimes, e, \tau)$ be a small symmetric monoidal category. We denote by $C^{-1}C$ the category whose objects are pairs of objects of C. Morphisms in $C^{-1}C$ from (C_1, D_1) to (C_2, D_2) are equivalence classes of pairs of morphisms

$$(f: C_1 \otimes E \to C_2, g: D_1 \otimes E \to D_2),$$

where E is an object of C. Such a pair is equivalent to

$$(f': C_1 \otimes E' \to C_2, g': D_1 \otimes E' \to D_2)$$

if there is an isomorphism $h \in C(E, E')$, such that the diagram

$$(C_1 \otimes E, D_1 \otimes E) \xrightarrow{(1_{C_1} \otimes h, 1_{D_1} \otimes h)} (C_1 \otimes E, D_1 \otimes E)$$

with maps (f,g) and (f',g') to (C_2, D_2)

commutes.

The category $C^{-1}C$ is called the *Grayson–Quillen construction of C*.

Note that the special case $E = e$ guarantees that every pair of morphisms in C gives rise to a morphism in $C^{-1}C$.

Lemma 13.3.2 *The category $C^{-1}C$ is symmetric monoidal, there is a lax symmetric monoidal functor $j: C \to C^{-1}C$, and $\pi_0(C^{-1}C)$ is an abelian group.*

Proof You can check that $C^{-1}C$ is actually a category. We define its symmetric monoidal structure coordinatewise by declaring that

$$(C_1, D_1) \otimes (C_2, D_2) := (C_1 \otimes C_2, D_1 \otimes D_2).$$

Note that in order to establish the naturality of $\otimes$, we need the fact that C is *symmetric* monoidal.

We define $j: C \to C^{-1}C$ on objects as $j(C) = (C, e)$, and for a morphism $f: C \to C'$, we set $j(f)$ to be the composition

$$(C \otimes e, e \otimes e) \cong (C, e) \xrightarrow{(f, 1_e)} (C', e)$$

and we abuse notation and denote this morphism by $[(f, 1_e)]$.

In $\mathcal{C}^{-1}\mathcal{C}$, there is a chain of morphisms

$$(C, D) \otimes (D, C) = (C \otimes D, D \otimes C) \xrightarrow{(1_{C \otimes D}, \tau_{D,C})} (C \otimes D, C \otimes D) \longleftarrow (e, e) \cdot$$

and thus, the set of path components $\pi_0(B\mathcal{C}^{-1}\mathcal{C})$ is an abelian group, because the inverse of the equivalence class of (C, D) is the class of (D, C). $\qquad\square$

Hence, we get that $Bj \colon B\mathcal{C} \to B\mathcal{C}^{-1}\mathcal{C}$ is a morphism of associative and commutative H-spaces and the induced map

$$\pi_0(Bj) \colon \pi_0(B\mathcal{C}) \to \pi_0(B\mathcal{C}^{-1}\mathcal{C})$$

is a map of abelian monoids.

Definition 13.3.3 Let $\mathcal{C}$ be a small symmetric monoidal category. Then, the *K-theory space of* $\mathcal{C}$, $K\mathcal{C}$, is defined as $K\mathcal{C} := B(\mathcal{C}^{-1}\mathcal{C})$, and the *nth K-group of* $\mathcal{C}$, $K_n\mathcal{C}$, is its nth homotopy group:

$$K_n\mathcal{C} := \pi_n B(\mathcal{C}^{-1}\mathcal{C}).$$

On the level of π_0, we obtain a group completion.

Lemma 13.3.4 *Let $\mathcal{C}$ be a small symmetric monoidal category. Then,*

$$K_0(\mathcal{C}) = \pi_0(K\mathcal{C}) \cong G(\pi_0(\mathcal{C})).$$

Proof   We consider the map from the set of objects of $\mathcal{C}^{-1}\mathcal{C}$ into the Grothendieck group on the abelian monoid $\pi_0(\mathcal{C})$, which is given by

$$\phi \colon (C, D) \mapsto [C] - [D].$$

For a fixed object E of $\mathcal{C}$, we obtain

$$\phi(C \otimes E, D \otimes E) = [C \otimes E] - [D \otimes E] = [C] + [E] - [D] - [E]$$
$$= [C] + [D] = \phi(C, D).$$

If $[(f, g)] \colon (C_1, D_1) \to (C_2, D_2)$ is a morphism in $\mathcal{C}^{-1}\mathcal{C}$, that is, there is an object E of $\mathcal{C}$ and morphisms $f \colon C_1 \otimes E \to C_2$ and $g \colon D_1 \otimes E \to D_2$ in $\mathcal{C}$, then

$$\phi(C_1, D_1) = [C_1] - [D_1] = [C_2] - [D_2] = \phi(C_2, D_2).$$

Thus, ϕ factors through $\pi_0(\mathcal{C}^{-1}\mathcal{C})$ and we get an induced map

$$\bar{\phi} \colon \pi_0(\mathcal{C}^{-1}\mathcal{C}) \to G(\pi_0(\mathcal{C})).$$

This map is surjective because elements of the form $[C] - [D]$ generate the group $G(\pi_0(\mathcal{C}))$. If $\phi(C, D) = 0$, then the classes $[C]$ and $[D]$ are equal in $\pi_0(\mathcal{C})$, and therefore, there is a finite zigzag of morphisms in $\mathcal{C}$ between C

and D. We can recycle this zigzag in order to get a zigzag of morphisms in the category $\mathcal{C}^{-1}\mathcal{C}$ between (C, D) and (D, D), and thus, we obtain that the class of (C, D) in $\pi_0(\mathcal{C}^{-1}\mathcal{C})$ is trivial.   □

Remark 13.3.5 Note that the map $\bar{\phi}$ constructed in the proof is inverse to the universal map from the group completion of $\pi_0(\mathcal{C})$ to $\pi_0(\mathcal{C}^{-1}\mathcal{C})$.

Example 13.3.6 There is a small version of the category of finitely generated projective left R-modules, $\mathcal{P}(R)$, for an associative ring R. As for each such module P, there is a module Q with $P \oplus Q \cong R^n$, we can think of such projective modules as the image of a projection map, and by slight abuse of notation, we call this P, so $P \colon R^n \to R^n$ is an R-linear map with $P^2 = P$.

The objects of $\mathcal{P}(R)$ are of the form (n, P), where $n \in \mathbb{N}_0$ and $P \in M_n(R)$ is a projection matrix, that is, $P^2 = P$. The morphisms in $\mathcal{P}(R)((n, P), (m, Q))$ are trivial for $n \neq m$. For $n = m$, the set of morphisms $\mathcal{P}(R)((n, P), (n, Q))$ consists of R-linear isomorphisms $f \colon \mathrm{im}(P) \to \mathrm{im}(Q)$. Then, two objects are equivalent in $\pi_0(\mathcal{P}(R))$ if and only if they are connected by a finite zigzag of R-isomorphism of the projective modules that are the images of the projection; hence,

$$\pi_0(\mathcal{P}(R)) \cong \mathrm{Proj}(R)$$

and $\pi_0(\mathcal{P}(R)^{-1}\mathcal{P}(R)) \cong K_0(R)$, the classical K_0-group of the ring R, which is the group completion of the monoid of isomorphism classes of finitely generated projective R-modules.

Example 13.3.7 Alternatively, we can use the small permutative category of finitely generated free R-modules $F(R)$ that we saw earlier. In this case, we set $GL_n(R)$ as the set of morphisms. Then, $\pi_0 F(R) \cong \mathbb{N}_0$, and the group completion is just $\mathbb{Z}$.

Example 13.3.8 One can identify the group completion of the category Σ explicitly. Sagave and Schlichtkrull [**SaSc12**] define a category $\mathcal{J}$, whose objects are pairs of objects of $\mathcal{I}$, the category of finite sets and injections from Example 1.2.3. A morphism in $\mathcal{J}((\mathbf{n}_1, \mathbf{n}_2), (\mathbf{m}_1, \mathbf{m}_2))$ is a triple (α, β, σ), where $\alpha \in \mathcal{I}(\mathbf{n}_1, \mathbf{m}_1)$, $\beta \in \mathcal{I}(\mathbf{n}_2, \mathbf{m}_2)$, and σ is a bijection

$$\sigma \colon \mathbf{m}_1 \setminus \alpha(\mathbf{n}_1) \to \mathbf{m}_2 \setminus \beta(\mathbf{n}_2).$$

For another morphism $(\gamma, \delta, \xi) \in \mathcal{J}((\mathbf{m}_1, \mathbf{m}_2), (\mathbf{l}_1, \mathbf{l}_2))$, the composition is the morphism $(\gamma \circ \alpha, \delta \circ \beta, \tau(\xi, \sigma))$, where $\tau(\xi, \sigma)$ is the permutation

$$\tau(\xi, \sigma)(s) = \begin{cases} \xi(s), & \text{if } s \in \mathbf{l}_1 \setminus \gamma(\mathbf{m}_1), \\ \delta(\sigma(t)), & \text{if } s = \gamma(t) \in \gamma(\mathbf{m}_1 \setminus \alpha(\mathbf{n}_1)). \end{cases}$$

Note that $\mathbf{l}_1 \setminus \gamma(\alpha(\mathbf{n}_1))$ is the disjoint union of $\mathbf{l}_1 \setminus \gamma(\mathbf{m}_1)$ and $\gamma(\mathbf{m}_1 \setminus \alpha(\mathbf{n}_1))$.

With these definitions, $\mathcal{J}$ is actually a category, and it inherits a symmetric monoidal structure from $\mathcal{I}$ via componentwise disjoint union [**SaSc12**, Proposition 4.3]. In particular, the category of $\mathcal{J}$-spaces is symmetric monoidal with the Day convolution product. Note that the unit for the monoidal structure $\boxtimes_{\mathcal{J}}$ is $\mathcal{J}((\mathbf{0}, \mathbf{0}), (-, -))$, and this is *not* a constant functor, but $\mathcal{J}((\mathbf{0}, \mathbf{0}), (\mathbf{n}, \mathbf{n}))$ can be identified with the symmetric group Σ_n.

There is an isomorphism of categories $\Sigma^{-1}\Sigma \cong \mathcal{J}$, which is the identity on objects and which sends a morphism in $\Sigma^{-1}\Sigma$, represented by (f, g) with $f : \mathbf{n}_1 \sqcup \mathbf{l} \to \mathbf{m}_1$ and $g : \mathbf{n}_2 \sqcup \mathbf{l} \to \mathbf{m}_2$, to the morphism $(f|_{\mathbf{n}_1}, g|_{\mathbf{n}_2}, (f|_{\mathbf{l}}) \circ (g|_{\mathbf{l}})^{-1})$ in $\mathcal{J}$.

13.4 Group Completion of H-Spaces

Definition 13.4.1 Let X be an associative H-space of the homotopy type of a CW complex in which left translation by any element of X is homotopic to right translation by the same element. The *group completion of X* is an associative H-space Y of the homotopy type of a CW complex that satisfies the same left versus right translation property, together with a morphisms of H-spaces $f : X \to Y$, such that the map $\pi_0(f) : \pi_0(X) \to \pi_0(Y)$ identifies the abelian monoid $\pi_0(Y)$ as the Grothendieck group of $\pi_0(X)$ and such that $H_*(f; k)$ induces an isomorphism

$$H_*(X; k)[\pi_0(X)^{-1}] \cong H_*(Y; k) \tag{13.4.1}$$

for all commutative rings k.

For an H-space X as in Definition 13.4.1, the singular homology of X is a graded ring with unit, where the multiplication

$$H_p(X; k) \otimes_k H_q(X; k) \to H_{p+q}(X \times X; k) \to H_{p+q}(X; k)$$

is given as the composite of the Künneth map and the map that is induced by the H-space structure on X, $\mu : X \times X \to X$. In particular, $H_*(X; k)$ is a graded $H_0(X; k)$-module and $H_0(X; k)$ is the group algebra $k[\pi_0(X)]$, such that the elements of $\pi_0(X)$ are central. Therefore, it makes sense to localize the ring $H_*(X; k)$ at $\pi_0(X)$. The homological requirement in the definition of a group completion of X implies that this localization is isomorphic to the homology of Y.

Remark 13.4.2 Quillen [**Q94**, Remark 1.4] showed that it suffices to check the isomorphism from (13.4.1) for $k = \mathbb{Q}$ and $k = \mathbb{F}_p$ for all primes p.

If X is a group-like H-space, we want to compare it to its group completion. To this end, consider the following version of the Whitehead theorem for homology with local coefficients (see, for instance, [**DK01**, Theorem 6.71] for a proof):

Theorem 13.4.3 *Let $f : X \to Y$ be an arbitrary continuous map between connected CW spaces X and Y. If $\pi_1(f)$ is an isomorphism and if*

$$H_n(X; \mathcal{A}) \to H_n(Y; \mathcal{A})$$

is an isomorphism for all local coefficient systems $\mathcal{A}$ and for all n, then f is a homotopy equivalence.

If X is a group-like H-space, then

$$X \simeq \pi_0(X) \times X_0,$$

where X_0 is the path component of the homotopy unit – all the path components are homeomorphic and every $x \in X$ can be sent to X_0 with the help of a correction term in $\pi_0(X)$, because the latter is a group.

Let Y be a group completion of X, and assume that X and Y are CW complexes. For every H-space X, any local coefficient system is simple (in the sense of Example 2.1.3); hence, it suffices to check that the map $f : X_0 \to Y_0$ induces an isomorphism on homology groups.

The map $X_0 \to Y_0$ induces an isomorphism on homology by the very definition of a group completion and the fact that $\pi_0(X)$ is a group. Hence, by Theorem 13.4.3, X is homotopy equivalent to its group completion.

We mention one important identification of a group completion. Note that for a topological group G, a standard argument (see for example, [**Ha02**, Proposition 4.66]) yields that there is a weak homotopy equivalence $G \to \Omega BG$.

Proposition 13.4.4 [**May74**, Theorem 1.6] *If M is a topological monoid such that for M and ΩBM, left translation by any element is homotopic to right translation by the same element, then the map*

$$\alpha : M \to \Omega BM$$

is a group completion.

Here, α is the adjoint of the map

$$\Sigma M \to BM, \quad [t, m] \mapsto [m, (t, 1 - t)].$$

Proposition 11.3.6 implies that for such topological monoids M, we obtain that $\pi_1(BM) \cong G(M)$. Examples include

$$M = \bigsqcup_{n=0}^{\infty} B\Sigma_n, \quad M = \bigsqcup_{n=0}^{\infty} BGL_n(R).$$

An analog of α for iterated and infinite loop spaces is as follows:

Theorem 13.4.5

- [**May74**, Lemma 2.1] *Let $\mathcal{O}$ be an E_∞-operad in topological spaces. There is a functor G from $\mathcal{O}$-spaces to spaces, together with a natural transformation $\eta\colon \mathrm{Id} \Rightarrow G$, such that η_X is a group completion for all $\mathcal{O}$-spaces X.*
- [**Se73**] [**CLM76**, III.3.3] *For all $n > 1$, $C_n X \to \Omega^n \Sigma^n X$ is a group completion.*

Theorem 13.4.6 [**Gr76**] *If $\mathcal{C}$ is a small groupoid that is symmetric monoidal and such that for every object C of $\mathcal{C}$, the functor*

$$(-) \otimes C\colon \mathcal{C} \to \mathcal{C}$$

is faithful, then $B\mathcal{C}^{-1}\mathcal{C}$ is a group completion of $B\mathcal{C}$.

In the examples Σ and $F(R)$, we obtain that the group completion has π_0 isomorphic to $\mathbb{Z}$ and that

$$B(\Sigma^{-1}\Sigma) \simeq \Omega B \left(\bigsqcup_{n \geq 0} B\Sigma_n \right)$$

and

$$B(F(R)^{-1}F(R)) \simeq \Omega B \left(\bigsqcup_{n \geq 0} BGL_n(R) \right).$$

The latter is one version of the classical definition of the K-theory space of a ring R due to Quillen.

Remark 13.4.7 Kan and Thurston show [**KT76**] that for every connected space X, there is a discrete group G_X, such that there is a homology isomorphism from the classifying space of G_X, BG_X, to X. As BG_X has only one nontrivial homotopy group, there cannot be an analogous result about homotopy types. However, Dusa McDuff [**McD79**] extended the Kan–Thurston result and proved that for every connected space X, there is a discrete monoid M, such that X has the same weak homotopy type as BM.

14

Approaches to Iterated Loop Spaces
via Diagram Categories

Since the 1970s, diagram categories have been used to model loop spaces, so spaces of the form $\Omega^n X$, mostly for $n = 1, 2$, and ∞. During the last decades, new approaches, in particular for modelling iterated loop spaces $\Omega^n X$ for all n, have been developed. This chapter intends to give an overview, and I mostly refer to the literature for proofs and technical details.

Before we start with the more complicated topic of iterated loop spaces, we explain how diagram categories can be used to detect algebraic structures in the discrete and in the k-linear setting, where k is a commutative ring with unit.

14.1 Diagram Categories Determine Algebraic Structure

Some categories are suitable for encoding algebraic properties. For instance, if you consider the category Δ^o, then there is the equality $d_1 d_2 = d_1 d_1 \colon [3] \to [1]$ of morphisms, and if you assign M^n to $[n]$ in a functorial manner, then this can correspond to an associativity condition for a multiplication $d_1 \colon M^2 \to M$. In contrast, when you consider the category Fin, then the extra symmetry coming from $\Sigma_n \subset \text{Fin}(\mathbf{n}, \mathbf{n})$ will encode commutativity, and you have to desymmetrize Fin if you want to avoid that.

Some of the examples of diagram categories in Chapter 15 are actually categories of operators. The following is a discrete version of the one in [**MayTh78**, Construction 4.1]:

Definition 14.1.1 Let $\mathcal{O}$ be an operad in the category of sets with $\mathcal{O}(0) = *$. The *category of operators with respect to* $\mathcal{O}$, $\Gamma(\mathcal{O})$, has the same objects as the category of finite pointed sets, Γ, and a morphism in $\Gamma(\mathcal{O})([n], [m])$ is a tuple $(f; \omega_0, \ldots, \omega_m)$ with $f \in \Gamma([n], [m])$ and $\omega_i \in \mathcal{O}(|f^{-1}(i)|)$, where $|f^{-1}(i)|$

denotes the cardinality of the fiber of f over $i \in [m]$. The identity morphism $1_{[n]}$ in $\Gamma(\mathcal{O})([n], [n])$ is $(1_{[n]} \in \Gamma([n], [n]); \eta(e), \dots, \eta(e))$. Composition of morphisms in $\Gamma(\mathcal{O})$ is induced by the composition of morphisms in Γ and the composition in the operad $\mathcal{O}$.

Example 14.1.2 For the operad Com in Sets, we obtain $\Gamma(\text{Com}) \cong \Gamma$. For the operad As in Sets, the category of operators $\Gamma(\text{As})$ is precisely the category of finite pointed associative sets $\Gamma(as)$ from Definition 15.3.2. A morphism $f \in \Gamma(as)([m], [n])$ is a morphism $f \in \Gamma([m], [n])$, together with a total ordering on every fiber $f^{-1}(i)$. This is a desymmetrized version of Γ.

May and Thomason use categories of operators in [**MayTh78**] for comparing infinite loop space machines. Loop spaces need base points, so in their context, it is natural to work with finite *pointed* sets. In an algebraic setting, basepoints can be used for dealing with coefficients, for instance, when one wants to consider Hochschild homology of an algebra with coefficients in a bimodule (see Definition 15.3.1). If we are interested in purely algebraic structures, we might use the category of (unbased) finite sets, Fin.

A prototypical example of the description of an algebraic structure via a diagram category is the following basic result, which we already saw in disguise in Proposition 10.14.3.

Proposition 14.1.3 *Let M be a set. Then, M is a monoid if and only if the assigment*

$$[n] \mapsto M^n$$

gives rise to a functor from Δ^o to Sets.

Proof In the proof of Proposition 10.14.3, we already saw that the fact that $[n] \mapsto M^n$ defines a functor from Δ^o to Sets ensures that M is a monoid. For the converse, if $(M, \cdot, 1)$ is a monoid, then we set

$$d_i \colon M^n \to M^{n-1}, d_i(m_1, \dots, m_n)$$

$$:= \begin{cases} (m_2, \dots, m_n), & \text{for } i = 0, \\ (m_1, \dots, m_i \cdot m_{i+1}, \dots, m_n), & \text{for } 1 \leq i \leq n-1, \\ (m_1, \dots, m_{n-1}), & \text{for } i = n. \end{cases}$$

The degeneracy map s_i inserts 1 in slot number $i + 1$. This gives rise to a simplicial object, as we know from the definition of the nerve (Definition 11.1.1) of $\mathcal{C}_M$. $\qquad\square$

The following is a typical k-linear example:

Proposition 14.1.4 *Let k be a commutative ring and let A be a k-module. Then, A is a commutative k-algebra if and only if the assignment $\{1, \ldots, n\} = \mathbf{n} \mapsto A^{\otimes_k n}$ is a functor $\mathcal{L}^k(A) \colon \mathsf{Fin} \to k\text{-mod}$.*

Proof If A is a commutative k-algebra and $f \in \mathsf{Fin}(\mathbf{n}, \mathbf{m})$, then we define the morphism $\mathcal{L}^k(f) \colon \mathcal{L}^k(A)(\mathbf{n}) \to \mathcal{L}^k(A)(\mathbf{m})$ by sending a generator $a_1 \otimes \cdots \otimes a_n$ to $b_1 \otimes \cdots \otimes b_m$, where

$$b_i = \prod_{j \in f^{-1}(i)} a_j,$$

and if the fiber of i is empty, then we set $\prod_\varnothing = 1_A$. As A is commutative, it does not matter in which order we multiply the element. For $g \in \mathsf{Fin}(\mathbf{m}, \ell)$, the equality of fibers $(g \circ f)^{-1}(i) = f^{-1}(g^{-1}(i))$ guarantees that $\mathcal{L}^k(g \circ f) = \mathcal{L}^k(g) \circ \mathcal{L}^k(f)$

Conversely, if we just know that A is a k-module, but, in addition, we know that $\mathcal{L}^k(A) \colon \mathsf{Fin} \to k\text{-mod}$ is a functor, then we define a multiplication μ on A as $\mathcal{L}^k(A)(v)$, where v is the unique map from $\mathbf{2}$ to $\mathbf{1}$ in Fin. Note that μ is bilinear because $\mathcal{L}^k(A)(v)$ is a map of k-modules.

The object $\mathbf{0}$ is initial in Fin, and let u denote the unique map $u \in \mathsf{Fin}(\mathbf{0}, \mathbf{1})$. We define the unit $\eta \colon k = A^{\otimes_k 0} \to A^{\otimes_k 1} = A$ as $\mathcal{L}^k(A)(u)$.

Twisting 1 and 2 in $\mathbf{2}$ first and then applying v is equal to v; therefore, μ is commutative. Associativity follows from the equality $v \circ (v \oplus \mathrm{id}) = v \circ (\mathrm{id} \oplus v)$:

in Fin. The identity on $\mathbf{1}$ is equal to the composite that identifies $\mathbf{1}$ with $\varnothing \oplus \mathbf{1}$, sends this to $\mathbf{1} \oplus \mathbf{1} \cong \mathbf{2}$, and applies v. Similarly, we can first use $\mathbf{1} \cong \mathbf{1} \oplus \varnothing$ and compose with the same string of morphisms and get the identity on $\mathbf{1}$. This implies that η satisfies the unit axioms. □

A probably less obvious identification relates a skeletal version of the category of finitely generated free groups with the category of commutative Hopf algebras. This result is crucial, for instance, for the work of [**BRY**∞, §5].

Definition 14.1.5 Let Free be the category whose objects are $\mathbf{n}$ for $n \in \mathbb{N}_0$. Morphisms in Free from $\mathbf{n}$ to $\mathbf{m}$ are group homomorphisms from the free group on n generators to the free group on m generators.

Note that in Free, the object **0** corresponds to the free group on 0 generators and hence to the trivial group. Therefore, it is a zero object in Free.

Proposition 14.1.6 (see [**P02**, Theorem 5.2]) *Let H be a k-module. Then, H is a commutative Hopf algebra over k if and only if the assignment $\mathbf{n} \mapsto H^{\otimes n}$ is a functor $\mathcal{L}^k(H)$: Free $\to k$-mod.*

Proof We only give a proof of the direction that H is a commutative Hopf algebra over k if $\mathcal{L}^k(H)$: Free $\to k$-mod is a functor. The fact that **0** is a zero object in Free gives rise to a unit map for H, $\eta\colon k \to H$ via $\mathcal{L}^k(H)(\mathbf{0} \to \mathbf{1})$ and to a counit map, $\varepsilon\colon H \to k$, via $\mathcal{L}^k(H)(\mathbf{1} \to \mathbf{0})$. The multiplication in H, $\mu\colon H \otimes_k H \to H$, corresponds to the map $\mathcal{L}^k(H)(v)$, where $v\colon \mathbf{2} \to \mathbf{1}$ is the morphism that sends the generators of the free group on two generators x_1, x_2 to the generator of the free group $\langle x_1 \rangle$. As the map that exchanges the generators composed with v is v, we get that the multiplication is commutative. As in the proof of Proposition 14.1.4, we see that the unit map, together with the multiplication map, turns H into a commutative k-algebra.

For the diagonal map $\Delta\colon H \to H \otimes_k H$, we take $\mathcal{L}^k(\psi)$, where $\psi\colon \langle x_1 \rangle \to \langle x_1, x_2 \rangle$ is the map that sends x_1 to $x_1 x_2$. As the free group on two generators is *not* commutative, this comultiplication will not be cocommutative in general. As we have the equality $(x_1 x_2)x_3 = x_1(x_2 x_3)$ in the free group on three generators, this diagonal is coassociative.

The counit ε corresponds to the map that sends the generator x_1 of $\langle x_1 \rangle$ to the empty word. Therefore, $(\varepsilon \otimes 1_H) \circ \Delta = 1_H = (1_H \otimes \varepsilon) \circ \Delta$.

The morphism i of free groups that sends x_1 to x_1^{-1} induces a self-map $\mathcal{L}^k(i)\colon H \to H$. The fact that the composites $x_1 \mapsto x_1 x_2 \mapsto x_1 x_2^{-1} \mapsto x_1 x_1^{-1}$ and $x_1 \mapsto x_1 x_2 \mapsto x_1^{-1} x_2 \mapsto x_1^{-1} x_1$ give the empty word shows that $S = \mathcal{L}^k(i)$ is an antipode for H.

We need to show that the diagram

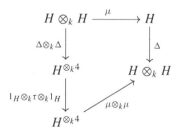

commutes (where τ denotes the twist morphism). If we chase the diagram of generators around in the same diagram, we obtain that both $\psi \circ v$ and $(v, v) \circ \tau \circ (\psi, \psi)$ send (x_1, x_2) to $(x_1 x_2, x_1 x_2)$. $\qquad\square$

14.2 Reduced Simplicial Spaces and Loop Spaces

We consider simplicial topological spaces.

Definition 14.2.1 A simplicial set or simplicial topological space X is *reduced* if X_0 is a point.

In analogy to the definition of Segal sets (see Definition 10.14.1), one can consider Segal spaces.

Definition 14.2.2 A simplicial topological space X or a bisimplicial set viewed as a simplicial object in simplicial sets is a *Segal space* if the canonical map

$$(i_1, \ldots, i_n) \colon X_n \to X_1 \times_{X_0} \cdots \times_{X_0} X_1$$

is a weak equivalence for all $n \geq 2$.

Note that for a reduced Segal space X, $\pi_0(X_1)$ carries the structure of a pointed monoid.

Every simplicial space has an *underlying topological space* by forming X_1. Segal proved the following result that identifies certain spaces as based loop spaces:

Theorem 14.2.3 [**Se74**, Proposition 1.5] *A reduced Segal space X has $X_1 \simeq \Omega|X|$ if and only if X_1 has homotopy inverses and if X is good.*

Here, *good* is the same technical assumption on a simplicial topological space that ensures that the fat realization is equivalent to the geometric realization (compare 10.9, [**Se74**, Appendix A]). If a simplicial topological space X arises as the partial geometric realization of a bisimplicial set, then X is good.

14.3 Gamma-Spaces

Recall the definition of the category Γ. The objects of Γ are the finite pointed sets $[n] = \{0, \ldots, n\}$ for $n \geq 0$, with zero as basepoint, and morphisms are functions preserving the basepoint. There is a canonical functor from Δ^o to Γ.

Definition 14.3.1 The *simplicial circle* is the functor $C \colon \Delta^o \to \Gamma$ with $C[n] = [n]$, and $C(s_i) \colon [n] \to [n+1]$ is the strict monotone injection missing the value $i + 1$. For $i < n$, the ith face map is

$$C(d_i)(j) = \begin{cases} j, & j \leq i, \\ j - 1, & j > i. \end{cases}$$

For $i = n$, we have

$$C(d_n)(j) = \begin{cases} j, & j \le n-1, \\ 0, & j = n. \end{cases}$$

It is easy to see that this actually defines a functor. Furthermore, there are only two nondegenerate simplices, the one in degree 0 and $1 \in C[1]$ and $d_0(1) = d_1(1) = 0$, and hence, the geometric realization of C is $\mathbb{S}^1$.

Segal's original definition in [Se74] uses the opposite category of Γ, Γ^o. It can be described as having as objects the sets $\mathbf{n} = \{1, \ldots, n\}$ with $\mathbf{0} = \varnothing$. A morphism in Γ^o from $\mathbf{m}$ to $\mathbf{n}$ is an m-tuple $(S_1, \ldots, S_m)$ of pairwise disjoint subsets of $\mathbf{n}$. The composite of $(T_1, \ldots, T_\ell): \ell \to \mathbf{m}$ and $(S_1, \ldots, S_m): \mathbf{m} \to \mathbf{n}$ is the ℓ-tuple

$$\left(\bigcup_{j_1 \in T_1} S_{j_1}, \ldots, \bigcup_{j_\ell \in T_\ell} S_{j_\ell} \right).$$

In order to see that this category is the opposite of Γ, just think of a morphism $f: [n] \to [m]$ as being determined by the preimages $f^{-1}(1), \ldots, f^{-1}(m)$.

Exercise 14.3.2 The symmetric group on n letters, Σ_n, is contained in $\Gamma([n], [n])$. Determine the morphism f_σ in $\Gamma^o(\mathbf{n}, \mathbf{n})$ that corresponds to a permutation $\sigma_n \in \Gamma([n], [n])$.

Definition 14.3.3

- Let $\mathcal{C}$ be any category with zero object $*$. Then, a Γ-*object in* $\mathcal{C}$ is an $F \in$ $\mathsf{Fun}(\Gamma, \mathcal{C})$ with $F[0] = *$.
- A Γ-*space* is a Γ-object in the category of pointed simplicial sets, that is, a $Y \in \mathsf{Fun}(\Gamma, s\mathsf{Sets}_*)$ with $Y[0] = \Delta_0$.

For two pointed simplicial sets X_1, X_2 with basepoints $*_{X_1}$ and $*_{X_2}$, their *smash product* is

$$X_1 \wedge X_2 = X_1 \times X_2 / X_1 \vee X_2,$$

where $X_1 \vee X_2 = X_1 \sqcup X_2 / *_{X_1} \sim *_{X_2}$.

Examples 14.3.4

(1) Every finite pointed set can be turned into a finite pointed simplicial set, using the constant functor. Thus, we have a canonical Γ-space

$$\mathbb{S}: \Gamma \to s\mathsf{Sets}_*.$$

(2) [**Se74**, §1] Let M be an abelian monoid. Then, we define the Γ-space HM as $HM[n] = M^n$, such that an $f \in \Gamma([n], [\ell])$ induces

$$HM(f) \colon M^n \to M^\ell, \quad HM(f)(m_1, \dots, m_\ell)$$

$$= \left(\sum_{i \in f^{-1}(1)} m_i, \dots, \sum_{i \in f^{-1}(\ell)} m_i \right).$$

Here, we place the unit 0 of M in the ith coordinate if $f^{-1}(i) = \varnothing$.

(3) If X is a simplicial set and Y is a Γ-space, then $X \wedge Y$ is the Γ-space

$$[n] \mapsto X \wedge (Y[n]).$$

(4) Consider a commutative ring k, a commutative k-algebra A, and a symmetric A-bimodule M. By forgetting structure, M is also a k-module. We denote $\otimes_k$ by $\otimes$. We consider as $\mathcal{C}$ the category of k-modules under and over M. The Loday functor

$$\mathcal{L}^k(A; M) \colon \Gamma \to \mathcal{C}, \mathcal{L}^k(A; M)[n] = M \otimes A^{\otimes n}$$

defines a Γ-object in $\mathcal{C}$ by declaring that for any $f \in \Gamma([n], [\ell])$, we send $a_0 \otimes a_1 \otimes \cdots \otimes a_n$ (with $a_0 \in M$, $a_i \in A$ for $i > 0$) to $b_0 \otimes b_1 \otimes \cdots \otimes b_\ell$, with

$$b_i = \prod_{j \in f^{-1}(i)} a_j,$$

where $b_i = 1_A$ if the preimage of i is empty.

Exercise 14.3.5 Show that $\mathbb{S}$ is isomorphic to the representable Γ-space $\Gamma([1], -)$.

Definition 14.3.6

(1) If Y is a Γ-space and X is a finite pointed simplicial set, then the evaluation of Y on X is the diagonal of the bisimplicial set

$$Y(X)_{p,q} = (Y(X_p))_q, \quad 0 \le p, q,$$

that is, $Y(X)_p = Y(X_p)_p$.

(2) If X_1 and X_2 are finite pointed simplicial sets and Y is a Γ-space, then there is an *assembly map*

$$X_1 \wedge Y(X_2) \to Y(X_1 \wedge X_2),$$

defined as follows: For a fixed $x \in X_1$, we get a function f_x sending X_2 to $X_1 \wedge X_2$:

$$f_x \colon X_2 \to X_1 \wedge X_2, \quad x_2 \mapsto [(x, x_2)],$$

where $[(x, x_2)]$ denotes the equivalence class of (x, x_2) in the smash product. For varying x, this gives a well-defined map

$$\sigma_{X_1, X_2} : X_1 \wedge Y(X_2) \to Y(X_1 \wedge X_2)$$

for every Γ space Y.

(3) The *spectrum associated with a Γ space* Y is the sequence of simplicial sets $(Y(S^n))$, where S^n is the simplicial model of the n-sphere $S^n = (S^1)^{\wedge n}$. The assembly maps induce structure maps

$$\sigma_{S^1, S^n} : S^1 \wedge Y(S^n) \to Y(S^{n+1}).$$

The spectrum associated with a Γ-space gives a notion of homotopy groups.

Definition 14.3.7 Let Y be a Γ-space. Its nth homotopy group is

$$\pi_n(Y) = \operatorname{colim}_i \pi_{n+i} Y(S^i),$$

where the structure maps σ_{S^1, S^n} induce the maps for the filtered colimit.

Note that for any Γ-space Y, $Y[1] = Y[S^0]$, the zeroth simplicial set in the spectrum associated with Y.

Examples 14.3.8

(1) The spectrum associated with the Γ-space $\mathbb{S}$ (Example (1)) has

$$\mathbb{S}(S^n) = S^n.$$

This is the *sphere spectrum*. Its homotopy groups are the stable homotopy groups of spheres.

(2) If A is an abelian group, then the Γ-space HA (Example (2)) has

$$HA(S^n) \cong A \otimes \bar{\mathbb{Z}}\{S^n\},$$

and this is an Eilenberg–Mac Lane space of type (A, n) because $\pi_j HA(S^n)$ is trivial for all j but n, where we get

$$\pi_n HA(S^n) \cong \pi_n A \otimes \bar{\mathbb{Z}}\{S^n\} \cong \bar{H}_n(\mathbb{S}^n; A) = A.$$

The spectrum $(HA(S^n))n \geq 0$ is the *Eilenberg–Mac Lane spectrum of A*.

(3) For a simplicial set X, the associated spectrum of the Γ-space $X \wedge \mathbb{S}$ is the *suspension spectrum of X*.

(4) The homotopy groups of the functor $\mathcal{L}^k(A; M): \Gamma \to k$-mod are isomorphic to the Gamma homology of A with coefficients in M [**PR00**, Theorem 1], a homology theory for commutative algebras introduced by Alan Robinson and Sarah Whitehouse in [**RoWh02**]; see Section 15.5.

Remark 14.3.9 The first two examples should look familiar. In Example 10.15.2, we already encountered them in the context of symmetric spectra. This is no accident. Every spectrum associated with a Gamma space carries the structure of a symmetric spectrum by using the action of Σ_n on $(S^1)^{\wedge n}$ that permutes the smash factors.

A Γ space gives rise to a spectrum, but we can also forget down to simplicial sets.

Definition 14.3.10 The *underlying pointed simplicial set of a Γ-space Y* is $U(Y) = Y[1]$.

Exercise 14.3.11 Show that the functor $U \colon \Gamma$ spaces $\to s$Sets has a left adjoint given by $X \mapsto X \wedge \mathbb{S}$, where $\mathbb{S}$ is the Γ-space from Example (1).

Segal posed extra conditions on Γ-spaces to single out those Γ-spaces that give rise to infinite loop spaces.

Definition 14.3.12
(1) A Γ-space Y is *special* if the projection maps from $[k + \ell] \cong [k] \vee [\ell]$ to $[k]$ and $[\ell]$ induce weak equivalences of simplicial sets $Y[k + \ell] \to Y[k] \times Y[\ell]$.
(2) A special Γ-space Y is *very special* if $\pi_0 Y[1]$ is a group.

Note that the condition for special Γ-spaces is equivalent to requiring that the projection maps $p_i \colon [n] \to [1]$, $p_i(i) = 1$, $p_i(j) = 0$ for all $j \neq i$ induce weak equivalences of simplicial sets

$$Y[n] \to Y[1]^n. \tag{14.3.1}$$

In particular, $Y[2] \simeq Y[1] \times Y[1]$.

Compare this to the condition of a Segal set from Definition 10.14.1 and to Segal spaces from Definiton 14.2.2. As we require $Y[0] = *$, (14.3.1) is about products and not about actual fiber products.

There is a fold map $\xi \colon [2] \to [1]$ in Γ, sending 1 and 2 to 1. Thus, we get a zigzag

$$Y[1] \times Y[1] \xleftarrow{\ \simeq\ } Y[2] \xrightarrow{\ \xi\ } Y[1]$$

that turns $\pi_0 Y[1]$ into an abelian monoid. The extra condition of being very special demands that this abelian monoid actually be an abelian group.

Segal then shows the following:

Theorem 14.3.13 [Se74, Proposition 1.4] *If Y is a very special Γ-space, then $Y[1]$ is an infinite loop space.*

The concept of a Γ-space looks innocent, but we saw that it is strong enough for infinite loop spaces. Bousfield and Friedlander showed [**BF78**, Theorem 5.8] that there is a model structure on Γ-spaces, which gives rise to a Quillen equivalence with the model category of connective spectra, that is, spectra with homotopy groups concentrated in non-negative degrees.

As a diagram category, Γ-spaces possess a Day convolution product, as in Section 9.8.

Definition 14.3.14 Let Y_1 and Y_2 be two Γ-spaces. The *external smash product of Y_1 and Y_2* is the functor

$$Y_1 \tilde{\wedge} Y_2 \colon \Gamma \times \Gamma \to s\text{Sets}, \quad ([n], [m]) \mapsto Y_1[n] \wedge Y_2[m].$$

The *smash product of Y_1 and Y_2, $Y_1 \wedge Y_2$*, is the Day convolution product with respect to the external smash product and the symmetric monoidal structure of the smash product on Γ, that is, $(Y_1 \wedge Y_2)[n] = \text{colim}_{f \in \Gamma([m] \wedge [\ell], [n])} Y_1[m] \wedge Y_2[\ell]$:

$$
\begin{array}{ccc}
\Gamma \times \Gamma & \xrightarrow{Y_1 \tilde{\wedge} Y_2} & s\text{Sets.} \\
\wedge \downarrow & \nearrow Y_1 \wedge Y_2 & \\
\Gamma & &
\end{array}
$$

Remark 14.3.15 Note that $[m] \wedge [\ell]$ can be identified with $[m\ell]$. Manos Lydakis realized that the smash product turns the category of Γ-spaces into a symmetric monoidal category, and he used that to develop a model of connective spectra with a strict symmetric monoidal smash product [**Ly99**]. Schwede [**Schw99**] then investigated the corresponding world of spectra with a strict multiplication, ring spectra, via Γ-spaces. Many spectra that play a role in algebraic K-theory have nice models in Γ-spaces, and we will say a bit about some of those in the following section. If you want to learn more about this (and many more related topics), then [**DGM13**] is an excellent source.

A word of warning: Tyler Lawson showed [**La09**] that not all connective E_∞-ring spectra can be modelled by commutative Γ-rings.

14.4 Segal K-Theory of a Permutative Category

Although not all connective E_∞-spectra arise as Γ-spaces, there are particularly nice examples. Segal describes in [**Se74**, §2] a way of associating a spectrum with a symmetric monoidal category. For simplicity, we deal with the permutative case here, as in [**May78**, Construction 10] and follow the description in [**EM06**].

Definition 14.4.1 Let $(\mathcal{C}, \oplus, 0, \tau)$ be a small permutative category and let X be a finite pointed set with basepoint $*$. Let $\mathcal{C}(X)$ be the category whose objects are families $(C_S, \varrho_{S,T}; S, T \subset X \setminus \{*\}, S \cap T = \varnothing)$, where the C_Ss are objects of $\mathcal{C}$ and the $\varrho_{S,T}$s are isomorphisms $\varrho_{S,T} : C_S \oplus C_T \to C_{S \cup T}$ in $\mathcal{C}$. These objects satisfy the following conditions:

- $C_S = 0$ if $S = \varnothing$.
- $\varrho_{S,T}$ is the identity morphism if S or T is the empty subset.
- The $\varrho_{S,T}$s are commutative and associative in the sense that the following diagrams commute:

$$
\begin{array}{ccc}
C_S \oplus C_T & \xrightarrow{\varrho_{S,T}} & C_{S \cup T} \\
\downarrow{\tau_{S,T}} & & \| \\
C_T \oplus C_S & \xrightarrow{\varrho_{T,S}} & C_{T \cup S}
\end{array}
\quad \text{and} \quad
\begin{array}{ccc}
C_S \oplus C_T \oplus C_U & \xrightarrow{\varrho_{S,T} \oplus 1_{C_U}} & C_{S \cup T} \oplus C_U \\
\downarrow{1_{C_S} \oplus \varrho_{T,U}} & & \downarrow{\varrho_{S \cup T, U}} \\
C_S \oplus C_{T \cup U} & \xrightarrow{\varrho_{S, T \cup U}} & C_{S \cup T \cup U}.
\end{array}
$$

Morphisms in $\mathcal{C}(X)$ from $(C_S, \varrho_{S,T})$ to $(\tilde{C}_S, \tilde{\varrho}_{S,T})$ are families of morphisms $f_S \in \mathcal{C}(C_S, \tilde{C}_S)$ for $S \subset X \setminus \{*\}$, such that $f_\varnothing = 1_0$ and the f_Ss are compatible with the ϱs and $\tilde{\varrho}$s. The diagram

$$
\begin{array}{ccc}
C_S \oplus C_T & \xrightarrow{f_S \oplus f_T} & \tilde{C}_S \oplus \tilde{C}_T \\
\downarrow{\varrho_{S,T}} & & \downarrow{\tilde{\varrho}_{S,T}} \\
C_{S \cup T} & \xrightarrow{f_{S \cup T}} & \tilde{C}_{S \cup T}
\end{array}
$$

commutes for all $S, T \subset X \setminus \{*\}$ with $S \cap T = \varnothing$.

Example 14.4.2 If $X = [2]$, then an object in $\mathcal{C}([2])$ is a square of objects of $\mathcal{C}$:

$$
\begin{array}{cc}
C_\varnothing = 0 & C_{\{1\}} \\
C_{\{2\}} & C_{\{1,2\}},
\end{array}
$$

together with an isomorphism $\varrho_{\{1\},\{2\}} : C_{\{1\}} \oplus C_{\{2\}} \cong C_{\{1,2\}}$, so up to isomorphism, $C_{\{1,2\}}$ is determined by $C_{\{1\}} \oplus C_{\{2\}}$, but you remember an isomorphism as part of the data.

Similarly, for $X = [3]$, we get a cube of objects

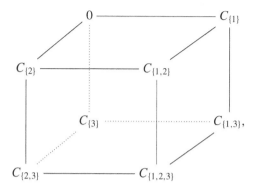

and you remember all possible ways of forming $C_{\{1,2,3\}}$ out of $C_{\{1\}}$, $C_{\{2\}}$, and $C_{\{3\}}$ by merging sums. The lines in the picture are not meant to be maps; they are just there to indicate the cube.

This construction gives a functor $\mathcal{C}(-) \colon \Gamma \to \mathrm{cat}$ as follows: If $g \colon X \to Y$ is a morphism of finite pointed sets, then an object $(C_S, \varrho_{S,T})$ is sent to $(C_{S'}^g, \varrho_{S',T'}^g)$, where $C_{S'}^g := C_{g^{-1}(S')}$ and $\varrho_{S',T'}^g := \varrho_{g^{-1}(S), g^{-1}(T)}$. As g preserves the basepoint, this is well-defined.

Definition 14.4.3 Let $\mathcal{C}$ be a small permutative category. Then, the Γ-space

$$N\mathcal{C}(-) \colon \Gamma \to s\mathrm{Sets}$$

is the *Segal K-theory of the permutative category* $\mathcal{C}$. We denote it by $K_S(\mathcal{C})$.

We can also form the corresponding spectrum $(N\mathcal{C}(S^n))_{n \geq 0}$. One can identify the homotopy type of the spectrum in many cases (see, for instance, [**May77**, VI.5]).

Examples 14.4.4

- The process of remembering splittings of projective modules into direct sums should sound familiar from Example 13.3.6, and it should not be a surprise that the Segal K-theory of the category of finitely generated projective R-modules for some ring R, $\mathcal{P}(R)$, gives the K-theory spectrum of that ring, $K(R)$.
- The Segal K-theory of the category of finite sets and bijections, Σ, gives rise to the sphere spectrum.
- The category of complex vector spaces, $\mathcal{V}_{\mathbb{C}}$, gives rise to a spectrum via $K_S(\mathcal{V}_{\mathbb{C}})$, which is a connective version of complex topological K-theory, ku. Its real analog, $\mathcal{V}_{\mathbb{R}}$, gives connective real topological K-theory, ko.

- If A is an abelian group, then the discrete category associated with A, with $a \in A$ as objects, is permutative (using the abelian group structure of A), and the Segal K-theory gives the Eilenberg–Mac Lane spectrum of A, HA.

14.5 Injections and Infinite Loop Spaces

Establishing E_∞-structures on spaces is usually not easy. Schlichtkrull gives in [**Sc09**, Proposition 6.5] an explicit action of the Barratt–Eccles operad on a model of the homotopy colimit of a commutative $\mathcal{I}$-space. In the simplicial setting, the action is as follows:

Proposition 14.5.1 *Let* $X \colon \mathcal{I} \to s\mathrm{Sets}$ *be a commutative $\mathcal{I}$-space monoid with multiplication* $\mu \colon X \square X \to X$. *Then,* $\mathrm{hocolim}_{\mathcal{I}} X$ *is a simplicial set with an action of the Barratt–Eccles operad* $(N\mathcal{E}_{\Sigma_m})_{m \geq 0}$, *and hence, its geometric realization is an E_∞-space.*

Proof   We define a functor $F_m \colon \mathcal{E}_{\Sigma_m} \times (X \backslash \mathcal{I})^m \to X \backslash \mathcal{I}$ for every m by

$$F_m(\sigma, (\mathbf{n}_1, x_1), \cdots, (\mathbf{n}_m, x_m))$$
$$= (\mathbf{n}_{\sigma^{-1}(1)} \sqcup \cdots \sqcup \mathbf{n}_{\sigma^{-1}(m)}, \sigma(n_1, \ldots, n_m)(\mu((\mathbf{n}_1, x_1), \ldots, (\mathbf{n}_m, x_m)))).$$

Here, the permutation $\sigma(n_1, \ldots, n_m) \in \Sigma_{n_1 + \cdots + n_m}$ permutes the blocks $\mathbf{n}_1$ up to $\mathbf{n}_m$, as $\sigma \in \Sigma_m$ permutes the numbers 1 to m.

On morphisms, we define F_m as follows: For $\tau \in \mathcal{E}_{\Sigma_m}(\sigma, \tau \circ \sigma)$, we set

$$F_m(\tau, 1_{(\mathbf{n}_1, x_1)}, \ldots, 1_{(\mathbf{n}_m, x_m)}) = \tau(n_{\sigma^{-1}(1)}, \ldots, n_{\sigma^{-1}(m)}).$$

For $f_i \colon (\mathbf{n}_i, x_i) \to (\mathbf{p}_i, y_i)$ morphisms in $X \backslash \mathcal{I}$ for $1 \leq i \leq m$, that is, $f_i \in \mathcal{I}(\mathbf{n}_i, \mathbf{p}_i)$, such that $X(f_i)(x_i) = y_i$, we define

$$F_m(1_{\mathbf{m}}, f_1, \ldots, f_m) = f_{\sigma^{-1}(1)} \sqcup \cdots \sqcup f_{\sigma^{-1}(m)}. \qquad \square$$

Note that in the case of the terminal $\mathcal{I}$-space, this reduces to May's result about actions of the Barratt–Eccles operad on the classifying space of any permutative category, here in the case $B\mathcal{I}$ (Theorem 13.1.5).

Sagave and Schlichtkrull discuss important examples of such structures in [**SaSc12**, §1].

Example 14.5.2   The families of groups $O(n)$, $U(n)$, and $GL_n(R)$ for R some discrete commutative ring give rise to commutative $\mathcal{I}$-spaces $\mathbf{n} \mapsto BO(n)$, $\mathbf{n} \mapsto BU(n)$, and $\mathbf{n} \mapsto BGL_n(R)$ and the corresponding Bousfield–Kan homotopy colimits are BO, BU, and $BGL(R)^+$.

Example 14.5.3 Any pointed simplicial set X and any well-based topological space give rise to the commutative $\mathcal{I}$-space $\mathsf{Sym}(X)$ with $\mathsf{Sym}(X)(\mathbf{n}) = X^n$. Schlichtkrull showed in [**Sc08**] that $\mathrm{hocolim}_{\mathcal{I}}\mathsf{Sym}(X)$ is equivalent to the Barratt–Eccles construction [**BE74**]. For connected X, this is a model for $Q(X) = \Omega^\infty \Sigma^\infty X$.

14.6 Braided Injections and Double Loop Spaces

One can use a different diagram category in order to identify double loop spaces. It is stated in Stasheff [**Sta92**, pp. 122–123] that one can deloop braided monoidal categories twice. By a result of Fiedorowicz [**Fi∞**, Theorem 2], we know that the classifying space of a braided monoidal category is a double loop space after group completion. Berger [**Be99**] proves this fact by using simplicial operads.

The idea of [**ScSo16**] is to desymmetrize the category of finite sets and injections by allowing braid groups as automorphisms and not symmetric groups. This can be viewed as a *crossed semisimplicial group* in analogy to the crossed simplicial groups of Fiedorowicz and Loday [**FiLo91**]. Using the corresponding diagram category of braided injections, they produce an explicit twofold delooping.

Definition 14.6.1

- Let $\mathcal{MI}$ denote the *category of order-preserving injections*. The objects of $\mathcal{MI}$ are the sets $\mathbf{n} = \{1, \ldots, n\}$ for $n \geq 0$ and the morphisms are order-preserving functions of finite sets.
- As in Example 8.6.6, we denote by Br_n the braid group on n strands.
- Let $\mathfrak{B}$ be the *category of braided injections*. The objects of $\mathfrak{B}$ are the same as the objects of $\mathcal{MI}$, but the morphisms $f \in \mathfrak{B}(\mathbf{n}, \mathbf{m})$ are composites of morphisms $f = i \circ \sigma$, where $\sigma \in \mathrm{Br}_n$ and $i \in \mathcal{MI}(\mathbf{n}, \mathbf{m})$. Composition in $\mathfrak{B}$ is defined by concatenation and combing. If $g = j \circ \tau$ with $j \in \mathcal{MI}(\mathbf{m}, \ell)$, then $g \circ f = j \circ \tau \circ i \circ \sigma$ can be uniquely written as $(j \circ \tau_*(i)) \circ (i^*(\tau) \circ \sigma)$ with $j \circ \tau_*(i) \in \mathcal{MI}(\mathbf{n}, \ell)$ and $i^*(\tau) \circ \sigma \in \mathrm{Br}_n$, as in [**FiLo91**, Theorem 3.8].

Remark 14.6.2 Note that the decomposition of morphisms given earlier is analogous to the one in the category $\mathcal{I}$. Every $f \in \mathcal{I}(\mathbf{n}, \mathbf{m})$ can be uniquely written as a composite $f = i \circ \sigma$, where $i \in \mathcal{MI}(\mathbf{n}, \mathbf{m})$ and $\sigma \in \Sigma_n$. Thus, in the category $\mathfrak{B}$, you just replace the family of symmetric groups by the family of braid groups.

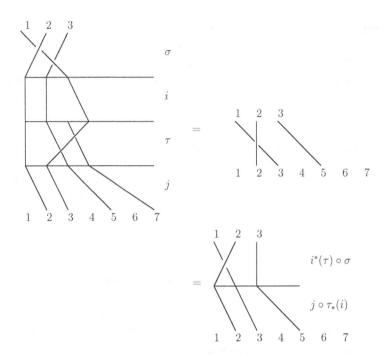

The category $\mathfrak{B}$ is a braided monoidal category [**ScSo16**, §2], with monoidal product given by disjoint ordered union of sets:

$$(\mathbf{n}, \mathbf{m}) \mapsto \mathbf{n} \sqcup \mathbf{m},$$

where we declare the elements of $\mathbf{n}$ to be smaller than the elements of $\mathbf{m}$, and thus, $\mathbf{n} \sqcup \mathbf{m}$ identifies with $\mathbf{n} + \mathbf{m}$. There are two choices for the braiding: using the braid $\tilde{\chi}(n, m) \in \mathrm{Br}_{n+m}$ that pulls the first n strands in front and the last m strands to the back and then moves them around each other, or we can take its inverse. We use $\tilde{\chi}(n, m)$.

Proposition 14.6.3 ([**ScSo16**, Proposition 3.12]) *The Day convolution product from Definition 9.8.1 turns the category of functors from $\mathfrak{B}$ to the category of simplicial sets into a braided monoidal category.*

Following Schlichtkrull and Solberg, we call such functors $\mathfrak{B}$-*spaces.* Explicitly, for two $\mathfrak{B}$-spaces X and Y, the value of the Day convolution product $X \boxtimes Y$ on an object $\mathbf{n}$ is given by

$$(X \boxtimes Y)(\mathbf{n}) = \mathrm{colim}_{\ell \sqcup \mathbf{m} \to \mathbf{n}} X(\ell) \times Y(\mathbf{m}).$$

The morphisms $\tilde{\chi}(n, m)$ define a braiding, that is, an isomorphism $\beta : X \boxtimes Y \to Y \boxtimes X$. The unit for the Day convolution product is the terminal $\mathfrak{B}$-space $\mathfrak{B}(0, -)$.

Definition 14.6.4 A $\mathfrak{B}$-space A is a *commutative $\mathfrak{B}$-space monoid* if it is a monoid with multiplication μ and unit and if the diagram

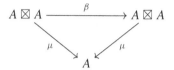

commutes.

Given a monoidal structure, we can form a two-sided bar construction of any monoid, as in Definition 10.4.2. As the unit $\mathfrak{B}(0, -)$ is the terminal object in the category of $\mathfrak{B}$-spaces, it is also a bimodule over any monoid. In particular, we can construct the reduced bar construction $B_\bullet^\boxtimes(\mathfrak{B}(0, -), A, \mathfrak{B}(0, -))$ of any commutative $\mathfrak{B}$-space monoid A. Note that this is a bisimplicial object.

Definition 14.6.5 The *reduced bar construction, $B^\boxtimes(A)$, of a $\mathfrak{B}$-space monoid A* is the diagonal simplicial $\mathfrak{B}$-space associated with $B_\bullet^\boxtimes(\mathfrak{B}(0, -), A, \mathfrak{B}(0, -))$.

Schlichtkrull and Solberg show [**ScSo16**, Lemma 6.1] that the reduced bar construction of a commutative $\mathfrak{B}$-space monoid is a $\mathfrak{B}$-space monoid, and hence, one can apply the reduced bar construction once more. This yields the following result:

Proposition 14.6.6 [**ScSo16**, Proposition 6.4] *If A is a commutative $\mathfrak{B}$-space monoid (with underlying flat $\mathfrak{B}$-space), then $B^\boxtimes(B^\boxtimes(A))_{h\mathfrak{B}}$ is a double delooping of the group completion of the Bousfield–Kan homotopy colimit $\mathrm{hocolim}_\mathfrak{B} A$.* □

Here, the flatness condition ensures correct homotopical behavior.

14.7 Iterated Monoidal Categories as Models for Iterated Loop Spaces

Our understanding of single loop spaces and infinite loop spaces is rather good. For double loop spaces, we saw that we have a fair understanding in terms of braided monoidal categories or braided operads. It is harder to find explicit models for iterated loop spaces of the form $\Omega^n X$, with $2 < n < \infty$. An early approach can be found in Cobb's paper [**Co74**]. He uses a diagram category P_n and relates connected special P_n-spaces to n-fold loop spaces [**Co74**, Corollary 3.1] via an iterated delooping.

We discuss two different approaches: one via iterated monoidal categories due to [**BFSV03**] and a different one using the category Θ_n in Section 14.8.

If you have a double loop space $\Omega^2 X = k\underline{\mathrm{Top}}_*(\mathbb{S}^2, X)$ on a based space X, then you can reexpress $\Omega^2 X$ as $\Omega(\Omega X)$, and in fact, one can use this reformulation and the Eckmann–Hilton argument from (8.1.1) to get that $\pi_2(X)$ is abelian, by considering the two H-space structures coming from the two loop coordinates.

As monoidal categories model loop spaces, the rough idea is that double loop spaces should be modelled by categories with two monoidal structures that are compatible. However, the compatibility must not be too good, because otherwise, you could apply the Eckmann–Hilton trick and you would get a commutative structure. This idea is worked out in [**BFSV03**].

Definition 14.7.1 Let n be bigger than 1. An *n-fold monoidal category* is a category $\mathcal{C}$ with the following data:

- There are n strict monoidal structures

$$\square_1, \ldots, \square_n : \mathcal{C} \times \mathcal{C} \to \mathcal{C}$$

on $\mathcal{C}$ that share a unit object; that is, there is an object e in $\mathcal{C}$ that is the common strict unit for all of these strict monoidal structures.
- For all $1 \le i < j \le n$, there is a natural transformation φ^{ij}, whose components are

$$\varphi^{ij}_{A,B,C,D} : (A\square_j B)\square_i(C\square_j D) \to (A\square_i C)\square_j(B\square_i D).$$

These natural transformations $(\varphi^{ij})_{1 \le i < j \le n}$ satisfy the following conditions:

(1) (unit conditions) For all objects A, B of $\mathcal{C}$:

$$\varphi^{ij}_{A,B,e,e} = 1_{A\square_j B} = \varphi^{ij}_{e,e,A,B} \text{ and } \varphi^{ij}_{A,e,B,e} = 1_{A\square_i B} = \varphi^{ij}_{e,A,e,B}.$$

(2) (associativity conditions) The diagrams

$$(A\square_jB)\square_i(C\square_jD)\square_i(E\square_jF) \xrightarrow{\varphi^{ij}_{A,B,C,D}\square_i 1_{E\square_jF}} ((A\square_iC)\square_j(B\square_iD))\square_i(E\square_jF)$$

$$\left\downarrow {\scriptstyle 1_{A\square_jB}\square_i\varphi^{ij}_{C,D,E,F}} \qquad\qquad\qquad\qquad \right\downarrow {\scriptstyle \varphi^{ij}_{A\square_iC,B\square_iD,E,F}}$$

$$(A\square_jB)\square_i((C\square_iE)\square_j(D\square_iF)) \xrightarrow{\varphi^{ij}_{A,B,C\square_iE,D\square_iF}} (A\square_iC\square_iE)\square_j(B\square_iD\square_iF)$$

and

$$(A\square_jB\square_jC)\square_i(D\square_jE\square_jF) \xrightarrow{\varphi^{ij}_{A\square_jB,C,D\square_jE,D}} ((A\square_jB)\square_i(D\square_jE))\square_j(C\square_iF)$$

$$\left\downarrow {\scriptstyle \varphi^{ij}_{A,B\square_jC,D,E\square_jF}} \qquad\qquad\qquad\qquad \right\downarrow {\scriptstyle \varphi^{ij}_{A,B,D,E}\square_j 1_{C\square_iF}}$$

$$(A\square_iD)\square_j((B\square_jC)\square_i(E\square_jF)) \xrightarrow{1_{A\square_iD}\square_j\varphi^{ij}_{B,C,E,F}} (A\square_iD)\square_j(B\square_iE)\square_j(C\square_iF)$$

commute for all objects A, B, C, D, E, F of $\mathcal{C}$ and for all $i < j$.

(3) For all $1 \le i < j < k \le n$, the following hexagon diagram commutes. (We omit the objects at the φs to ease notation.)

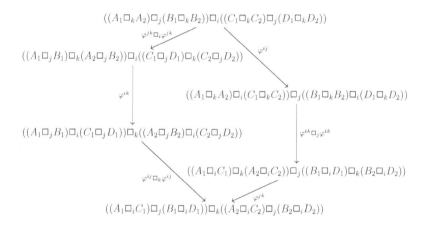

Note that the φ^{ij}s are *not* required to be isomorphisms. In the associativity diagrams, the second diagram looks like a backward version of the first, but note the asymmetric condition that $i < j$. The flow of the associativity diagrams and the hexagon diagram is upward from a $\square_i$-product as the central product to a $\square_j$- or $\square_k$-product for $i < j$ or $i < j < k$.

Theorem 14.7.2 [BFSV03, Theorem 2.2] *The group completion of the classifying space of an n-fold monoidal category is an n-fold based loop space.*

14.8 The Category Θ_n

André Joyal defined in [**Jo-a∞**] a category Θ_n for every $n \geq 1$, in order to define weak versions of n-categories. Michael Batanin [**Ba08**] and Clemens Berger [**Be07**] used the category Θ_n for understanding n-fold based loop spaces. The categories Θ_n also play an important role in higher category theory for the understanding of (∞, n)-categories [**BRez13b, BRez13a, BRez∞**].

Definition 14.8.1 Let $\mathcal{C}$ be an arbitrary small category.

- The category $\Theta(\mathcal{C})$ has objects $([m]; C_1, \ldots, C_m)$, where $[m]$ is an object of Δ and the C_is are objects of $\mathcal{C}$. For $m = 0$, there is an object $([0])$ of $\Theta(\mathcal{C})$, which we also denote by $([0]; -)$. A morphism in $\Theta(\mathcal{C})$ from $([m]; C_1, \ldots, C_m)$ to $([n]; D_1, \ldots, D_n)$ consists of an $f \in \Delta([m], [n])$ and morphisms $g_{ij} \in \mathcal{C}(C_i, D_j)$ if $f(i-1) < j \leq f(i)$. For $m = 0$, a morphism from $([0], -)$ to $([n]; D_1, \ldots, D_n)$ is just a morphism $f \in \Delta([0], [n])$.
- The same category $\Theta(\mathcal{C})$ is also denoted by $\Delta \wr \mathcal{C}$ and is called the *wreath construction of the category Δ with the category $\mathcal{C}$*.

Remark 14.8.2 Wreath constructions can be with other categories than Δ, and we will see another example later in Definition 14.8.12. Precursors of this idea can be found as early as in [**Co74**].

A nice way of visualizing the objects $([m]; C_1, \ldots, C_m)$ of the category $\Theta \wr \mathcal{C}$ is to write the object C_i of $\mathcal{C}$ between the elements $i - 1$ and i of $[m]$. A morphism in $\Theta \wr \mathcal{C}$ from $([m]; C_1, \ldots, C_m)$ to $([n]; D_1, \ldots, D_n)$ consists of a morphism $f \in \Delta([m], [n])$ and morphisms in $\mathcal{C}$ from C_i to D_j whenever the object C_i can see the object D_j, that is, when its view is not blocked by the arrows of f.

Example 14.8.3 Consider the map $f : [4] \to [4]$ in Δ, which is given by $f(0) = 1$, $f(1) = 2$, and $f(2) = f(3) = f(4) = 4$. A morphism from $([4]; C_1, C_2, C_3, C_4)$ to $([4]; D_1, D_2, D_3, D_4)$ with f as the first coordinate, is then of the form

$$(f, g_{12} : C_1 \to D_2, g_{23} : C_2 \to D_3, g_{24} : C_2 \to D_4).$$

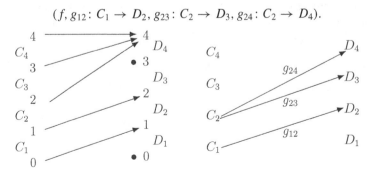

Exercise 14.8.4 What are possible morphisms in $\Theta(\mathcal{C})$ whose first component is a δ_i or a σ_j?

There is a canonical functor $\Delta \times \mathcal{C} \to \Theta(\mathcal{C})$.

Definition 14.8.5 We define a functor $\delta \colon \Delta \times \mathcal{C} \to \Theta(\mathcal{C})$ on objects by setting $\delta([m], C) := ([m]; \underbrace{C, \ldots, C}_{m})$ for $m > 0$ and $\delta([0], C) = ([0]; -)$. A pair of morphisms (f, g) with $f \in \Delta([m], [n])$ and $g \in \mathcal{C}(C, C')$ is sent to the morphism in $\Theta(\mathcal{C})$ that has f as a first component and $g_{ij} = g$ whenever g_{ij} is defined.

Recall that we denote the terminal category by $[0]$.

Definition 14.8.6 The categories Θ_n are defined in an iterative way. We set $\Theta_0 = [0]$ and $\Theta_{n+1} := \Theta(\Theta_n)$.

Note that Θ_1 is isomorphic to Δ; thus, if you prefer the wreath notation,

$$\Theta_n = \underbrace{\Delta \wr \cdots \wr \Delta}_{n}.$$

Objects in Θ_2 are of the form $([n]; [k_1], \ldots, [k_n])$ with objects $[n]$, $[k_1], \ldots, [k_n]$ of Δ. For Θ_3, we get an additional copy of Δ on the left, and for $[n]$ in Δ, we need n objects $([m_1]; [k_1^1], \ldots, [k_{m_1}^1]), \ldots,$ $([m_n]; [k_1^n], \ldots, [k_{m_n}^n])$ to form an object

$$([n]; ([m_1]; [k_1^1], \ldots, [k_{m_1}^1]), \ldots, ([m_n]; [k_1^n], \ldots, [k_{m_n}^n]))$$

in Θ_3. For sake of clarity, we reorder these as

$$([n]; [m_1], \ldots, [m_n]; [k_1^1], \ldots, [k_{m_1}^1], \ldots, [k_1^n], \ldots, [k_{m_n}^n])$$

and call $[n]$ the left-most object and $([k_j^n])_{j=1}^{m_n}$ the right-most sequence of objects of the object.

The iterative nature of the definition of Θ_n allows us to connect the categories Θ_n for different n via suspension and inclusion functors (see [**Rez10**, 11.3]).

Definition 14.8.7

- The *suspension functor* $\sigma_n \colon \Theta_n \to \Theta_{n+1}$ sends an object x of Θ_n to the object $([1]; x)$ of Θ_n. A morphism $f \in \Theta_n(x, y)$ is sent to the morphism

$$\sigma_n(f) = (1_{[1]}; f) \in \Theta_{n+1}(\sigma_n(x), \sigma_n(y)).$$

- The *inclusion functor* $i_n \colon \Theta_n \to \Theta_{n+1}$ sends an object $x \in \Theta_n$ to the object $(x; [0], \ldots, [0])$ of Θ_{n+1}, where we add as many instances of $[0]$ as prescribed by the right-most sequence of objects in x. On morphisms, i_n adds the identity on $[0]$ everywhere where it is needed.

Remark 14.8.8 Note that the functor $\delta \colon \Delta \times \Delta \to \Theta(\Delta) = \Delta \wr \Delta$ from Definition 14.8.5 can be iterated to a functor

$$\delta_n \colon \underbrace{\Delta \times \cdots \times \Delta}_{n} \to \Theta_n. \tag{14.8.1}$$

Exercise 14.8.9 Convince yourself that for $n > 1$, the functor δ_n is neither full nor faithful, and, of course, it is not essentially surjective.

14.8.1 Trees with n-Levels, Finite Combinatorial n-Disks and Θ_n

As an alternative, you can think about objects of Θ_n as planar trees with up to n levels. For $\Theta_1 = \Delta$, the object $[m]$ for $m \geq 0$ corresponds to the planar corolla tree with m leaves:

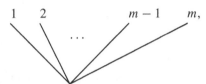

where the tree is empty for $m = 0$. As $\Theta_2 = \Delta \wr \Delta$, an object $([m]; [n_1], \ldots, [n_m])$ of Θ_2 corresponds to a planar tree with two levels, where level 1 is the tree above and the second level consists of corolla trees with n_i leaves, such that the n_i-corolla tree in level 2 is glued to the ith leave in the first level. For example, the object $([2]; [3], [2])$ corresponds to

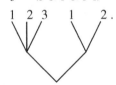

Exercise 14.8.10 Describe morphisms in Θ_n in terms of morphisms between planar n-level trees.

We formalize the description given earlier in order to obtain a model for the opposite category of Θ_n, Θ_n^o, and we use the category of finite combinatorial n-disks in the sense of [**Jo-a∞**] as an intermediate category for the comparison.

Recall that $\mathbf{n}$ denotes the set $\{1, \ldots, n\}$ with the convention that $\mathbf{0} = \varnothing$.

Definition 14.8.11 [**Jo-a∞**, p. 3] [**Ba08**, Def. 4.3] The category Ω_n has as objects chains of order-preserving map

$$T = (\mathbf{k_n} \xrightarrow{f_n} \mathbf{k_{n-1}} \xrightarrow{f_{n-1}} \cdots \xrightarrow{f_2} \mathbf{k_1} \xrightarrow{f_1} \mathbf{1}).$$

A morphism from T to

$$S = (\ell_n \xrightarrow{g_n} \ell_{n-1} \xrightarrow{g_{n-1}} \cdots \xrightarrow{g_2} \ell_1 \xrightarrow{g_1} \mathbf{1})$$

consists of functions $(\alpha_i : \mathbf{k_i} \to \ell_i, 1 \le i \le n)$, such that the diagram

$$
\begin{array}{ccccccccc}
\mathbf{k_n} & \xrightarrow{f_n} & \mathbf{k_{n-1}} & \xrightarrow{f_{n-1}} & \cdots & \xrightarrow{f_2} & \mathbf{k_1} & \xrightarrow{f_1} & \mathbf{1} \\
\downarrow{\scriptstyle \alpha_n} & & \downarrow{\scriptstyle \alpha_{n-1}} & & & & \downarrow{\scriptstyle \alpha_1} & & \| \\
\ell_n & \xrightarrow{g_n} & \ell_{n-1} & \xrightarrow{g_{n-1}} & \cdots & \xrightarrow{g_2} & \ell_1 & \xrightarrow{g_1} & \mathbf{1}
\end{array}
$$

commutes and such that the restriction of each α_i to every fiber $f_{i-1}^{-1}(j)$ is order-preserving.

As for Θ_n, the objects of Ω_n correspond to planar trees with less or equal to n levels. For $n = 1$, we just have objects $\mathbf{k_1} \to \mathbf{1}$ of Ω_1, which correspond to corollas with k_1 leaves.

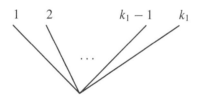

If $k_1 = 0$, then we get the empty corolla.

In Ω_2, there are corollas stacked on every $i \in \mathbf{k_1}$, but as we do not only consider surjective order-preserving functions, we might also get empty preimages. A typical example of an object in Ω_2 is

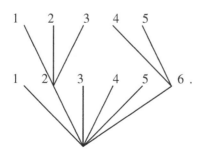

We will establish Ω_n as a model for Θ_n^o up to homotopy equivalence (see [**Jo-a∞**] and [**Be07**, §3.6]). We obtain an immediate model of Θ_n^o via the category of intervals, **I**. To this end, we define the wreath product of the interval category of Definition 10.3.1 with a small category.

Definition 14.8.12 Let $\mathcal{C}$ be a small category.

- The objects of the category $\mathbf{I} \wr \mathcal{C}$ are tuples $([n + 1]; C_1, \ldots, C_n)$, where $[n + 1]$ is an object of **I** and the C_is are objects of $\mathcal{C}$.
- A morphism in $\mathbf{I} \wr \mathcal{C}$ from $([n + 1]; C_1, \ldots, C_n)$ to $([m + 1]; D_1, \ldots, D_m)$ consists of a $g \in \mathbf{I}([n + 1], [m + 1])$ and morphisms $f_{ij} \colon C_i \to D_j$ in $\mathcal{C}$ for all i and j with $f(j) = i$.

In contrast to $\Delta \wr \mathcal{C}$, one can view the objects C_i as being glued to $i \in [n+1]$, and they move with f.

Example 14.8.13 Observe that $(\Delta \wr \Delta)^o = \Theta_2^o$ is isomorphic to $\mathbf{I} \wr \mathbf{I}$: We know that $\mathbf{I} \cong \Delta^o$; hence, the crucial thing is to observe that the morphisms really move in the opposite direction.

Consider the morphism

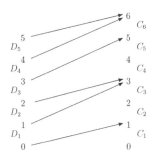

in $\Delta \wr \Delta$ with $g_{12} \colon D_1 \to C_2$, $g_{13} \colon D_1 \to C_3$, $g_{34} \colon D_3 \to C_4$, $g_{35} \colon D_3 \to C_5$, and $g_{46} \colon D_4 \to C_6$.

The corresponding morphism $g = D(f) \colon [7] \to [6]$ of proper intervals is

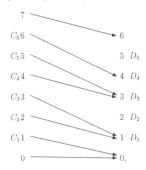

and we obtain $f_{12}\colon C_2 \to D_1$, $f_{13}\colon C_3 \to D_1$, $f_{34}\colon C_4 \to D_3$, $f_{35}\colon C_5 \to D_3$, and $f_{46}\colon C_6 \to D_4$ in $\mathbf{I} \wr \mathbf{I}$.

An iteration of the preceding argument gives that the n-fold iterated wreath product of $\mathbf{I}$ with itself is isomorphic to Θ_n^o. For brevity, we set

$$\Psi_1 := \mathbf{I}, \text{ and } \Psi_n := \mathbf{I} \wr \Psi_{n-1} \text{ for } n \geq 2.$$

Proposition 14.8.14 *For all $n \geq 1$,*

$$\Psi_n \cong \Theta_n^o.$$ $\square$

There is a beautiful description in [**Jo-a∞**] of the category that you can build by using balls of different dimensions and their canonical inclusion and projection maps. For simplicity, consider the standard balls in Euclidean space

$$B_n := \{x \in \mathbb{R}^n, \|x\| \leq 1\},$$

with the convention that B_0 is a point. We can embed B_0 into the boundary of B_1 in two different ways, as the starting point or as the endpoint of the interval B_1. In higher dimensions, you express ∂B_n as the union of the two hemispheres and get two canonical inclusions $B_{n-1} \to B_n$. We also have projection maps $q_n\colon B_n \to B_{n-1}$ collapsing the last dimension. The corresponding small category is as follows: We use a variant of the definition given in [**Jo-a∞**] and [**Be07**, Proof of Theorem 3.10], where we require proper intervals as fibers.

Definition 14.8.15 The *category* $\mathcal{D}_n$ *of finite combinatorial n-disks* has as objects sequences of finite sets and functions

$$D_n \underset{\longleftarrow}{\overset{\longleftarrow}{\rightrightarrows}} D_{n-1} \underset{\longleftarrow}{\overset{\longleftarrow}{\rightrightarrows}} \ldots \underset{\longleftarrow}{\overset{\longleftarrow}{\rightrightarrows}} D_1 \underset{\longleftarrow}{\overset{\longleftarrow}{\rightrightarrows}} D_0 = \{*\}, \qquad (14.8.2)$$

with functions $q_i\colon D_i \to D_{i-1}$ for $1 \leq i \leq n$ and $s_i, t_i\colon D_{i-1} \to D_i$ for $1 \leq i \leq n$. These have to satisfy the following conditions:

- For each $x \in D_{i-1}$, the fiber $q_i^{-1}(x)$ is a proper interval with minimal element $s_i(x)$ and maximal element $t_i(x)$.
- For all i,

$$\begin{aligned} q_i s_i &= q_i t_i = 1_{D_{i-1}}, \\ t_{i+1} t_i &= s_{i+1} t_i, \text{ and} \\ t_{i+1} s_i &= s_{i+1} s_i. \end{aligned}$$

- For $i \geq 2$, the equalizer of t_i and s_i is $s_{i-1}(D_{i-2}) \cup t_{i-1}(D_{i-2})$.

A morphism in $\mathcal{D}_n$ is a commutative diagram

$$
\begin{array}{ccccccccc}
D_n & \rightleftarrows & D_{n-1} & \rightleftarrows & \cdots & \rightleftarrows & D_1 & \rightleftarrows & \{*\} \\
\downarrow{\scriptstyle f_n} & & \downarrow{\scriptstyle f_{n-1}} & & & & \downarrow{\scriptstyle f_1} & & \| \\
D'_n & \rightleftarrows & D'_{n-1} & \rightleftarrows & \cdots & \rightleftarrows & D'_1 & \rightleftarrows & \{*\},
\end{array}
$$

such that the f_i respect the order of the fibers and the minimal and maximal elements.

Theorem 14.8.16 *For all $n \geq 1$, the n-fold wreath product of $\mathbf{I}$, Ψ_n, is equivalent to $\mathcal{D}_n$.*

Proof We have $\mathbf{I} = \Psi_1 \cong \mathcal{D}_1$. By induction, we assume that the claim is shown for all $k \leq n - 1$, and we show that $\mathbf{I} \wr \mathcal{D}_{n-1} \cong \mathcal{D}_n$. This is a proof by cutting and stacking.

We define a functor $F : \mathbf{I} \wr \mathcal{D}_{n-1} \to \mathcal{D}_n$ by taking an object

$$
\left([m+1], D_{n-1}^1 \rightleftarrows \cdots \rightleftarrows D_1^1 \rightleftarrows D_0^1 = \{*\}, \cdots, D_{n-1}^m \rightleftarrows \cdots \rightleftarrows D_1^m \rightleftarrows D_0^m = \{*\} \right)
$$

of $\mathbf{I} \wr \mathcal{D}_{n-1}$ and stacking it together, sending it to the object

$$
D_{n-1}^1 \sqcup \cdots \sqcup D_{n-1}^m \to \cdots \to D_1^1 \sqcup \cdots \sqcup D_1^m \to [m+1] \to \{*\}
$$

in $\mathcal{D}_n$, where we have ordered the m endpoints of the objects and have added a minimal and a maximal element to obtain $[m+1]$.

The functor $G : \mathcal{D}_n \to \mathbf{I} \wr \mathcal{D}_{n-1}$ cuts a finite combinatorial n-disk, as in (14.8.2), into $m = |D_1| - 2$ finite combinatorial $(n-1)$-disks but remembers the object $[m+1]$ of $\mathbf{I}$ from the cutting procedure. We can write $D_1 = \{\min, x_1, \ldots, x_m, \max\}$ and take the m combinatorial $(n-1)$-disks that lie over $x_1, \ldots, x_m$.

The functors F and G are defined on morphisms in the only possible way. As F and G are inverse to each other, this proves the claim. $\quad\square$

If we discard the part of the diagram (14.8.2) that involves the functions s_i and t_i, then we focus on the inner part of the disks.

Definition 14.8.17

- The *boundary of the set D_i* of a finite combinatorial n-disk, as in (14.8.2), is $\partial D_i = s_i(D_{i-1}) \cup t_i(D_{i-1})$. We set $\partial D_0 = \varnothing$.
- The *interior of the set D_i* is the complement $D_i \setminus \partial D_i$. We denote it by $\mathring{D}_i$.
- The *interior* of a finite combinatorial n-disk (14.8.2) is the diagram

$$\overset{\circ}{D}_n \xrightarrow{q_n} \overset{\circ}{D}_{n-1} \xrightarrow{q_{n-1}} \cdots \xrightarrow{q_1} \overset{\circ}{D}_1 \longrightarrow \{*\},$$

where we abuse the notation and reuse q_i for the restriction of q_i to the interior.

- Morphisms between such open finite combinatorial n-disks are commutative diagrams of functions

$$
\begin{array}{ccccccccc}
\overset{\circ}{D}_n & \xrightarrow{q_n} & \overset{\circ}{D}_{n-1} & \xrightarrow{q_{n-1}} & \cdots & \xrightarrow{q_1} & \overset{\circ}{D}_1 & \longrightarrow & \{*\} \\
\downarrow{f_n} & & \downarrow{f_{n-1}} & & & & \downarrow{f_1} & & \\
\overset{\circ}{D}'_n & \xrightarrow{q'_n} & \overset{\circ}{D}'_{n-1} & \xrightarrow{q_{n-1}} & \cdots & \xrightarrow{q_1} & \overset{\circ}{D}'_1 & \longrightarrow & \{*\},
\end{array}
$$

such that the f_is are order-preserving on the fibers of the q_is.

- We denote the *category of open finite combinatorial n-disks* by $\overset{\circ}{\mathcal{D}}_n$.

Proposition 14.8.18

(1) The categories $\overset{\circ}{\mathcal{D}}_n$ and Ω_n are isomorphic.

(2) There is a functor

$$(-) \colon \mathcal{D}_n \to \overset{\circ}{\mathcal{D}}_n,$$

and $(-)$ has a left adjoint.

Proof

(1) There is a total ordering on every fiber $q^{-1}(x)$ for $q_i \colon \overset{\circ}{D}_i \to \overset{\circ}{D}_{i-1}$. We can use the total ordering on $\overset{\circ}{D}_1$ in order to identify $\overset{\circ}{D}_1$ with the labels of the first layer of a tree in Ω_n, $|\mathbf{D_1}|$. The total ordering on $\overset{\circ}{D}_1$ induces a total ordering on $\overset{\circ}{D}_2$ if we order the fibers, such that the elements in $q_2^{-1}(x)$ are less than the elements in $q_2^{-1}(y)$ for $x < y$. This allows us to identify $\overset{\vee}{D}_2$ with the nodes in the second layer of a tree in Ω_n. Iteratively, the object

$$\overset{\circ}{D}_n \xrightarrow{q_n} \overset{\circ}{D}_{n-1} \xrightarrow{q_{n-1}} \cdots \xrightarrow{q_1} \overset{\circ}{D}_1 \longrightarrow \{*\}$$

corresponds to a unique tree in Ω_n. Morphisms in $\overset{\circ}{\mathcal{D}}_n$ correspond to morphisms in Ω_n.

(2) Taking an object in $\mathcal{D}_n$ to its interior $\overset{\circ}{\mathcal{D}}_n$ defines a functor $(-) \colon \mathcal{D}_n \to \overset{\circ}{\mathcal{D}}_n$. We can take an object

$$\overset{\circ}{D}_n \xrightarrow{q_n} \overset{\circ}{D}_{n-1} \xrightarrow{q_{n-1}} \cdots \xrightarrow{q_1} \overset{\circ}{D}_1 \longrightarrow \{*\}$$

and set $D'_0 = D_0 = \{*\}$. We define the D_is for $0 < i \leq n$ by adding a minimal and a maximal element to each fiber of q_i. This defines D_i

and also a function $q_i' : D_i' \to D_{i-1}'$. Then, we define $s_i : D_{i-1} \to D_i$ by sending an $x \in D_{i-1}'$ to the minimal element in the fiber $(q_i')^{-1}(x)$, and t_i sends x to the maximal element in $(q_i')^{-1}(x)$. This defines the left adjoint L to $(\overset{\circ}{-})$ on the level of objects.

For a morphism f in $\overset{\circ}{\mathcal{D}}_n$, we define $L(f)$ by requiring that $L(f)$ is equal to f on the subsets $\overset{\circ}{D}_i \subset D_i$ and that it respects minimal and maximal elements in the fibers. This defines a morphism in $\mathcal{D}_n$.

As $(\overset{\circ}{-})$ forgets structure, it is easy to see that L is left adjoint to $(\overset{\circ}{-})$. $\square$

Corollary 14.8.19 *There is a homotopy equivalence between $B\Omega_n$ and $B\Theta_n^o$.*

Proof   As we have a left-adjoint L to $(\overset{\circ}{-})$, we get that $B\Omega_n \simeq B\mathcal{D}_n$, and we know from Theorem 14.8.16 that $B\mathcal{D}_n \cong B(\Psi_n)$ and from Proposition 14.8.14 that $B(\Psi_n) \cong B(\Theta_n^o)$. $\square$

14.8.2 Θ_n and n-Fold Loop Spaces

Explaining Berger's result [**Be07**] in full detail would go well beyond the scope of this book, so we only sketch his approach. For functors $X : \Theta_n^o \to$ cg, he has a notion of being reduced, which means that $X(i_{n-1}(x)) = *$ for every object $x \in \Theta_{n-1}$ and that $X([0], -) = *$. Here, $i_{n-1} : \Theta_{n-1} \to \Theta_n$ is the inclusion functor from Definition 14.8.7. Berger defines a topological realization functor $| - |_{\Theta_n}$ [**Be07**, Corollary 3.11] that sends a functor $X : \Theta_n^o \to$ Sets (or $X : \Theta_n^o \to$ cg) to a topological space. There is also a model categorical requirement that ensures the correct homotopical behavior. For a functor $X : \Theta_n^o \to$ cg, one denotes by $U(X)$ the evaluation of X on the object $([1]; [1]; \ldots ; [1])$, of Θ_n. Note that this is the object $\sigma_{n-1} \circ \cdots \sigma_1[1]$, where $\sigma_i : \Theta_i \to \Theta_{i+1}$ is the suspension functor from Definition 14.8.7.

Theorem 14.8.20 [**Be07**, Theorem 4.5]   *Let $X : \Theta_n^o \to$ cg be reduced and cofibrant-fibrant. Then, $U(X)$ is weakly equivalent to $\Omega^n |X|_{\Theta_n}$, and hence, in this case, $|X|_{\Theta_n}$ is an n-fold delooping of X.*

Remark 14.8.21   The case $n = 1$ of the preceding result recovers Segal's Theorem 14.2.3 that identifies suitable reduced Segal spaces as based loop spaces.

Exercise 14.8.22   Prove that precomposition with δ_n sends reduced Θ_n-spaces to reduced n-fold simplicial spaces. Here, an n-fold simplicial space Y is reduced when $Y([k_1], \ldots, [k_n]) = *$ if $k_i = 0$ for one $1 \le i \le n$.

In the algebraic setting, n-fold deloopings are realized by so-called E_n-homology. Let k be a commutative ring and let A be a commutative augmented k-algebra. Note that k is a left and right A-module. We consider the two-sided bar construction of A, $B(k, A, k)$, as in Definition 10.4.2, which is an algebraic analog of a delooping. If G is a group, then $B(\{e\}, G, \{e\})$ is a model for the classifying space BG, and there is a weak homotopy equivalence $G \to \Omega BG$.

As A is commutative, the chain complex associated with the simplicial k-module $B(k, A, k)$ carries a multiplication. Let us denote elements in $B_p(k, A, k)$ by $[a_1 \mid \cdots \mid a_p]$, omitting the scalars in k. Then,

$$[a_1 \mid \cdots \mid a_p] \cdot [a_{p+1} \mid \cdots \mid a_{p+q}] = \sum_{\sigma \in Sh(p,q)} \text{sign}(\sigma)[a_{\sigma^{-1}(1)} \mid \cdots \mid a_{\sigma^{-1}(p+q)}],$$

$$(14.8.3)$$

where the sum runs over all (p, q)-shuffles, that is, all permutations in the symmetric group on $p + q$ letters, which respect the ordering of the first p and the last q elements.

The shuffle multiplication in the bar construction can be encoded by the category Ω_2. Consider the morphism

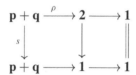

in Ω_2, where ρ maps every $i \in \{1, \dots, p\}$ to 1 and the complement to 2. The unlabelled morphisms are uniquely determined because **1** is terminal. The function s has to be order-preserving on the fibers of ρ, so it has to preserve the ordering $1 < \cdots < p$ and $p + 1 < \cdots < p + q$. If s is bijective, then s is a shuffle in $Sh(p, q)$.

The bar construction can be turned into a functor $\mathcal{L}^k(A; k)$ from Ω_2 to the category of k-modules by sending an object $\mathbf{p + q} \xrightarrow{\rho} \mathbf{2} \longrightarrow \mathbf{1}$ to $A^{\otimes p+q}$. A morphism, as considered earlier, induces the shuffle multiplication from (14.8.3). This idea is used in [**LR11**] in order to show that E_n-homology of commutative augmented algebras possesses a functor homology description. Instead of Ω_n, we work with a variant, Epi_n, that has as objects chains of order-preserving surjections [**LR11**, Definition 3.1], whereas in Definition 14.8.11, one allows chains of arbitrary order-preserving maps.

14.8.3 Θ_n and Higher Categories

The objects in Θ_n correspond to certain full subcategories of the category of strict n-categories. An object in Θ_1 is of the form $[n]$ for some $n \geq 0$, and $[n]$ corresponds to the strict 1-category (which is nothing but a category)

$$0 \to 1 \to \cdots \to n.$$

An object $([n]; [k_1], \ldots, [k_n])$ in Θ_2 encodes the strict 2-category, whose set of objects is $[n] = \{0, \ldots, n\}$ and where we have k_i 1-morphisms between the object $i - 1$ and i of $[n]$ and $(k_i - 1)$ 2-morphisms. For example, the object $([5]; [0], [2], [1], [4], [3])$ represents the strict 2-category with 1-morphisms

$$0 \longrightarrow 1 \rightrightarrows 2 \Longrightarrow 3 \rightrightarrows 4 \rightrightarrows 5.$$

In an n-category, you have k-morphisms for all $1 \leq k \leq n$.

The suspension functor $\sigma_n \colon \Theta_n \to \Theta_{n+1}$ upgrades objects to 1-morphisms and i-morphisms to $(i + 1)$-morphisms for $i \leq n$.

For instance, the suspension of the object $([5]; [0], [2], [1], [4], [3])$ of Θ_2 is the object $([1]; [5]; [0], [2], [1], [4], [3])$ of Θ_3.

Rezk [**Rez10**, p. 523] considers particular objects O_i in Θ_i: $O_0 = [0]$ is the only object of Θ_0. The suspension of O_0 is $O_1 = ([1]; O_0) = ([1], [0])$. Iteratively, we set

$$O_{n+1} = \sigma_n O_n \text{ for all } n \geq 0.$$

In the interpretation of strict n-categories, O_n corresponds to an n-cell:

$$O_0 = 0; \quad O_1 = 0 \xrightarrow{\ 0\ } 1; \quad O_2 = 0 \overset{0}{\underset{1}{\Downarrow 0}} 1, \ldots$$

The term ∞-category denotes a higher version of a category, where one has k-morphisms for all $k \geq 1$. If you take a topological space X, you can take its points as objects. A 1-morphism between x and y in X is a path from x to y. If you have two such paths, then a 2-morphism between them could be a continuous map from a unit square that is a basepoint preserving homotopy between the two paths. But nobody can prevent you at this point from defining k-morphisms for arbitrary $k \geq 1$ as homotopies between the $(k - 1)$-morphisms that you had. This is the ∞-groupoid of the space X, and Grothendieck's homotopy hypothesis demands that everything that wants to be called an ∞-groupoid should be equivalent to the ∞-groupoid of a space. An ∞-category then gets rid of the assumption of the invertibility of morphisms.

An (∞, n)-category is an ∞-category such that the k-morphisms are weakly invertible for all $k > n$. Lurie's book [**Lu09**] is the standard reference for the model of quasi-categories as a model for $(\infty, 1)$-categories. For an overview about other models, see [**B18**].

Charles Rezk used the category Θ_n for a model of (∞, n)-categories for $n \geq 1$ via Θ_n-spaces [**Rez10**]. Together with Julie Bergner, they compare the model of Θ_n-spaces to other models of (∞, n)-categories [**BRez13a, BRez∞**]. See [**B20**] for an overview.

15

Functor Homology

Let $\mathcal{C}$ be a small category and let $\mathcal{A}$ be an abelian category. Functor homology assigns to a functor $F : \mathcal{C} \to \mathcal{A}$ the groups $\mathrm{Tor}_*^{\mathcal{C}}(G, F)$ for some fixed functor $G : \mathcal{C}^o \to \mathcal{A}$. If you have your favorite homology theory for some kind of algebraic objects, say associative algebras over a fixed commutative ground ring k, then a functor homology interpretation would say that your homology theory applied to some algebra A is of the form $\mathrm{Tor}_*^{\mathcal{C}}(G, F_A)$ for some functor F_A depending on A.

Why do we want functor homology interpretations?

- Combinatorial features of the parametrizing category $\mathcal{C}$ can be used in order to get extra structure, additional spectral sequences and more, for instance the Hodge decomposition of Hochschild homology [**Lo98**, Theorem 6.4.5] and higher Hochschild homology [**P00a**] can be expressed in terms of functor homology.
- Tor and Ext functors have universal properties, and this helps to obtain uniqueness and comparison results: You might want to compare two (co)homology theories by comparing both of them to functor (co)homology. For instance, the functor homology description of Gamma homology from [**PR00**] combined with [**P00a**, Proposition 2.2] allows a comparison of Gamma homology with stable homotopy.
- In order to get functor homology interpretations we have to understand what something *really* is.

15.1 Tensor Products

We defined coends in Section 4.4.6 and saw special cases of coends in the form of tensor products. In all examples, the target category was a cocomplete symmetric monoidal category, and this is, in fact, the suitable generality for considering such tensor products.

Definition 15.1.1 Let $\mathcal{D}$ be a small category, $(\mathcal{C}, \otimes, 1)$ be a cocomplete symmetric monoidal category, and $F \colon \mathcal{D}^o \to \mathcal{C}$ and $G \colon \mathcal{D} \to \mathcal{C}$ be functors. Then, their *tensor product*, $F \otimes_{\mathcal{D}} G$, is defined as the coend of the functor

$$H \colon \mathcal{D}^o \times \mathcal{D} \to \mathcal{C},$$

where a pair of objects (D_1, D_2) is mapped to $F(D_1) \otimes G(D_2)$.

Explicitly, $F \otimes_{\mathcal{D}} G$ is given as the coequalizer of

$$\bigsqcup_{f \in \mathcal{D}(D_1, D_2)} F(D_2) \otimes G(D_1) \underset{\bigsqcup \mathcal{D}(1_{D_2}, f)}{\overset{\bigsqcup \mathcal{D}(f, 1_{D_1})}{\rightrightarrows}} \bigsqcup_{\text{objects } D \text{ of } \mathcal{D}} F(D) \otimes G(D).$$

Example 15.1.2 Let R be a ring, M be a right R-module, and N be a left R-module. Let $\mathcal{R}$ be the category with one object $*$ and with R as endomorphisms. We define $F_M \colon \mathcal{R}^o \to \mathsf{Ab}$ as $F_M(*) = M$ and $G_N \colon \mathcal{R} \to \mathsf{Ab}$ as $G_N(*) = N$. Then, the tensor product of the additive functors F_M and G_N can be identified with the tensor product of the right R-module M with the left R-module N:

$$F_M \otimes_{\mathcal{R}} G_N \cong M \otimes_R N.$$

For rings and modules, we know that the tensor product is adjoint to a homfunctor. The analogous statement is true for tensor products of functors if $\mathcal{C}$ is closed and bicomplete. Let $\mathcal{D}$ be a small category and let C be an object of $\mathcal{C}$. For a functor $G \colon \mathcal{D} \to \mathcal{C}$, we consider the functor

$$\mathsf{Sets}(G(-), C) \colon \mathcal{D}^o \to \mathcal{C}, \quad D \mapsto \underline{\mathcal{C}}(G(D), C), \quad g \in \mathcal{D}(D, D') \mapsto \underline{\mathcal{C}}(G(g), C).$$

This functor is adjoint to the tensor product.

Theorem 15.1.3 *Let $\mathcal{C}$ be a closed symmetric monoidal and bicomplete category. For all functors $F \colon \mathcal{D}^o \to \mathcal{C}$, $G \colon \mathcal{D} \to \mathcal{C}$ and for all objects C of $\mathcal{C}$, there is an isomorphism in $\mathcal{C}$*

$$\underline{\mathcal{C}}(F \otimes_{\mathcal{D}} G, C) \cong \mathsf{nat}_{\mathcal{C}}(F, \underline{\mathcal{C}}(G(-), C)),$$

where $\mathsf{nat}_{\mathcal{C}}$ denotes the enriched end of natural transformations from F to $\underline{\mathcal{C}}(G(-), C)$.

Proof The functor $\underline{C}(-, C)$ sends colimits to limits, so we can rewrite

$$\underline{C}(F \otimes_D G, C) \cong \lim_{\mathcal{D}} \underline{C}(F(D) \otimes G(D), C).$$

Adjunction tells us that the latter is isomorphic to $\lim_{\mathcal{D}} \underline{C}(F(D), \underline{C}(G(D), C))$, and this limit is precisely the enriched end of natural transformations, as claimed. $\qquad\square$

Remark 15.1.4 As we will mostly work with the target category of k-modules for some commutative ring with unit k, we will make Theorem 15.1.3 explicit in this setting. For all functors $F \colon \mathcal{D}^o \to k\text{-mod}$ and $G \colon \mathcal{D} \to k\text{-mod}$ and for all k-modules M, the k-module of k-linear maps from $F \otimes_D G$ to M is isomorphic to the k-module of natural transformations from F to $k\text{-mod}(G(-), M)$:

$$k\text{-mod}(F \otimes_D G, M) \cong \text{nat}(F, k\text{-mod}(G(-), M)). \tag{15.1.1}$$

The following result is a special case of Proposition 9.3.10. We will mostly be interested in the case where $\mathcal{D}$ has the standard enrichment in $\mathcal{C}$, that is, $\mathcal{D}_{\mathcal{C}}(D_1, D_2) = \bigsqcup_{f \in \mathcal{D}(D_1, D_2)} e$.

Corollary 15.1.5 *Assume that $\mathcal{C}$ is a closed symmetric monoidal and bicomplete category. Let $G \colon \mathcal{D} \to \mathcal{C}$ be a $\mathcal{C}$-enriched representable functor $G = \mathcal{D}_{\mathcal{C}}(D, -)$ and let $F \colon \mathcal{D}^o \to \mathcal{C}$ be an arbitrary functor. Then, there is an isomorphism in $\mathcal{C}$*

$$F \otimes_{\mathcal{D}} \mathcal{D}_{\mathcal{C}}(D, -) \cong F(D)$$

that is natural in F and D. Similarly,

$$\mathcal{D}_{\mathcal{C}}(-, D) \otimes_{\mathcal{D}} G \cong G(D)$$

for every $G \colon \mathcal{D} \to \mathcal{C}$.

15.2 Tor and Ext

Let $(\mathcal{A}, \otimes)$ be an abelian bicomplete category. If $\mathcal{D}$ is a small category and $\mathcal{A}$ has enough projectives, then the functor categories $\text{Fun}(\mathcal{D}, \mathcal{A})$ and $\text{Fun}(\mathcal{D}^o, \mathcal{A})$ are again abelian categories (see Proposition 7.3.7), and they have enough projectives.

Lemma 15.2.1 *Assume that $\mathcal{A}$ has enough projectives. If P is a projective object in $\mathcal{A}$, then for all objects D of $\mathcal{D}$,*

$$F_D(P) = \bigoplus_{f \in \mathcal{D}(D, -)} P \tag{15.2.1}$$

is a projective object in $\text{Fun}(\mathcal{D}, \mathcal{A})$, and $\text{Fun}(\mathcal{D}, \mathcal{A})$ has enough projectives.

Proof   The evaluation functor

$$\varepsilon_D \colon \mathsf{Fun}(\mathcal{D}, \mathcal{A}) \to \mathcal{A}, \quad G \mapsto G(D)$$

has

$$F_D \colon \mathcal{A} \to \mathsf{Fun}(\mathcal{D}, \mathcal{A}), \quad A \mapsto \bigoplus_{f \in \mathcal{D}(D, -)} A$$

as a left adjoint, and hence, for P projective, $F_D(P)$ is projective.

Let G be an arbitrary functor $G \colon \mathcal{D} \to \mathcal{A}$ and let $\rho_D \colon P_D \to G(D)$ be an epimorphism with P_D projective in $\mathcal{A}$. Then, there is an epimorphism

$$\bigoplus_{D \text{ an object of } \mathcal{D}} \bigoplus_{\mathcal{D}(D, -)} F_D(P_D) \to G,$$

whose restriction to a (D, f)-component in $\bigoplus_{D \text{ an object of } \mathcal{D}} \bigoplus_{f \in \mathcal{D}(D, D')}$ $F_D(P_D)$ is given by the composite of the epimorphism ρ_D with $G(f)$,

$$P_D \xrightarrow{\rho_D} G(D) \xrightarrow{G(f)} G(D'). \qquad \square$$

Hence, we can do homological algebra in these functor categories.

Definition 15.2.2 Let $\mathcal{A}$ be a bicomplete abelian category with enough projectives, and assume that $\mathcal{A}$ is closed symmetric monoidal. Let $F \colon \mathcal{D}^o \to \mathcal{A}$ and $G \colon \mathcal{D} \to \mathcal{A}$ be functors. Then,

$$\mathsf{Tor}^{\mathcal{D}}_*(F, G) := H_*(P_* \otimes_{\mathcal{D}} G),$$

where P_* is a projective resolution of F in $\mathsf{Fun}(\mathcal{D}^o, \mathcal{A})$.

Similarly, we can define Ext groups.

Definition 15.2.3 Let $F, G \colon \mathcal{D} \to \mathcal{A}$ be functors. Then,

$$\mathsf{Ext}^{\mathcal{D}}_*(F, G) := H_*(\mathsf{nat}(P_*, G)),$$

where P_* is a projective resolution of F in $\mathsf{Fun}(\mathcal{D}, \mathcal{A})$.

Theorem 15.2.4 (Axiomatic description of Tor and Ext) [**CE56**, III.5], [**FFPS03**, p. 110, Proposition 2.1] *If H_* is a functor from $\mathsf{Fun}(\mathcal{C}, k\text{-mod})$ to the category of graded k-modules, such that*

- *$H_0(F)$ is canonically isomorphic to $G \otimes_{\mathcal{C}} F$ for all $F \in \mathsf{Fun}(\mathcal{C}, k\text{-mod})$,*
- *$H_*(-)$ maps short exact sequences of functors in $\mathsf{Fun}(\mathcal{C}, k\text{-mod})$ to long exact sequences in a way that is natural in short exact sequences, and*
- *$H_i(F) = 0$ for all projective F and $i > 0$,*

then $H_i(F) \cong \mathsf{Tor}^{\mathcal{C}}_i(G, F)$ for all F in $\mathsf{Fun}(\mathcal{C}, k\text{-mod})$.

15.3 How Does One Obtain a Functor Homology Description?

The following is a common situation. You have a homology theory and you want to describe it as functor homology, for instance, because you hope that combinatorial features of the diagram category might give you some additional means for calculations (*e.g.* via a spectral sequence) or because you hope to understand your homology theory better by extracting what diagrams you need, in order to describe it.

There are several functor homology interpretations of homology theories and their applications in the literature (see, for instance, [**Dj∞, FP98, FFPS03, LR11, PR02, PR00**] and [**Z16**] for a tiny and very biased selection of some of them). But how does one actually find such a thing? How do you guess the diagram category, and how do you find the "right" functors?

Well, it's not rocket science, and I will explain the procedure by using the example of Hochschild homology.

First of all, one homology theory can have different descriptions as functor homology. Hochschild homology has one using the simplicial category [**Lo98**]. I'll explain the one from [**PR02**].

So, what is Hochschild homology? You start with a commutative ring with unit k, an k-algebra A, and an A-bimodule M.

Definition 15.3.1 The *ith Hochschild homology group of A over k with coefficients in M*, $\mathrm{HH}_i^k(A; M)$ is defined as

$$H_i\left(\cdots \xrightarrow{\ b\ } M \otimes A^{\otimes 2} \xrightarrow{\ b\ } M \otimes A \xrightarrow{\ b\ } M \right).$$

Here, the tensor products are over k and $b = \sum_{i=0}^{n}(-1)^i d_i$, where

$$d_i(a_0 \otimes \cdots \otimes a_n) = \begin{cases} a_0 \otimes \cdots \otimes a_i a_{i+1} \otimes \cdots a_n, & \text{for } i < n \text{ and} \\ a_n a_0 \otimes \cdots \otimes a_{n-1}, & \text{for } i = n. \end{cases} \quad (15.3.1)$$

for $a_0 \in M$ and $a_i \in A$ for $0 < i < n$.

A nice way to visualize this is to draw elements in the Hochschild complex in a cyclic manner:

$$\begin{matrix} & & & a_1 & & \\ & & \otimes & & \otimes & \\ & a_0 & & & & \ddots \\ & \otimes & & & & \cdot \\ & a_n & & & & \\ & & \otimes & & \cdots & \end{matrix}$$

Then, the ith face map in the Hochschild complex just multiplies the elements a_i and a_{i+1} together, where now, the indices have to be read modulo $n+1$. If we take A as an A-module, then this gives rise to the important cyclic structure on the Hochschild complex [**Lo98**]; see also (15.4.1).

15.3.1 How Do We Guess the Diagram Category?

First of all, we do not assume that A is commutative and that M is a symmetric bimodule, so one should think of the elements of being ordered, and it is important that we get $a_i \cdot a_{i+1}$ in the ith face map and not $a_{i+1} \cdot a_i$. The other important feature is that the coordinate of the bimodule M always stays on the left. It is acted upon with the face maps d_0 and d_n, but the leftmost entry is always in M, and we think of M as being glued to a basepoint. Therefore, if we want to have a diagram category modelling the Hochschild complex, then it should have basepoint preserving maps as morphisms.

In every chain degree, we have something of the form $M \otimes A^{\otimes n}$ for $n \geq 0$, so we do not need more than finite sets to model this complex, and the rough idea is to assign $M \otimes A^{\otimes n}$ to a finite set with $n+1$ elements. That's it. So, we have

- finite sets,
- with a basepoint,
- an ordering.

The corresponding category is as follows:

Definition 15.3.2 Let $\Gamma(as)$ be the *category of finite pointed associative sets*. Its objects are the finite pointed sets $[n] = \{0, 1, \ldots, n\}$ for $n \geq 0$ with 0 as a basepoint. A morphism in $\Gamma(as)([n], [m])$ is a basepoint preserving function of finite sets $f \colon [n] \to [m]$ together with a total ordering on each fiber $f^{-1}(j)$ for all $j \in [m]$.

Example 15.3.3 An $f \in \Gamma(as)([5], [0])$ is a function from $[5]$ to $[0]$ (it is automatically preserving the basepoint), together with a total ordering of the fiber, but here, the only fiber is $[5]$, so we can identify f with a total ordering of the set $\{0, \ldots, 5\}$, and this is nothing but a permutation in Σ_6.

In order to model the Hochschild complex, we have to define a functor $\mathcal{L}^k(A; M) \colon \Gamma(as) \to k\text{-mod}$, such that

$$\mathsf{HH}_*(A; M) \cong \mathsf{Tor}_*^{\Gamma(as)}\big(\bar{\mathsf{b}}, \mathcal{L}^k(A; M)\big),$$

where $\bar{\mathsf{b}} \colon \Gamma(as)^o \to k\text{-mod}$ is a functor that we will specify later.

Definition 15.3.4 The functor $\mathcal{L}^k(A; M)\colon \Gamma(as) \to k$-mod is defined as

$$\mathcal{L}^k(A; M)[n] := M \otimes A^{\otimes n},$$

and for an $f \in \Gamma(as)([n], [m])$, we set

$$\mathcal{L}^k(A; M)(f)(a_0 \otimes \cdots \otimes a_n) = b_0 \otimes \cdots \otimes b_m$$

where $b_i = \prod^<_{f(j)=i} a_j$ is the product of the a_js, according to the total order of the fiber $f^{-1}(i)$.

Example 15.3.5 Let $f \in \Gamma(as)([4], [2])$ be the morphism with fibers $f^{-1}(0) = \{0\}$, $f^{-1}(1) = \{4\}$, and $f^{-1}(2) = \{2 < 1 < 3\}$.

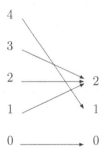

Then,

$$\mathcal{L}^k(A; M)(f)(a_0 \otimes \cdots \otimes a_4) = a_0 \otimes a_4 \otimes a_2 \cdot a_1 \cdot a_3.$$

If your algebras are not associative algebras but are, maybe, algebras over an operad in the category of sets (see Definition 12.2.1), then a diagram category that you can try is the category of operators for that operad (see Definition 14.1.1).

So far, we have an educated guess for the diagram category and for the covariant functor. What we need now is the contravariant functor $\bar{b}$, which is global; that is, for every $\mathcal{L}^k(A; M)$ for varying A and M, the groups $\mathrm{Tor}_*^{\Gamma(as)}(\bar{b}, \mathcal{L}^k(A; M))$ should give Hochschild homology.

15.3.2 How to Guess the Functor?

In order to guess the correct contravariant functor $\bar{b}\colon \Gamma(as)^o \to k$-mod, we have a look at the bottom part of the Hochschild complex, because the axiomatic description from Theorem 15.2.4 of the Tor functor tells us that

$$\mathrm{Tor}_0^{\Gamma(as)}(\bar{b}, \mathcal{L}^k(A; M)) \cong \bar{b} \otimes_{\Gamma(as)} \mathcal{L}^k(A; M) = \mathsf{HH}_0(A; M),$$

and we know that $HH_0(A; M) \cong M/\langle am - ma, m \in M, a \in A\rangle$. Thus, we want a quotient of $M = \mathcal{L}^k(A; M)[0]$, but thanks to Corollary 15.1.5, we can express this as

$$\mathcal{L}^k(A; M)[0] \cong k\{\Gamma(as)(-, [0])\} \otimes_{\Gamma(as)} \mathcal{L}^k(A; M).$$

The Hochschild complex in chain degree one is

$$M \otimes A = \mathcal{L}^k(A; M)[1] \cong k\{\Gamma(as)(-, [1])\} \otimes_{\Gamma(as)} \mathcal{L}^k(A; M).$$

Therefore, we don't have a choice but to define $\bar{b}$ as the cokernel of a transformation

$$d \colon k\{\Gamma(as)(-, [1])\} \to k\{\Gamma(as)(-, [0])\}.$$

This transformation is forced on us by the boundary map $M \otimes A \to M$, which takes $m \otimes a$ to $ma - am$, which is an alternating sum of two face maps. The first one takes $m \otimes a$ to ma (in that order), and the second one switches the elements first and then applies the module structure map. So, both maps must be induced by morphisms in $\Gamma(as)$ from [1] to [0], and there are only two such maps. Both send 0 and 1 to 0, but one has the order $0 < 1$ on the fiber and the other map specifies that the order on the fiber is $1 < 0$. We call the first one $d_{0<1}$ and the second one $d_{1<0}$. That's it:

$$k\{\Gamma(as)(-, [1])\} \xrightarrow{d_{0<1}-d_{1<0}} k\{\Gamma(as)(-, [0])\} \longrightarrow \bar{b} \longrightarrow 0.$$

15.3.3 How to Prove that This Is Actually Correct?

The steps so far were rather easy in this case. You didn't need much creativity to come up with the correct diagram category and the correct functor. However, guessing an adequate diagram category can be more involved in other examples (see, for instance, [**LR11**] or [**HV15**]). What you have to do now is to extend the definition of Hochschild homology to functors from $\Gamma(as)$ to the category of k-modules. That is straightforward, because here also, you don't really have a choice.

Definition 15.3.6 Let $F \colon \Gamma(as) \to k$-mod be a functor. Then, *the Hochschild homology of F, $HH_*(F)$*, is the homology

$$H_*\left(\cdots \xrightarrow{b} F[2] \xrightarrow{b} F[1] \xrightarrow{b} F[0] \right).$$

Here, $b \colon F[n] \to F[n-1]$ is $b = \sum_{i=0}^{n}(-1)^i d_i$, where $d_i \colon F[n] \to F[n-1]$ is $d_i = F(\partial_i)$, and for $i \neq n$,

$$\partial_i : [n] \to [n-1], \quad \partial_i^{-1}(j) = \begin{cases} \{j\}, & \text{for } i \neq j, \\ \{i < i+1\}, & \text{for } i = j. \end{cases}$$

However, for $i = n$,

$$\partial_n : [n] \to [n-1], \quad \partial_n^{-1}(j) = \begin{cases} \{j\}, & \text{for } j \neq 0, \\ \{n < 0\}, & \text{for } j = n. \end{cases}$$

With these definitions, you will get that $\mathrm{Tor}_0^{\Gamma(as)}(\bar{b}, F) \cong \mathrm{HH}_0(F)$.

The bad news is that what you have done so far might not work! So what can go wrong? You have to show that this newly defined Hochschild homology vanishes on projectives in positive degrees. That is the part of the argument where you actually have to work.

Proving something for *all* projective objects might be too involved. Often, one can simplify things.

Definition 15.3.7 A set of projective objects $\{P^i | i \in I\}$ in an abelian category $\mathcal{A}$ is called a *family of projective generators* if every object of $\mathcal{A}$ can be written as the cokernel of a morphism from a direct sum of P^is.

Lemma 15.3.8 *If $\mathcal{D}$ is a small category, then the functor category* $\mathrm{Fun}(\mathcal{D}, k\text{-mod})$ *always possesses a family of projective generators given by the family*

$$\{k\{\mathcal{D}(D, -)\}, D \text{ an object of } \mathcal{D}\}.$$

Proof Let $F : \mathcal{D} \to k$-mod be a functor. The category of k-modules has the module k as a projective generator. Using (15.2.1), we get an epimorphism

$$\bigoplus_{D \text{ and object of } \mathcal{D}} \bigoplus_{f \in \mathcal{D}(D, -)} P_D \to F,$$

where P_D is a direct sum of copies of k. Hence, we get an epimorphism

$$\bigoplus_X k\{\mathcal{D}(D, -)\} \to F,$$

where X is a suitable indexing set, and hence, the functors $k\{\mathcal{D}(D, -)\}$ are a family of projective generators. $\square$

In the case of finite pointed associative set $\Gamma(as)$, we have the projective generators $\Gamma(as)^n$ of $\mathrm{Fun}(\Gamma(as), k\text{-mod})$ with

$$\Gamma(as)^n([m]) = k\{\Gamma(as)([n], [m])\},$$

and thus, one has to show that

$$\mathrm{HH}_*(\Gamma(as)^n) \cong 0 \text{ for } * > 0 \text{ and for all } n \geq 0.$$

In [**PR02**, 2.2], this is done by showing that $\mathsf{HH}_*(\Gamma(as)^n)$ is the singular homology with coefficients in k of $n!$ copies of the standard n-simplex. Thus, we obtain the following:

Theorem 15.3.9 ([**PR02**, Theorem 1.3] *For all* $F : \Gamma(as) \to k$-mod,

$$\mathsf{HH}_*(F) \cong \mathsf{Tor}_*^{\Gamma(as)}(\bar{\mathsf{b}}, F).$$

Remark 15.3.10 In [**Lo98**, Corollary 6.2.3], you find a different functor homology description of Hochschild homology in terms of the simplicial category Δ (see Definition 10.1.1): For all associative k-algebras A and any A-bimodule M,

$$\mathsf{HH}_*(A; M) \cong \mathsf{Tor}_*^{\Delta^o}(k, \mathcal{L}^k(A; M)),$$

where k is the constant functor $k : (\Delta^o)^o = \Delta \to k$-mod with value k. We will show this later in Corollary 16.5.10. Jolanta Słomińska proved a direct comparison of both functor homology descriptions, using a decomposition of the category $\Gamma(as)$. See [**Sł03**] for details.

In other examples, the acyclicity of the projective generators might be proven using an explicit chain homotopy (*e.g.* in [**PR00**]) or a spectral sequence argument (*e.g.* in [**LR11**] and [**HV15**]).

15.4 Cyclic Homology as Functor Homology

Cyclic homology takes the cyclic symmetry of the Hochschild complex serious. For an associative k-algebra A, we consider A as an A-bimodule. Then, there is a visible action of the cyclic group $\mathbb{Z}/(n + 1)\mathbb{Z}$ on $A^{\otimes n+1}$ by rotating tensor factors:

$$\begin{matrix} & & a_1 & \\ & \otimes & & \otimes & \\ a_0 & & & & \ddots \\ \otimes & & & & \vdots \\ a_n & & & & \\ & \otimes & \cdots & \end{matrix}$$

(15.4.1)

So, in this situation, we do not consider basepoints, but any tensor factor behaves like any other one, so we work with finite sets instead of finite *pointed* sets. But, we still consider associative algebras A, so we need to fix an ordering of elements.

Definition 15.4.1 [**PR02**] Let $\mathcal{F}(as)$ denote the *category of finite associative sets*. Its objects are the finite sets $[n] = \{0, 1, \ldots, n\}$ for $n \geq 0$, and a morphism $f \in \mathcal{F}(as)([n], [m])$ is a function $f : [n] \to [m]$, together with a total ordering of the fibers $f^{-1}(i)$ for $0 \leq i \leq n$.

The Loday functor $\mathcal{L}^k(A) := \mathcal{L}^k(A; A)$ can be viewed as a functor $\mathcal{L}^k(A) : \mathcal{F}(as) \to k\text{-mod}$, sending $[n]$ to $A^{\otimes n+1}$ and $\mathcal{L}^k(A)(f)(a_0 \otimes \cdots \otimes a_n) = b_0 \otimes \cdots \otimes b_m$, where $b_i = \prod_{f(j)=i}^{<} a_j$ is the product of the a_js, according to the total order of the fiber $f^{-1}(i)$.

Remark 15.4.2 Similar to Proposition 14.1.4, one can show that a k-module A is an associative algebra if and only if $\mathcal{L}^k(A) : \mathcal{F}(as) \to k\text{-mod}$ is a functor.

Remark 15.4.3 Note that $\Gamma(as)$ is a subcategory of $\mathcal{F}(as)$ and that we can also embed the category Δ into $\mathcal{F}(as)$ by taking the identity on objects and by sending an order-preserving function to the same function, together with the standard ordering on each fiber. There are also forgetful functors that forget orderings, so we get a diagram of categories:

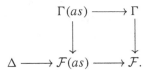

Actually, we prove in [**PR02**] that $\mathcal{F}(as)$ is isomorphic to the category ΔS from [**FiLo91**]: Every morphism f in $\mathcal{F}(as)$ has a unique composition as $f = g \circ h$, where g is order-preserving and h is a permutation.

For every functor $F : \mathcal{F}(as) \to k\text{-mod}$, you can define the cyclic homology of F, $HC_*(F)$ as the homology of the total complex of the cyclic bicomplex. We have to understand $HC_0(F)$ in terms of representable functors. But, $HC_0(F)$ is the zeroth homology of the total complex of

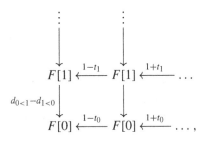

where $t_1 = F((0, 1))$ is the map induced by the transposition $(0, 1)$ and t_0 is the identity due to lack of space for permutations. Thus, $1 - t_0 = 0$ and

$$HC_0(F) = \text{coker}(d_{0<1} - d_{1<0} \colon F[1] \to F[0]).$$

As earlier, we rewrite that as

$$HC_0(F) = \text{coker}(d_{0<1} - d_{1<0} \colon \mathcal{F}(as)_1 \otimes_{\mathcal{F}(as)} F \to \mathcal{F}(as)_0 \otimes_{\mathcal{F}(as)} F),$$

with $\mathcal{F}(as)_n \colon \mathcal{F}(as)^o \to k\text{-mod}$, $\mathcal{F}(as)_n([m]) = k\{\mathcal{F}(as)([m], [n])\}$.

So, the (educated) guess is to define the functor b as

$$\mathsf{b} = \text{coker}(d_{0<1} - d_{1<0} \colon \mathcal{F}(as)_1 \to \mathcal{F}(as)_0),$$

and the content of [**PR02**, Theorem 1.3] is that this actually works.

Theorem 15.4.4 *For every* $F \colon \mathcal{F}(as) \to k\text{-mod}$,

$$\text{Tor}_*^{\mathcal{F}(as)}(\mathsf{b}, F) \cong HC_*(F),$$

in particular, $HC_*(A) \cong \text{Tor}_*^{\mathcal{F}(as)}(\mathsf{b}, \mathcal{L}^k(A))$ *for every associative* k*-algebra* A.

15.5 The Case of Gamma Homology

Gamma homology of a commutative k-algebra A with coefficients in an A-module M is defined in [**RoWh02**] in terms of an explicit chain complex. It is a homology theory that views A as an E_∞-algebra and takes its homology in that setting. Robinson develops an obstruction theory for E_∞-structures on ring spectra [**Ro03**], whose obstruction groups are the Gamma cohomology groups of the corresponding algebra of cooperations. Gamma cohomology groups can actually be described as André–Quillen cohomology in the category of simplicial (or non-negatively graded differential) E_∞-algebras [**BasRi04**, §2]. The obstruction groups in Robinson's approach [**Ro03**] are therefore isomorphic to the obstruction groups in the setting of Goerss–Hopkins [**GH04**], and both approaches have been used in important applications for establishing E_∞-structures on ring spectra.

As we take commutative algebras and as we want a homology theory with coefficients, our diagram category is Γ, the category of finite pointed sets.

The Loday functor for A and M is defined as earlier:

$$\mathcal{L}^k(A; M) \colon \Gamma \to k\text{-mod}, \quad [n] \mapsto M \otimes A^{\otimes n},$$

but as we have commutativity, $\mathcal{L}^k(A; M)$ is well-defined as a functor on Γ. We don't have to specify orderings of the fibers of morphisms. We denote by Γ_n the functor

$$\Gamma_n \colon \Gamma^o \to k\text{-mod}, \quad [m] \mapsto k\{\Gamma([m], [n])\}.$$

The zeroth Gamma homology group of A over k with coefficients in M, $\mathrm{H}\Gamma_0^k(A; M)$, is isomorphic to the first Hochschild homology group of A over k with coefficients in M, $\mathrm{HH}_1^k(A; M) \cong M \otimes A/b(M \otimes A^{\otimes 2})$, where the boundary is

$$b(m \otimes a_1 \otimes a_2) = ma_1 \otimes a_2 - m \otimes a_1 a_2 + a_2 m \otimes a_1 \text{ for } m \in M, a_1, a_2 \in A.$$

(For those of you who know about Kähler differentials, $\mathrm{HH}_1^k(A; M)$, and therefore $\mathrm{H}\Gamma_0^k(A; M)$, is isomorphic to $M \otimes_A \Omega_{A|k}^1$.) Thus, $\mathrm{H}\Gamma_0(A; M)$ is a quotient of $\mathcal{L}^k(A; M)[1]$ by a submodule generated by an image of $\mathcal{L}^k(A; M)[2]$ under the b-differential.

Definition 15.5.1 Let $t \colon \Gamma^o \to k\text{-mod}$ be the functor

$$t = coker(\chi \colon \Gamma_2 \to \Gamma_1),$$

with $\chi = d_0 - d_1 + d_2$.

Again, for every $F \colon \Gamma \to k\text{-mod}$, we come up with a suitable definition of $\mathrm{H}\Gamma_*(F)$, and we show the following in [**PR00**]:

Theorem 15.5.2 *For all* $F \colon \Gamma \to k\text{-mod}$,

$$\mathrm{H}\Gamma_*(F) \cong \mathrm{Tor}_*^\Gamma(t, F),$$

in particular,

$$\mathrm{H}\Gamma_*(A; M) \cong \mathrm{Tor}_*^\Gamma(t, \mathcal{L}^k(A; M)).$$

This functor homology description, together with the identification of $\mathrm{Tor}_*^\Gamma(t, F)$ as the stable homotopy groups of F, made it possible to calculate Gamma homology in some crucial classes of examples, such as abelian group algebras [**RiRo04**].

Example 15.5.3 [**RiRo04**, Proposition 3.2] A nice sample example is the Gamma homology of a polynomial ring in one variable $k[x]$ with coefficients in k with $k[x]$ acting on k via the augmentation that sends x to zero. Although $k[x]$ is free as a commutative k-algebra, it is far from being free as an E_∞-algebra, and Gamma homology detects this:

$$\mathrm{H}\Gamma_*(k[x]; k) \cong Hk_*H\mathbb{Z},$$

where $Hk_* H\mathbb{Z}$ is the stable k-homology of the Eilenberg–Mac Lane spectrum of the integers. This is huge. For instance, for $k = \mathbb{F}_p$, we get "half" of the dual of the Steenrod algebra:

$$(H\mathbb{F}_p)_* H\mathbb{Z} \cong \begin{cases} \mathbb{F}_2[\xi_1^2, \xi_2, \xi_3, \ldots], & \text{for } p = 2, \\ \mathbb{F}_p[\xi_1, \xi_2, \ldots] \otimes \Lambda_{\mathbb{F}_p}(\tau_1, \tau_2, \ldots), & \text{for } p \text{ odd.} \end{cases}$$

Here, as usual, ξ_i is of degree $2^i - 1$ for $p = 2$ and $2p^i - 2$ for p odd. The τ_is are of degree $2p^i - 1$.

Exercise 15.5.4 Take $k = A = M$, and show that $\mathrm{H}\Gamma_n(k; k) = 0$ for all $n \geq 0$ by identifying $\mathcal{L}^k(k; k)$.

15.6 Adjoint Base-Change

For a morphism of rings $f \colon R_1 \to R_2$, there is a flat base-change result. If R_2 is flat as an R_1-module via f, then there is an isomorphism of Tor groups

$$\mathrm{Tor}_n^{R_2}(M \otimes_{R_1} R_2, N) \cong \mathrm{Tor}_n^{R_1}(M, N_f)$$

for all right R_1-modules M, all left R_2-modules N, and all $n \geq 0$. Here, N_f is the R_1-module whose underlying abelian group is N, such that $r_1.n = f(r_1)n$ for all $r_1 \in R_1$, $n \in N$.

There is a general base-change result for Tor groups in functor categories. Let $\mathcal{C}$ and $\mathcal{D}$ be two small categories, and assume that there is an adjoint pair of functors $L \colon \mathcal{C} \to \mathcal{D}$ and $R \colon \mathcal{D} \to \mathcal{C}$. Then, we can precompose any functor $F \colon \mathcal{D}^o \to k\text{-mod}$ with $L^o \colon \mathcal{C}^o \to \mathcal{D}^o$ and obtain a functor $L^*(F) \colon \mathcal{C}^o \to k\text{-mod}$ and precomposition with R maps functors $G \colon \mathcal{C} \to k\text{-mod}$ to functors $R^*(G) \colon \mathcal{D} \to k\text{-mod}$. These changes of categories always preserve exactness.

Lemma 15.6.1 *Let $L \colon \mathcal{C} \to \mathcal{D}$ be a functor between small categories and let*

$$0 \Longrightarrow G' \overset{\tau}{\Longrightarrow} G \overset{\xi}{\Longrightarrow} G'' \Longrightarrow 0$$

be an exact sequence of functors. Then,

$$0 \Longrightarrow L^* G' \overset{\tau}{\Longrightarrow} L^* G \overset{\xi}{\Longrightarrow} L^* G'' \Longrightarrow 0$$

is exact.

Proof Exactness is checked objectwise, so the first sequence is exact if and only if

$$0 \longrightarrow G'(D) \overset{\tau_D}{\longrightarrow} G(D) \overset{\xi_D}{\longrightarrow} G''(D) \longrightarrow 0$$

is exact for every object D of $\mathcal{D}$. In particular, the sequence is exact for every object $L(C)$ of $\mathcal{D}$, and hence, the second sequence is exact. $\qquad\square$

So, we do not need flatness for preserving exactness, but we need that the corresponding functors build an adjoint pair in order to get isomorphisms on Tor groups.

Theorem 15.6.2 [**FFPS03**, p. 113, Lemma 2.7] *For any adjoint pair of functors* $\mathcal{C} \underset{R}{\overset{L}{\rightleftarrows}} \mathcal{D}$ *and any* $F: \mathcal{D}^o \to k$-mod, $G: \mathcal{C} \to k$-mod, *there are isomorphisms for all* $n \geq 0$:

$$\mathrm{Tor}_n^{\mathcal{C}}(L^*(F), G) \cong \mathrm{Tor}_n^{\mathcal{D}}(F, R^*(G)).$$

Proof　We first show the claim for functors of the form $F = k\{\mathcal{D}(-, D)\}$ and $G = k\{\mathcal{C}(C, -)\}$. In these cases, we do not get higher Tor groups, and from Corollary 15.1.5, we obtain

$$\begin{aligned} F \otimes_{\mathcal{D}} R^* G &\cong R^* G(D) \\ &= k\{\mathcal{C}(C, RD)\} \\ &\cong k\{\mathcal{C}(-, RD)\} \otimes_{\mathcal{C}} k\{\mathcal{C}(C, -)\} \\ &\cong k\{\mathcal{D}(L(-), D)\} \otimes_{\mathcal{C}} k\{\mathcal{C}(C, -)\} \\ &= L^* F \otimes_{\mathcal{C}} G. \end{aligned}$$

This proves the claim for the projective generators and hence for projectives. As the precompositions with L and R are exact functors, the general claim follows. $\qquad\square$

Exercise 15.6.3 Assume that $\mathcal{C}$ is a small category and $\mathcal{D}$ is a small category with an initial object $\varnothing$. Show that the functor $i_1: \mathcal{C} \to \mathcal{C} \times \mathcal{D}$, $i_1(C) = (C, \varnothing)$ is left adjoint to the projection functor $p_1: \mathcal{C} \times \mathcal{D} \to \mathcal{C}$ with $p_1(C, D) = C$. Formulate the consequence of the adjoint base-change result of Theorem 15.6.2.

16

Homology and Cohomology of Small Categories

There are many variants of the (co)homology of a small category with coefficients in a suitable system of coefficients. We start with a description of the one that is probably the most general one and then focus on some important special cases and examples. We close with a comparison result between functor homology and homology of small categories.

16.1 Thomason Cohomology and Homology of Categories

We follow [**GCNT13**] for the description of the Thomason (co)homology of categories and its properties.

Recall that the objects of the category Δ are of the form $[n] = \{0 < \cdots < n\}$, and thus, each of these objects is a category, with objects $\{0, \ldots, n\}$ and morphism coming from the poset structure. A morphism $h \in \Delta([n], [m])$ can be viewed as a functor from the poset category $[n]$ to $[m]$.

For a small category, the nerve of $\mathcal{C}$ is a functor $N(\mathcal{C}) \colon \Delta^o \to$ Sets, $[n] \mapsto N_n(\mathcal{C})$. As earlier, we use the category $N(\mathcal{C}) \backslash \Delta^o$, whose objects are of the form $([n], [f_n| \cdots |f_1])$, with $[n]$ an object of Δ and $[f_n| \cdots |f_1] \in N_n(\mathcal{C})$. A morphism

$$h \in N(\mathcal{C}) \backslash \Delta^o(([n], [f_n| \cdots |f_1]), ([m], [g_m| \cdots |g_1]))$$

is a morphism $h \in \Delta([m], [n])$, such that $N(\mathcal{C})(h)[f_n| \cdots |f_1] = [g_m| \cdots |g_1]$. This is called the *simplex category of* $\mathcal{C}$ in [**GCNT13**] and corresponds to the category of elements of a simplicial sets from Definition 10.2.7. We abbreviate $[f_n| \cdots |f_1]$ to F and $[g_m| \cdots |g_1]$ to G with $F(i - 1 < i) = f_i$ and $G(j - 1 < j) = g_j$. Then, the condition on a morphism h is that $G \circ h$ is equal to F. We need suitable coefficients for the (co)homology of the category $\mathcal{C}$.

Definition 16.1.1

- A *Thomason natural system with values in a category* $\mathcal{E}$ is a functor $M \colon N(\mathcal{C})\backslash\Delta^o \to \mathcal{E}$.
- A *contravariant Thomason natural system with values in a category* $\mathcal{E}$ is a functor $L \colon (N(\mathcal{C})\backslash\Delta^o)^o \to \mathcal{E}$.

If $h \in N(\mathcal{C})\backslash\Delta^o(([n], F), ([m], G))$ (hence $G \circ h = F$), then we get induced morphisms $M(h) \colon M([n], G \circ h) \to M(([m], G))$ and $L(h) \colon L(([m], G)) \to L([n], G \circ h)$.

Definition 16.1.2 Let $\mathcal{A}$ be an abelian category that is complete and cocomplete and has exact products and coproducts.

- The *Thomason cochain complex of* $\mathcal{C}$ *with coefficients in* $M \colon N(\mathcal{C})\backslash\Delta^o \to \mathcal{A}$ has as nth cochain group

$$C_T^n(\mathcal{C}; M) := \prod_{([n], F) \text{ object of } N(\mathcal{C})\backslash\Delta^o} M([n], F).$$

The coboundary map $\delta \colon C_T^n(\mathcal{C}; M) \to C_T^{n+1}(\mathcal{C}; M)$ is the alternating sum of the maps induced by the $\delta_i \in \Delta([n], [n+1])$:

$$\delta = \sum_{i=0}^{n} (-1)^i M(\delta_i).$$

The *nth Thomason cohomology group of* $\mathcal{C}$ *with coefficients in* M is the nth cohomology of the Thomason cochain complex. We denote it by $H_T^n(\mathcal{C}; M)$.

- Dually, the *nth Thomason chain group of* $\mathcal{C}$ *with coefficients in a contravariant Thomason natural system* L is

$$C_n^T(\mathcal{C}; L) = \bigoplus_{([n], F) \text{ object of } N(\mathcal{C})\backslash\Delta^o} L([n], F).$$

The differential is induced by $d = \sum_{i=0}^{n} (-1)^i L(\delta_i)$. The *$n$th Thomason homology group of* $\mathcal{C}$ *with coefficients in* L is the nth homology of the Thomason chain complex. We denote it by $H_n^T(\mathcal{C}; L)$.

Remark 16.1.3 Note that the Thomason chain complex actually is a chain complex associated with a simplicial object in $\mathcal{A}$, and similarly, the Thomason cochain complex is the cochain complex associated with a cosimplicial object in $\mathcal{A}$.

16.2 Quillen's Definition

There is a projection functor $P \colon N(\mathcal{C}) \backslash \Delta^o \to \mathcal{C}$ sending an object $([n], F)$ to $F(n)$. If $h \in \Delta([n], [m])$ with $F = G \circ h$, then $h(n) \leq m$, and, in particular, $C_n = D_{h(n)}$. The composite $g_m \circ \cdots \circ g_{h(n)+1}$ is a morphism from $F(n) = C_n$ to $D_m = G(m)$, and this is what we define as $P(h) \colon F(n) \to G(m)$.

Dually, there is a projection $P' \colon (N(\mathcal{C}) \backslash \Delta^o)^o \to \mathcal{C}$ defined by $P'([n], F) = F(0) = C_0$ and $P'(h) = g_{h(0)} \circ \cdots \circ g_1$, where

$$g_{h(0)} \circ \cdots \circ g_1 \colon D_0 \to D_{h(0)} = C_0.$$

Definition 16.2.1 Let $\mathcal{A}$ be again complete and cocomplete, with exact products and coproducts. Let $K \colon \mathcal{C} \to \mathcal{A}$ be a functor. Then, the *Thomason natural system associated with K* is the functor

$$K \circ P \colon N(\mathcal{C}) \backslash \Delta^o \to \mathcal{C} \to \mathcal{A}.$$

Dually, the precomposition with P' defines a *contravariant Thomason natural system associated with K*.

Therefore, Thomason homology and cohomology of categories are defined in this context. Explicitly, we get as a chain complex

$$C_n^T(\mathcal{C}, K \circ P) = \bigoplus_{[f_n|\cdots|f_1] \in N_n(\mathcal{C})} K(C_0),$$

and the corresponding nth homology group coincides with Quillen's definition of $H_n(\mathcal{C}, K)$ [**Q73**, p. 91].

Definition 16.2.2 The *homology of a small category $\mathcal{C}$ with coefficients in a functor $K \colon \mathcal{C} \to \mathcal{A}$* is

$$H_n(\mathcal{C}, K) := H_n^T(\mathcal{C}, K \circ P).$$

If L is a functor $L \colon \mathcal{C} \to k\text{-mod}$ that sends every morphism in $\mathcal{C}$ to an isomorphism, then Quillen's homology of $\mathcal{C}$ with coefficients in L has a topological interpretation. Recall from Theorem 11.5.5 that there is an equivalence of categories between covering spaces of $B\mathcal{C}$ and morphism-inverting functors $L \colon \mathcal{C} \to \text{Sets}$. If L has as a target the category of abelian groups, $L \colon \mathcal{C} \to \text{Ab}$, and is morphism-inverting, then its corresponding functor $L \colon \mathcal{C}[\text{Mor}(\mathcal{C})^{-1}] \to \text{Ab}$ is a *bundle of groups* on $B\mathcal{C}$ in the sense of [**Wh78**, VI.1], and hence, such Ls give rise to local coefficient systems for the homology of $B\mathcal{C}$ [**Wh78**, VI.2].

Theorem 16.2.3 [Q73] *Denote by $H_*(B\mathcal{C}; L)$ the singular homology of the topological space $B\mathcal{C}$, with coefficients in the local coefficient system associated with L. Then, there is an isomorphism*

$$H_*(\mathcal{C}; L) \cong H_*(B\mathcal{C}; L).$$

Proof   We consider the skeleton filtration

$$B\mathcal{C}^{(0)} \subset B\mathcal{C}^{(1)} \subset \cdots \subset B\mathcal{C}^{(n)} \subset B\mathcal{C}^{(n+1)} \subset \cdots$$

The associated spectral sequence has as E^1 term

$$E^1_{p,q} = H_{p+q}(B\mathcal{C}^{(p)}, B\mathcal{C}^{(p-1)}; L),$$

and this is trivial for $q > 0$. For $q = 0$, we get precisely the homology of the normalized chain complex associated with the simplicial k-module

$$[p] \mapsto \bigoplus_{[f_p| \cdots |f_1] \in N_p(\mathcal{C})} L(C_0),$$

and this calculates $H_p(\mathcal{C}; L)$. As the spectral sequence converges to the singular homology of $B\mathcal{C}$ with coefficients in L, we get the result. $\square$

Example 16.2.4   Let $\mathcal{C}_G$ be the category associated with a discrete group G and let k be a commutative ring with unit. Then, every morphism in $\mathcal{C}_G$ is an isomorphism, and every functor $L: \mathcal{C}_G \to k$-mod satisfies the requirement that $L(g)$ is an isomorphism; hence, we can apply Quillen's result and get that $H_*(\mathcal{C}_G; L)$ is isomorphic to the singular homology groups of the space BG, with coefficients in the local coefficient system defined by L. Let M denote $L(*)$. Then, in addition to being a k-module, M carries a G-action by k-linear maps via the isomorphisms $L(g)$ for $g \in G$. Hence, M is a module over the group algebra $k[G]$. The homology $H_*(\mathcal{C}_G; L)$ is nothing but the *group homology of G, with coefficients in the $k[G]$-module M*.

16.3 Spectral Sequence for Homotopy Colimits in Chain Complexes

Let k be a commutative ring, $\mathcal{D}$ be a small category, and $F: \mathcal{D} \to \mathrm{Ch}(k)_{\geq 0}$ be a functor. We saw in Remark 11.4.7 (11.4.1) that the homotopy colimit of F is the total complex associated with the bicomplex

$$\cdots \qquad\qquad \cdots \qquad\qquad \cdots$$

$$\downarrow \delta \qquad\qquad \downarrow -\delta$$

$$\bigoplus_{[f_2|f_1]\in N(\mathcal{D})_2} F(s(f_1))_0 \xleftarrow{\ \oplus d\ } \bigoplus_{[f_2|f_1]\in N(\mathcal{D})_2} F(s(f_1))_1 \xleftarrow{\ \oplus d\ } \cdots$$

$$\downarrow \delta \qquad\qquad \downarrow -\delta$$

$$\bigoplus_{[f_1]\in N(\mathcal{D})_1} F(s(f_1))_0 \xleftarrow{\ \oplus d\ } \bigoplus_{[f_1]\in N(\mathcal{D})_1} F(s(f_1))_1 \xleftarrow{\ \oplus d\ } \cdots$$

$$\downarrow \delta \qquad\qquad \downarrow -\delta$$

$$\bigoplus_{D\in\mathcal{D}} F(D)_0 \xleftarrow{\ \oplus d\ } \bigoplus_{D\in\mathcal{D}} F(D)_1 \xleftarrow{\ \oplus d\ } \cdots,$$

where $s(g)$ denotes the source of a morphism g, δ is the differential coming from the nerve, and d is the internal differential of the values of F.

To any such total complex, there are two standard spectral sequences [**W94**, 5.6] converging to the homology of the total complex. Filtration by columns gives a spectral sequence, whose E^1 term is

$$E^1_{p,q} = H^v_p\Big(\bigoplus_{[f_*|\cdots|f_1]\in N_*(\mathcal{D})} F(s(f_1))_q \Big),$$

and these groups are the homology groups of the category $\mathcal{D}$, with coefficients in the functor F_q, and thus, we obtain the following:

Theorem 16.3.1 *For every* $F\colon \mathcal{D} \to \mathrm{Ch}_{\geq 0}$, *there is a spectral sequence, with*

$$E^1_{p,q} = H_p(\mathcal{D}; F_q) \Rightarrow H_{p+q}\,\mathrm{hocolim}_{\mathcal{D}}\, F,$$

where $F_q\colon \mathcal{D} \to \mathrm{Ab}$ *is the functor given by* $F_q(D) = F(D)_q$.

16.4 Baues–Wirsching Cohomology and Homology

Recall the definition of a twisted arrow category of $\mathcal{C}$, $\mathcal{C}^\tau$, from Definition 4.5.1. Objects are the morphisms of $\mathcal{C}$, and a morphism in $\mathcal{C}^\tau$ from $f\colon C_1 \to C_2$ to $g\colon C_3 \to C_4$ is a pair of morphisms $(\alpha\colon C_3 \to C_1, \beta\colon C_2 \to C_4)$, such that $g = \beta \circ f \circ \alpha$:

$$\begin{array}{ccc} C_1 & \xleftarrow{\ \alpha\ } & C_3 \\ {\scriptstyle f}\downarrow & & \downarrow{\scriptstyle g} \\ C_2 & \xrightarrow{\ \beta\ } & C_4. \end{array}$$

Baues and Wirsching define in [**BW85**] the (co)homology of small categories with coefficients that are functors from the twisted arrow category of $\mathcal{C}$, $\mathcal{C}^\tau$, to the category of abelian groups. They call $\mathcal{C}^\tau$ the *category of factorizations of $\mathcal{C}$*.

Definition 16.4.1 Let $\mathcal{A}$ be an abelian category. A *natural system (in the sense of Baues and Wirsching)* on $\mathcal{C}$ is a functor M from $\mathcal{C}^\tau$ to $\mathcal{A}$.

Lemma 16.4.2 [GCNT13] *There is a projection functor* $\nu\colon N(\mathcal{C})\backslash\Delta^o \to \mathcal{C}^\tau$ *given by*

$$\nu([n], F) = f_n \circ \cdots \circ f_1 \colon C_0 \to C_n,$$

$$\nu(h\colon ([n], F) \to ([m], G)) = (g_{h(0)} \circ \cdots \circ g_1 \colon D_0 \to C_0, g_m \circ \cdots \circ g_{h(n)+1} \colon$$
$$C_n \to D_m)$$

Definition 16.4.3 Let $\mathcal{C}$ be a small category and let M be a natural system on $\mathcal{C}$. The *Baues–Wirsching homology of $\mathcal{C}$ with coefficients in M* is

$$H_*^{BW}(\mathcal{C}; M) = H_*^T(\mathcal{C}; M \circ \nu).$$

Example 16.4.4 An important class of examples of natural systems in the sense of Baues–Wirsching is given by functors

$$D\colon \mathcal{C}^o \times \mathcal{C} \to k\text{-mod}.$$

We can view D as a functor $D\colon \mathcal{C}^\tau \to k\text{-mod}$, by defining

$$D(f\colon C_1 \to C_2) := D(C_1, C_2),$$

and a morphism $(\alpha\colon C_3 \to C_1, \beta\colon C_2 \to C_4)$ induces a morphism

$$\alpha^* \circ \beta_* = \beta_* \circ \alpha^* = D(\alpha, \beta)\colon D(C_1, C_2) \to D(C_3, C_4):$$

(1) For a commutative ring k, let $F\colon \mathcal{C}^o \to k\text{-mod}$ and $G\colon \mathcal{C} \to k\text{-mod}$ be functors. Then,

$$F \otimes G\colon \mathcal{C}^o \times \mathcal{C} \to k\text{-mod}, \quad (C, C') \mapsto F(C) \otimes_k G(C')$$

gives rise to a Baues–Wirsching natural system.

(2) For an associative ring R, let $G_1, G_2 \colon \mathcal{C} \to R$-mod. Then,

$$R\text{-mod}(G_1, G_2) \colon \mathcal{C}^o \times \mathcal{C} \to R\text{-mod}, \quad (C, C') \mapsto R\text{-mod}(G_1(C), G_2(C'))$$

is a natural system.

(3) Let $G \colon \mathcal{C} \to k$-mod be an arbitrary functor and let $\underline{k} \colon \mathcal{C}^o \to k$-mod be the constant functor with value k. Then, $(\underline{k} \otimes G)(C_1, C_2) \cong G(C_2)$, and we can identify Baues–Wirsching homology with Quillen's homology:

$$H_*^{BW}(\mathcal{C}; \underline{k} \otimes G) \cong H_*(\mathcal{C}; G).$$

Exercise 16.4.5 Prove that

$$H_0^{BW}(\mathcal{C}; F \otimes G) = F \otimes_{\mathcal{C}} G \tag{16.4.1}$$

and that $H_{BW}^0(\mathcal{C}; R\text{-mod}(G_1, G_2))$ is the R-module of natural transformations from G_1 to G_2, and hence, H_0^{BW} calculates coends and H_{BW}^0 calculates ends.

16.5 Comparison of Functor Homology and Homology of Small Categories

Mamuka Jibladze and Teimuraz Pirashvili showed in [**JP91**] that under a mild projectivity assumption, functor homology can be expressed as Baues–Wirsching homology of categories with coefficients in a suitable tensor functor. They actually chose to work in the setting of cohomology. We do not claim any originality for the following result; it is a mere dualization of their work, and its proof can also be found in [**FFPS03**, p. 115, Proposition 2.10].

We prove the following result, extending the result from (16.4.1) to positive homological degree.

Theorem 16.5.1 ([**JP91**, Corollary 3.11]) *Let $\mathcal{C}$ be a small category, let k be a commutative ring, and let $F \colon \mathcal{C}^o \to k$-mod and $G \colon \mathcal{C} \to k$-mod be functors. If F or G has values in projective k-modules, then there is an isomorphism*

$$H_*^{BW}(\mathcal{C}; F \otimes G) \cong \mathrm{Tor}_*^{\mathcal{C}}(F, G).$$

The following lemma needs a bit of homological algebra.

Lemma 16.5.2 ([**JP91**, Proposition 3.7]) *Let $\varepsilon \colon C_\bullet \to C_{-1}$ be an augmented simplicial object in* cat. *Assume that all categories C_i are small and have the same set of objects, and assume that the $d_j s$ and $s_j s$ are the identities on objects. Also assume that for every pair of objects C, C' of the $C_i s$, the augmented simplicial set*

$$\varepsilon(C, C') \colon C_\bullet \to C_{-1}(C, C')$$

has a simplicial contraction h. Then, for any functor $D: \mathcal{C}_{-1}^o \times \mathcal{C}_{-1} \to k\text{-mod}$, there is a spectral sequence

$$E_{p,q}^1 = H_q^{BW}(\mathcal{C}_p; D) \Rightarrow H_{p+q}^{BW}(\mathcal{C}_{-1}; D).$$

Proof The Baues–Wirsching chain complex of the augmented simplicial object in cat gives rise to a double complex whose chain group in bidegree (p, q) is $C_p^{BW}(\mathcal{C}_q; D)$. To this bicomplex, we can associate two spectral sequences [**W94**, Section 5.6], both of which converge to the homology of the total complex. The first one arises from the filtration by columns. It takes vertical homology of entries in bidegree $(p, *)$, and this gives

$$^IE_{p,q}^1 = H_q^{BW}(\mathcal{C}_p; D) \Rightarrow H_{p+q} \text{Tot}(C_*^{BW}(\mathcal{C}_*; D)).$$

Filtration by rows gives the second spectral sequence, with

$$^{II}E_{p,q}^1 = H_q(C_p^{BW}(\mathcal{C}_*; D)) \Rightarrow H_{p+q} \text{Tot}(C_*^{BW}(\mathcal{C}_*; D)).$$

The simplicial contraction h gives rise to a chain contraction, and therefore, the $^{II}E^1$ term is concentrated in the $(q = 0)$-row with value $C_p^{BW}(\mathcal{C}_{-1}; D)$. Hence, this spectral sequence shows that the abutment is

$$H_p \text{Tot}(C_*^{BW}(\mathcal{C}_*; D)) \cong H_p^{BW}(\mathcal{C}_{-1}; D)$$

and that the first spectral sequence converges to $H_*^{BW}(\mathcal{C}_{-1}; D)$. $\square$

The next result investigates the homology of categories with coefficients in a tensor product, where the contravariant tensor factor is of the form $C' \mapsto k\{\mathcal{C}(C', C)\}$.

Lemma 16.5.3 ([**JP91**, Lemma 3.9]) *For every small category $\mathcal{C}$, every object C of $\mathcal{C}$, and every functor $G: \mathcal{C} \to k\text{-mod}$, we get*

$$H_*(\mathcal{C}; k\{\mathcal{C}(-, C)\} \otimes G) \cong \begin{cases} 0, & \text{if } * > 0, \\ G(C), & \text{for } * = 0. \end{cases}$$

Proof The value of the zeroth homology group is a consequence of Corollary 15.1.5 and Exercise 16.4.5:

$$k\{\mathcal{C}(-, C)\} \otimes_{\mathcal{C}} G \cong G(C).$$

We define a chain homotopy

$$(H_p: C_p^{BW}(\mathcal{C}; k\{\mathcal{C}(-, C)\} \otimes G) \to C_{p+1}^{BW}(\mathcal{C}; k\{\mathcal{C}(-, C)\} \otimes G))_p$$

as follows: We send $G(C)$ to $C_0^{BW}(\mathcal{C}; k\{\mathcal{C}(-, C)\} \otimes G)$ by mapping any element $m \in G(C)$ to $1_C \otimes m \in k\{\mathcal{C}(C, C)\} \otimes G(C)$. In higher degrees, we send an element $g \otimes m$ in component $[f_p | \cdots | f_1]$ in $C_p^{BW}(\mathcal{C}; k\{\mathcal{C}(-, C)\} \otimes G)$ to $1_C \otimes m$

in component $[g|f_p|\cdots|f_1]$, which is an element in $C_{p+1}^{BW}(\mathcal{C}; k\{\mathcal{C}(-, C)\} \otimes G)$. This defines a chain contraction, as claimed. □

As functor homology $\mathrm{Tor}_*^{\mathcal{C}}(F, G)$ vanishes for projective F, the same has to be true for the homology of categories with tensor coefficients.

Remark 16.5.4 ([**JP91**, Lemma 3.10]) We showed in Lemma 15.3.8 that the functors of the form $k\{\mathcal{C}(-, C)\}$ are a family of projective generators of the category $\mathrm{Fun}(\mathcal{C}^o, k\text{-mod})$; thus, every projective object of the category $\mathrm{Fun}(\mathcal{C}^o, k\text{-mod})$ receives an epimorphism from a direct sum of such functors. As we proved in Lemma 16.5.3 that the functors $k\{\mathcal{C}(-, C)\}$ are acyclic, we get the same result for any projective object F in $\mathrm{Fun}(\mathcal{C}^o, k\text{-mod})$:

$$H_*^{BW}(\mathcal{C}; F \otimes G) \cong \begin{cases} F \otimes_{\mathcal{C}} G, & \text{for } * = 0, \\ 0, & \text{for } * > 0. \end{cases}$$

The results so far suffice to establish an important spectral sequence that calculates functor homology via the homology of categories.

Proposition 16.5.5 [**JP91**, Theorem B] *Let $\mathcal{C}$ be a small category and let $F \colon \mathcal{C}^o \to k\text{-mod}$ and $G \colon \mathcal{C} \to k\text{-mod}$ be functors. Then, there is a spectral sequence*

$$E_{p,q}^2 = H_p^{BW}(\mathcal{C}; \mathrm{Tor}_q^k(F(-), G(-))) \Rightarrow \mathrm{Tor}_{p+q}^{\mathcal{C}}(F, G).$$

Here, $\mathrm{Tor}_q^k(F(-), G(-)) \colon \mathcal{C}^o \times \mathcal{C} \to k\text{-mod}$ *is the functor*

$$(C', C) \mapsto \mathrm{Tor}_q^k(F(C'), G(C)).$$

Proof We calculate $\mathrm{Tor}_*^{\mathcal{C}}(F, G)$ by choosing a projective resolution P_* in $\mathrm{Fun}(\mathcal{C}^o, k\text{-mod})$ of F and by taking the homology of $P_* \otimes_{\mathcal{C}} G$. The resolution P_* gives a functor

$$P_* \otimes G \colon \mathcal{C}^o \times \mathcal{C} \to \mathrm{Ch}_{\geq 0}(k),$$

where in chain degree ℓ, we take

$$(C', C) \mapsto P_\ell(C') \otimes G(C).$$

We know that for all ℓ, the homology $H_*^{BW}(\mathcal{C}; P_\ell \otimes G)$ is concentrated in degree $* = 0$. The hyperhomology spectral sequence [**W94**, 5.7.8] has

$$E_{p,q}^2 = H_p^{BW}(\mathcal{C}; H_q(P_* \otimes G))$$

and converges in our case to the $(p + q)$th homology of the complex $H_0^{BW}(\mathcal{C}; P_* \otimes G) = P_* \otimes_{\mathcal{C}} G$, which is nothing but $\mathrm{Tor}_{p+q}^{\mathcal{C}}(F, G)$.

For every object C' of $\mathcal{C}$, the chain complex $P_*(C')$ is a projective resolution of $F(C')$. Every $P_\ell(C')$ is a projective k-module because the projectivity of P_ℓ as a functor implies the projectivity on objects. Also, acyclicity is tested objectwise.

Thus, $H_q(P_* \otimes G)$ is nothing but $\mathrm{Tor}_q^k(F, G)$.   □

We can now return to the proof of the main result.

Proof of Theorem 16.5.1   By assumption, all values $F(C')$ are projective k-modules; hence, $\mathrm{Tor}_q^k(F, G)$ is trivial for $q \neq 0$ and the earlier spectral sequence collapses to the zeroth row

$$E_{p,0}^2 = H_p^{BW}(\mathcal{C}; F \otimes G),$$

and hence, by Proposition 16.5.5, the latter is isomorphic to $\mathrm{Tor}_p^{\mathcal{C}}(F, G)$.   □

Example 16.5.6 Gamma homology of a functor $F: \Gamma \to k\text{-mod}$ can be described as the homology of the category Γ,

$$\mathrm{Tor}_*^{\Gamma}(t, F) \cong H_*^{BW}(\Gamma; t \boxtimes F),$$

where $t \boxtimes F: \Gamma^o \times \Gamma \to k\text{-mod}$ is the bifunctor in the sense of Baues–Wirsching:

$$(t \boxtimes F)([m], [n]) = t[m] \otimes F[n].$$

An important application of Theorem 16.5.1 is the case where F is the constant functor with value k.

Corollary 16.5.7 *Let $\underline{k}: \mathcal{C}^o \to k\text{-mod}$ be the constant functor with value k. For every functor $G: \mathcal{C} \to k\text{-mod}$, Quillen's homology $H_*(\mathcal{C}; G)$ is isomorphic to $\mathrm{Tor}_*^{\mathcal{C}}(\underline{k}, G)$.*

Proof   By Theorem 16.5.1, we get

$$H_*(\mathcal{C}; G) \cong H_*^{BW}(\mathcal{C}; \underline{k} \otimes G) \cong \mathrm{Tor}_*^{\mathcal{C}}(\underline{k}, G).$$ □

Remark 16.5.8 We know that

$$H_0(\mathcal{C}, G) \cong \underline{k} \otimes_{\mathcal{C}} G,$$

and the latter is isomorphic to

$$\bigoplus_{C \text{ an object of } \mathcal{C}} G(C)/\sim,$$

where the relation $\sim$ takes into account the morphisms in $\mathcal{C}$. Hence,

$$H_0(\mathcal{C}, G) \cong \mathrm{colim}_{\mathcal{C}} G.$$

One can show that $\text{colim}_{\mathcal{C}}$ is a right-exact functor and that the higher homology groups $H_i(\mathcal{C}, G)$ are the left-derived functors of the colimit functor, often denoted by $\text{colim}_{\mathcal{C}}^i G$ [**GZ67**, Appendix II, Proposition 3.3]. This is compatible with Theorem 16.3.1, because here, we are considering functors with values in k-modules. You can view G as a functor to chain complexes, such that $G(C)$ is concentrated in chain degree zero. Therefore, the spectral sequence of Theorem 16.3.1 is concentrated in the zeroth column and $H_i(\mathcal{C}, G)$ coincides with the ith homology group of the homotopy colimit of G:

$$H_i \text{hocolim}_{\mathcal{C}} G \cong H_i(\mathcal{C}; G) = \text{colim}_{\mathcal{C}}^i G.$$

With $\mathcal{C} = \Delta^o$, the constant functor is the cosimplicial module $\underline{k}\colon (\Delta^o)^o = \Delta \to k\text{-mod}$, and Theorem 16.5.1 yields for every simplicial k-module G:

$$\text{Tor}_*^{\Delta^o}(\underline{k}, G) \cong H_*(\Delta^o, G).$$

In this particular case, we obtain an easy description of these groups.

Proposition 16.5.9 *For every simplicial k-module G,*

$$\text{Tor}_*^{\Delta^o}(\underline{k}, G) \cong H_*(C_*(G)).$$

Proof   We pick a particular resolution of the functor $\underline{k}$. We define $P_n = k\{\Delta([n], -)\}$ and take

$$d = \sum_{i=0}^n (-1)^i d_i : k\{\Delta([n], -)\} = P_n \to P_{n-1} = k\{\Delta([n-1], -)\}$$

as a boundary. We claim that P_* is a projective resolution of $\underline{k}$ in the category of cosimplicial k-modules. For a fixed object $[m]$ of Δ, we obtain a complex of k-modules:

$$\cdots \xrightarrow{d} k\{\Delta([n], [m])\} \xrightarrow{d} k\{\Delta([n-1], [m])\} \xrightarrow{d} \cdots \xrightarrow{d} k\{\Delta([0], [m])\}.$$

Its homology is isomorphic to the singular homology of Δ_m, and this is acyclic, with the zeroth homology being k, and thus, the complex $P_* \otimes_{\Delta^o} G$ calculates $\text{Tor}_*^{\Delta^o}(\underline{k}, G)$. As $P_n \otimes_{\Delta^o} G \cong G[n]$ and as the boundary gives precisely the boundary operator in the chain complex associated with G, the same complex also calculates $H_*(C_*(G))$. $\qquad\square$

Corollary 16.5.10 *Let A be an associative k-algebra and let M be an A-bimodule. Then,*

$$HH_*^k(A; M) \cong \text{Tor}_*^{\Delta^o}(\underline{k}, \mathcal{L}^k(A, M)) \cong H_*(\Delta^o, \mathcal{L}^k(A, M)).$$

References

[ARV10]   Jiří Adámek, Jiří Rosický, Enrico Maria Vitale, *What are sifted colimits?*
          Theory Appl. Categ. 23 (2010), 251–260. (2)

[Ad56]   J. Frank Adams, *On the cobar construction*, Proc. Nat. Acad. Sci. U.S.A.
          42 (1956), 409–412. 10.4.1

[Ad74]   J. Frank Adams, *Stable homotopy and generalised homology*, Reprint
          of the 1974 original. Chicago Lectures in Mathematics. University of
          Chicago Press, Chicago, IL (1995), x+373 pp. 2.2

[Ad78]   J. Frank Adams, *Infinite loop spaces*, Annals of Mathematics Studies,
          90. Princeton University Press, Princeton, NJ (1978), x+214 pp. 13.1

[AM10]   Marcelo Aguiar, Swapneel Mahajan, *Monoidal functors, species and
          Hopf algebras*, With forewords by Kenneth Brown and Stephen Chase
          and André Joyal. CRM Monograph Series, 29. American Mathematical
          Society, Providence, RI (2010), lii+784 pp. 9.7.1, 12.4.4

[BDR04]   Nils A. Baas, Bjørn Ian Dundas, John Rognes, *Two-vector bundles
          and forms of elliptic cohomology*, in: Topology, geometry and quantum
          field theory, London Mathematical Society Lecture Note Series, 308,
          Cambridge University Press, Cambridge (2004), 18–45. 9.1.2, 11.7

[BFSV03] Cornel Balteanu, Zbigniew Fiedorowicz, Roland Schwänzl, Rainer
          M. Vogt, *Iterated monoidal categories*, Adv. Math. 176 (2003), no. 2,
          277–349. 14.7, 14.7.2

[BaWe05] Michael Barr, Charles Wells, *Toposes, triples and theories*, Corrected
          reprint of the 1985 original, Repr. Theory Appl. Categ. 12 (2005), x+288
          pp. 6.2, 6.4, 6.6

[BE74]   Michael G. Barratt, Peter J. Eccles, Γ^+*-structures: I. A free group functor
          for stable homotopy theory*, Topology 13 (1974), 25–45. 12.3.9, 14.5.3

[BP72]   Michael Barratt, Steward Priddy, *On the homology of non-connected
          monoids and their associated groups*, Comment. Math. Helv. 47 (1972),
          1–14. 13.2.6

[BasRi04] Maria Basterra, Birgit Richter, *(Co-)homology theories for commuta-
          tive (S-)algebras*, in: Structured ring spectra, London Mathematical
          Society Lecture Note Series 315, Cambridge University Press (2004),
          115–131. 15.5

[Ba08]   Michael, A. Batanin, *The Eckmann–Hilton argument and higher oper-ads*, Adv. Math. 217 (2008), no. 1, 334–385. 14.8, 14.8.11

[BW85]   Hans Joachim Baues, Günther Wirsching, *Cohomology of small cate-gories*, J. Pure Appl. Algebra 38 (1985), no. 2–3, 187–211. 16.4

[Be69]   Jon Beck, *On H-spaces and infinite loop spaces*, in: Category theory, homology theory and their applications, III (Battelle Institute Con-ference, Seattle, Washington, 1968, Vol. 3), Springer, Berlin (1969), 139–153. 10.4.1, 10.5.4

[Ben67]   Jean Bénabou, *Introduction to bicategories*, Reports of the Midwest Category Seminar, Springer, Berlin (1967), 1–77. 9.6, 9.6.1

[BRY∞]   Yuri Berest, Ajay C. Ramadoss, Wai-Kit Yeung, *Representation homology of spaces and higher Hochschild homology*, preprint arXiv:1703.03505. 14.1

[Be97]   Clemens Berger, *Combinatorial models for real configuration spaces and E_n-operads*, Operads: Proceedings of Renaissance Conferences (Hartford, CT/Luminy, 1995), Contemporary Mathematics, 202. Ameri-can Mathematical Society, Providence, RI (1997), 37–52. 12.3.9

[Be99]   Clemens Berger, *Double loop spaces, braided monoidal categories and algebraic 3-type of space*, Higher homotopy structures in topology and mathematical physics (Poughkeepsie, NY, 1996). Contemporary Math-ematics, 227, American Mathematical Society, Providence, RI (1999), 49–66. 14.6

[Be07]   Clemens Berger, *Iterated wreath product of the simplex category and iterated loop spaces*, Adv. Math. 213 (2007), 230–270. 14.8, 14.8.1, 14.8.1, 14.8.2, 14.8.20

[B07]   Julia E. Bergner, *Simplicial monoids and Segal categories*, Categories in algebra, geometry and mathematical physics, Contemporary Mathemat-ics, 431. American Mathematical Society, Providence, RI (2007), 59–83. 10.14.4

[B18]   Julia E. Bergner, *The homotopy theory of $(\infty, 1)$-categories*, London Mathematical Society Student Texts, 90. Cambridge University Press, Cambridge (2018), xiv+273 pp. (document), 14.8.3

[B20]   Julia E. Bergner, *A survey of models for (∞, n)-categories*, in: Handbook of Homotopy Theory, edited by Haynes Miller, Chapman & Hall/CRC (2020), 263–295. 14.8.3

[BRez13b]   Julia E. Bergner, Charles Rezk, *Reedy categories and the Θ-construction*, Math. Z. 274 (2013), no. 1–2, 499–514. 14.8

[BRez13a]   Julia E. Bergner, Charles Rezk, *Comparison of models for (∞, n)-categories, I*, Geom. Topol. 17 (2013), no. 4, 2163–2202. 14.8, 14.8.3

[BRez∞]   Julia E. Bergner, Charles Rezk, *Comparison of models for (∞, n)-categories, II*, preprint arXiv:1406.4182. 14.8, 14.8.3

[BV68]   J. Michael Boardman, Rainer M. Vogt, *Homotopy-everything H-spaces*, Bull. Amer. Math. Soc. 74 (1968), 1117–1122. 12, 12.3.2

[BV73]   J. Michael Boardman, Rainer M. Vogt, *Homotopy invariant alge-braic structures on topological spaces*, Lecture Notes in Mathematics, Vol. 347, Springer-Verlag, Berlin and New York (1973), x+257 pp. 10.13, 12, 12.3.2

[Bö∞] Marcel Bökstedt, Topological Hochschild homology, preprint 1989. 9.7.1

[Bo94-1] Francis Borceux, *Handbook of Categorical Algebra 1, Basic Category Theory*, Encyclopedia of Mathematics and Its Applications, Cambridge University Press (1994), xvi+345 pp. (document), 1.3, 4, 4.1

[Bo94-2] Francis Borceux, *Handbook of Categorical Algebra 2, Categories and Structures*, Encyclopedia of Mathematics and Its Applications, Cambridge University Press (1994), xviii+443 pp. (document), 6.4, 6.6, 7.3, 9

[Bou87] Aldridge K. Bousfield, *On the homology spectral sequence of a cosimplicial space*, Amer. J. Math. 109 (1987), no. 2, 361–394. 10.4.1

[BF78] Aldridge K. Bousfield, Eric M. Friedlander, *Homotopy theory of Γ-spaces, spectra, and bisimplicial sets*, Geometric applications of homotopy theory (Proc. Conf., Evanston, Ill., 1977), II, pp. 80–130, Lecture Notes in Mathematics, 658, Springer, Berlin (1978). 14.3

[BK72] Aldridge K. Bousfield, Daniel M. Kan, *Homotopy limits, completions and localizations*, Lecture Notes in Mathematics, Vol. 304, Springer-Verlag, Berlin and New York (1972), v+348 pp. 10.8.6, 10.10, 10.10.4, 11.4, 11.4, 11.4, 11.4.1

[Br60] William Browder, *Homology operations and loop spaces*, Illinois J. Math. 4 1960, 347–357. 12.3.3

[CE56] Henri Cartan, Samuel Eilenberg, *Homological algebra*, With an appendix by David A. Buchsbaum. Reprint of the 1956 original. Princeton Landmarks in Mathematics. Princeton University Press, Princeton, NJ (1999), xvi+390 pp. 15.2.4

[CEF15] Thomas Church, Jordan S. Ellenberg, Benson Farb, *FI-modules and stability for representations of symmetric groups*, Duke Math. J. 164 (2015), no. 9, 1833–1910. 9.7.1

[Co74] Peter V. Z. Cobb, P_n-*spaces and n-fold loop spaces*, Bull. Amer. Math. Soc. 80 (1974), 910–914. 14.7, 14.8.2

[CLM76] Frederick R. Cohen, Thomas J. Lada, J. Peter May, *The homology of iterated loop spaces*, Lecture Notes in Mathematics, Vol. 533, Springer-Verlag, Berlin and New York (1976), vii+490 pp. 12.3.3, 12.3.3, 13.4.5

[DK01] James F. Davis, Paul Kirk, *Lecture notes in algebraic topology*, Graduate Studies in Mathematics, 35. American Mathematical Society, Providence, RI (2001), xvi+367 pp. 1.4.6, 13.4

[Day70a] Brian Day, *Construction of Biclosed Categories*, PhD Thesis, University of New South Wales (1970), available at www.math.mq.edu.au/~street/DayPhD.pdf. 9

[Day70b] Brian Day, *On closed categories of functors*, 1970 Reports of the Midwest Category Seminar, IV pp. 1–38, Lecture Notes in Mathematics, Vol. 137, Springer, Berlin (1970). 9, 9.8

[Dj∞] Aurélien Djament, *Décomposition de Hodge pour l'homologie stable des groupes d'automorphismes des groupes libres*, preprint arXiv:1510.03546. 15.3

[Do58] Albrecht Dold, *Homology of symmetric products and other functors of complexes*, Ann. of Math. (2) 68 (1958), 54–80. 10.11, 10.11.2, 10.11

[Du70]   Eduardo J. Dubuc, *Kan extensions in enriched category theory*, Lecture Notes in Mathematics, Vol. 145, Springer-Verlag, Berlin and New York (1970), xvi+173 pp. 4, 9

[D∞]   Daniel Dugger, *A primer on homotopy colimits*, notes available at pages.uoregon.edu/ddugger 11.4, (4), 11.4.6

[DS07]   Daniel Dugger, Brooke Shipley, *Topological equivalences for differential graded algebras*, Adv. Math. 212 (2007), no. 1, 37–61. 8.1.15

[DGM13]   Bjørn Ian Dundas, Thomas G. Goodwillie, Randy McCarthy, *The local structure of algebraic K-theory*, Algebra and Applications, 18. Springer-Verlag London, Ltd., London (2013), xvi+435 pp. (document), 14.3.15

[Dw74]   William G. Dwyer, *Strong convergence of the Eilenberg–Moore spectral sequence*, Topology 13 (1974), 255–265. 10.4.1

[DwHKS04]   William G. Dwyer, Philip S. Hirschhorn, Daniel M. Kan, Jeffrey H. Smith, *Homotopy limit functors on model categories and homotopical categories*, Mathematical Surveys and Monographs, 113. American Mathematical Society, Providence, RI (2004), viii+181 pp. 11.4.7

[DwSp95]   William G. Dwyer, Jan Spaliński, *Homotopy theories and model categories*, Handbook of algebraic topology, North-Holland, Amsterdam (1995), 73–126. (document)

[DL62]   Eldon Dyer, Richard K. Lashof, *Homology of iterated loop spaces*, Amer. J. Math. 84 (1962), 35–88. 12.3.3

[EH6162]   Benno Eckmann, Peter J. Hilton, *Group-like structures in general categories: I. Multiplications and comultiplications*, Math. Ann. 145 (1961/1962), 227–255. 8.1.3

[EZ50]   Samuel Eilenberg, Joseph A. Zilber, *Semi-simplicial complexes and singular homology*, Ann. of Math. (2) 51 (1950), 499–513. 10

[EM06]   Anthony D. Elmendorf, Michael A. Mandell, *Rings, modules, and algebras in infinite loop space theory*, Adv. Math. 205 (2006), no. 1, 163–228. 14.4

[Fi∞]   Zbigniew Fiedorowicz, *The symmetric bar construction*, preprint, available at https://people.math.osu.edu/fiedorowicz.1/ 14.6

[FiLo91]   Zbigniew Fiedorowicz, Jean-Louis Loday, *Crossed simplicial groups and their associated homology*, Trans. Amer. Math. Soc. 326 (1991), no. 1, 57–87. 14.6, 14.6.1, 15.4.3

[FP98]   Vincent Franjou, Teimuraz Pirashvili, *On the Mac Lane cohomology for the ring of integers*, Topology 37 (1998), no. 1, 109–114. 15.3

[FFPS03]   Vincent Franjou, Eric M. Friedlander, Teimuraz Pirashvili, Lionel Schwartz, *Rational representations, the Steenrod algebra and functor homology*, Panoramas et Synthèses, 16. Société Mathématique de France, Paris (2003), xxii+132 pp. 15.2.4, 15.3, 15.6.2, 16.5

[Fr09]   Benoit Fresse, *Modules over operads and functors*, Lecture Notes in Mathematics, 1967, Springer-Verlag, Berlin (2009), x+308 pp. 12

[GU71]   Peter Gabriel, Friedrich Ulmer, *Lokal Präsentierbare Kategorien*, Lecture Notes in Mathematics, Vol. 221. Springer-Verlag, Berlin and New York (1971), v+200 pp. 5.3

[GZ67]   Peter Gabriel, Michel Zisman, *Calculus of fractions and homotopy theory*, Ergebnisse der Mathematik und ihrer Grenzgebiete,

Band 35, Springer-Verlag New York, Inc., New York (1967), x+168 pp. 10, 10.6.9, 11.5, 11.5, 11.5.4, 11.5, 16.5.8

[GCNT13] Imma Galvez-Carrillo, Frank Neumann, Andrew Tonks, *Thomason cohomology of categories*, J. Pure Appl. Algebra 217 (2013), no. 11, 2163–2179. 16.1, 16.4.2

[GKZ94] Izrail M. Gelfand, Mikhail M. Kapranov, Andrey V. Zelevinsky, *Discriminants, resultants and multidimensional determinants*, Reprint of the 1994 edition. Modern Birkhäuser Classics. Birkhäuser Boston, Inc., Boston, MA (2008), x+523 pp. 12.3.4

[GH04] Paul G. Goerss, Michael J. Hopkins, *Moduli spaces of commutative ring spectra*, Structured ring spectra, London Math. Soc. Lecture Note Ser., 315, Cambridge Univ. Press, Cambridge (2004), 151–200. 10.10.4, 15.5

[GJ09] Paul G. Goerss, John F. Jardine, *Simplicial homotopy theory*, Reprint of the 1999 edition. Modern Birkhäuser Classics, Birkhäuser Verlag, Basel (2009), xvi+510 pp. 10, 10.4.1, 10.11, 10.12.12, 11.1, 11.7

[Gr76] Daniel Grayson, *Higher algebraic K-theory. II (after Daniel Quillen)*, Algebraic K-theory (Proc. Conf., Northwestern University, Evanston, Ill., 1976), Lecture Notes in Mathematics, Vol. 551, Springer, Berlin (1976), 217–240. (document), 13.3, 13.4.6

[Gro20] Moritz Groth, *A short course on ∞-categories*, in: Handbook of Homotopy Theory, edited by Haynes Miller, Chapman & Hall/CRC (2020), 549–617. (document), 11.8.2, 11.8.3, 11.8.13, 11.8.3

[G-SGA1] Alexander Grothendieck, *Revêtements étales et groupe fondamental (SGA 1)*, Séminaire de géométrie algébrique du Bois Marie 1960–61. Directed by A. Grothendieck. With two papers by M. Raynaud, Updated and annotated reprint of the 1971 original (Lecture Notes in Mathematics, 224, Springer, Berlin. Documents Mathématiques (Paris), 3. Société Mathématique de France, Paris (2003), xviii+327 pp. 9.6.5, 11.6

[Ha02] Allen Hatcher, *Algebraic topology*, Cambridge University Press, Cambridge (2002), xii+544 pp. 11.7, 13.1.2, 13.4

[HT10] Kathryn Hess, Andrew Tonks, *The loop group and the cobar construction*, Proc. Amer. Math. Soc. 138 (2010), no. 5, 1861–1876. 10.4.1

[Hi03] Philip S. Hirschhorn, *Model categories and their localizations*, Mathematical Surveys and Monographs, 99. American Mathematical Society, Providence, RI (2003), xvi+457. (document), 11.4.7, 11.4.1

[Hi∞] Philip S. Hirschhorn, *Notes on homotopy colimits and homotopy limits*, work in progress, available at www-math.mit.edu/~psh/notes/hocolim.pdf 11.4.10

[HV15] Eric Hoffbeck, Christine Vespa, *Leibniz homology of Lie algebras as functor homology*, J. Pure Appl. Algebra 219 (2015), no. 9, 3721–3742. 15.3.3, 15.3.3

[HM13] Karl H. Hofmann, Sidney A. Morris, *The structure of compact groups. A primer for the student – a handbook for the expert*, Third edition, revised and augmented. De Gruyter Studies in Mathematics, 25. De Gruyter, Berlin (2013), xxii+924 pp. 1.3

[Ho99] Mark Hovey, *Model categories*, Mathematical Surveys and Monographs, 63. American Mathematical Society, Providence, RI (1999), xii+209 pp. (document)

[Ho01] Mark Hovey, *Spectra and symmetric spectra in general model categories*, J. Pure Appl. Algebra 165 (2001), no. 1, 63–127. 10.15

[HoPS97] Mark Hovey, John H. Palmieri, Neil P. Strickland, *Axiomatic stable homotopy theory*, Mem. Amer. Math. Soc. 128 (1997), no. 610, x+114 pp. 8.4.7

[HoSS00] Mark Hovey, Brooke Shipley, Jeffrey H. Smith, *Symmetric spectra*, J. Amer. Math. Soc. 13 (2000), no. 1, 149–208. 10.15, 10.15, 10.15

[JP91] Mamuka Jibladze, Teimuraz Pirashvili, *Cohomology of algebraic theories*, J. Algebra 137 (1991), no. 2, 253–296. 16.5, 16.5.1, 16.5.2, 16.5.3, 16.5.4, 16.5.5

[Jo-a∞] André Joyal, *Disks, duality and Θ-categories*, preprint, available on https://ncatlab.org/nlab/files/JoyalThetaCategories.pdf. 10.3, 10.3, 14.8, 14.8.1, 14.8.11, 14.8.1, 14.8.1

[Jo-b∞] André Joyal, *Quasi-categories versus simplicial categories*, preprint, available at www.math.uchicago.edu/~may/IMA/Incoming/Joyal/QvsDJan9(2007).pdf 10.13

[Jo-c∞] André Joyal, *The Theory of Quasi-categories and its Applications*, book, available at www.mat.uab.cat/~kock/crm/hocat/advanced-course/Quadern45-2.pdf. 5.1.2, 10.13, 11.1, 11.8.3, 11.8.11, 11.8.12, 11.8.3

[Jo02] André Joyal, *Quasi-categories and Kan complexes*, Special volume celebrating the 70th birthday of Professor Max Kelly. J. Pure Appl. Algebra 175 (2002), no. 1-3, 207–222. 10.13

[JoSt93] André Joyal, Ross Street, *Braided tensor categories*, Adv. Math. 102 (1993), no. 1, 20–78. 8.6

[K57] Daniel M. Kan, *On c. s. s. complexes*, Amer. J. Math. 79 (1957), 449–476. 10

[K58a] Daniel M. Kan, *Adjoint functors*, Trans. Amer. Math. Soc. 87 (1958), 294–329. 2.4

[K58b] Daniel M. Kan, *A combinatorial definition of homotopy groups*, Ann. Math. 67 (1958), no. 2, 282–312. 10

[KT76] Daniel M. Kan, William Paul Thurston, *Every connected space has the homology of a $K(\pi, 1)$*, Topology 15 (1976), no. 3, 253–258. 13.4.7

[Ka93] Mikhail M. Kapranov, *The permutoassociahedron, Mac Lane's coherence theorem and asymptotic zones for the KZ equation*, J. Pure Appl. Algebra 85 (1993), no. 2, 119–142. 12.3.4

[Kash93] Takuji Kashiwabara, *On the homotopy type of configuration complexes. Algebraic topology*, (Oaxtepec, 1991), Contemp. Math., 146, Amer. Math. Soc., Providence, RI (1993), 159–170. 12.3.9

[K95] Christian Kassel, *Quantum groups*, Graduate Texts in Mathematics, 155. Springer-Verlag, New York (1995), xii+531 pp. 8.6, 8.6.9

[K64] Gregory Maxwell Kelly, *On MacLane's conditions for coherence of natural associativities, commutativities, etc*, J. Algebra 1 (1964), 397–402. 8.1

[K82]   Gregory Maxwell Kelly, *Basic concepts of enriched category theory*, London Mathematical Society Lecture Note Series, 64. Cambridge University Press, Cambridge-New York (1982), 245 pp. 8.1.13, 9

[KA56a]   Tatsuji Kudo; Shôrô Araki, *Topology of H_n-spaces and H-squaring operations*, Mem. Fac. Sci. Kyūsyū Univ. Ser. A. 10 (1956), 85–120. 12.3.3

[KA56b]   Tatsuji Kudo; Shôrô Araki, *On $H_*(\Omega^N(S^n); \mathbb{Z}_2)$*, Proc. Japan Acad. 32 (1956), 333–335. 12.3.3

[LTW79]   Dana May Latch, Robert W. Thomason, W. Stephen Wilson, *Simplicial sets from categories*, Math. Z. 164 (1979), no. 3, 195–214. 11.1.8

[La09]   Tyler Lawson, *Commutative Γ-rings do not model all commutative ring spectra*, Homology, Homotopy Appl. 11 (2009), no. 2, 189–194. 14.3.15

[Le∞]   Tom Leinster, Basic bicategories, notes available at the arxiv: math/9810017 9.6.5

[LR11]   Muriel Livernet, Birgit Richter, *An interpretation of E_n-homology as functor homology*, Math. Z. 269 (2011), no. 1-2, 193–219. 14.8.2, 15.3, 15.3.3, 15.3.3

[Lo94]   Jean-Louis Loday, *La renaissance des opérades*, Séminaire Bourbaki volume 1994/95, exposés 790–804, Astérisque no. 237 (1994–1995), Talk no. 792, p. 47–74. 12

[Lo98]   Jean-Louis Loday, *Cyclic homology*, Appendix E by María O. Ronco, Second edition. Chapter 13 by the author in collaboration with Teimuraz Pirashvili, Grundlehren der Mathematischen Wissenschaften 301, Springer-Verlag, Berlin (1998), xx+513 pp. 15, 15.3, 15.3, 15.3.10

[Lo04]   Jean-Louis Loday, *Realization of the Stasheff polytope*, Arch. Math. (Basel) 83 (2004), no. 3, 267–278. 12.3.4, 12.3.11, 12.3.4

[LoVa12]   Jean-Louis Loday, Bruno Vallette, *Algebraic operads*, Grundlehren der Mathematischen Wissenschaften 346, Springer, Heidelberg (2012), xxiv+634 pp. 12

[Lu09]   Jacob Lurie, *Higher topos theory*, Annals of Mathematics Studies, 170. Princeton University Press, Princeton, NJ (2009), xviii+925 pp. (document), 4.6.2, 5.1.2, 9, 10.13, 10.13.5, 11.1.3, 11.8.3, 11.8.11, 11.8.3, 11.8.3, 14.8.3

[Lu∞]   Jacob Lurie, *Higher Algebra*, book draft, available at www.math.harvard.edu/~lurie/. (document), 5.1.2, 11.6, 11.8.2, 11.8.2, 11.8.3, 12.3.4

[Ly99]   Manos Lydakis, *Smash products and Γ-spaces*, Math. Proc. Cambridge Philos. Soc. 126 (1999), no. 2, 311–328. 14.3.15

[MDS04]   John MacDonald, Manuela Sobral, *Aspects of monads*, in: Categorical foundations, Encyclopedia Math. Appl., 97, Cambridge Univ. Press, Cambridge (2004), 213–268.

[ML63]   Saunders Mac Lane, *Natural associativity and commutativity*, Rice Univ. Studies 49 (1963), no. 4, 28–46. 8.1, 8.1.8

[ML98]   Saunders Mac Lane, *Categories for the working mathematician*, Graduate Texts in Mathematics. 5, 2nd ed., Springer (1998), xii+314 pp. (document), 3.1, 4, 4.7, 6.6, 8.1.8, (1), 8.6, 8.6.7

[MMSS01] Michael A. Mandell, J. Peter May, Stefan Schwede, Brooke Shipley, Model categories of diagram spectra. Proc. London Math. Soc. (3) 82 (2001), no. 2, 441–512. 9.8

[MSS02] Martin Markl, Steve Shnider, Jim Stasheff, *Operads in algebra, topology and physics*, Mathematical Surveys and Monographs, 96, American Mathematical Society, Providence, RI (2002), x+349 pp. 12

[May67] J. Peter May, *Simplicial objects in algebraic topology*, Reprint of the 1967 original. Chicago Lectures in Mathematics, University of Chicago Press, Chicago, IL (1992), viii+161 pp. 10, 10.6.4, 10.12.12

[May72] J. Peter May, *The geometry of iterated loop spaces*, Lectures Notes in Mathematics, Vol. 271. Springer-Verlag, Berlin-New York (1972), viii+175 pp. 10.4.1, 10.4.1, 10.5.3, 10.5.4, 12, 12.3.2, 12.3.7

[May74] J. Peter May, E_∞ *spaces, group completions, and permutative categories*, New developments in topology (Proc. Sympos. Algebraic Topology, Oxford, 1972), London Math. Soc. Lecture Note Ser., No. 11, Cambridge Univ. Press, London (1974), pp. 61–93. 8.3.6, 13.1.5, 13.1, 13.4.4, 13.4.5

[May77] J. Peter May, E_∞ *ring spaces and* E_∞ *ring spectra*, with contributions by Frank Quinn, Nigel Ray, and Jørgen Tornehave, Lecture Notes in Mathematics, Vol. 577. Springer-Verlag, Berlin-New York (1977), 268 pp. 14.4

[May78] J. Peter May, *The spectra associated to permutative categories*, Topology 17 (1978), no. 3, 225–228. 14.4

[May99] J. Peter May, *A concise course in algebraic topology*, Chicago Lectures in Mathematics. University of Chicago Press, Chicago, IL, (1999), x+243 pp. 8.5

[MayTh78] J. Peter May, Robert W. Thomason, *The uniqueness of infinite loop space machines*, Topology 17 (1978), 205–224. 14.1, 14.1

[McCl01] John McCleary, *A user's guide to spectral sequences*, Second edition. Cambridge Studies in Advanced Mathematics, 58. Cambridge University Press, Cambridge (2001), xvi+561 pp. 10.4.1

[McCo69] Michael C. McCord, *Classifying spaces and infinite symmetric products*, Trans. Amer. Math. Soc. 146 (1969), 273–298. 8.5

[McD79] Dusa McDuff, *On the classifying spaces of discrete monoids*, Topology 18 (1979), no. 4, 313–320. 13.4.7

[M66] R. James Milgram, *Iterated loop spaces*, Ann. of Math. (2) 84 (1966), 386–403. 12.3.3

[Mi55] John Milnor, *On the construction FK*, notes from 1955, printed in J. Frank Adams, *Algebraic Topology – A Student's Guide*, London Mathematical Society Lecture Note Series (4), Cambridge University Press (1972), 118–136. 12.3.9

[Mi57] John Milnor, *The geometric realization of a semi-simplicial complex*, Ann. of Math. (2) 65 (1957), 357–362. 10.6, 10.6.4, 10.6

[Mo54/55] John C. Moore, *Homotopie des complexes monoïdaux*, I. Séminaire Henri Cartan, 7 no. 2 (1954-1955), Exp. No. 18, 8 p. 10.11, 10.12.10

[NS17] Thomas Nikolaus, Steffen Sagave, *Presentably symmetric monoidal ∞-categories are represented by symmetric monoidal model categories*, Algebr. Geom. Topol. 17 (2017), no. 5, 3189–3212. 11.8.3

[Ou10]   David Oury, *On the duality between trees and disks*, Theory Appl. Categ. 24 (2010), No. 16, 418–450. 10.3

[PS86]   Maria Cristina Pedicchio, Sergio Solimini, *On a "good" dense class of topological spaces*, J. Pure Appl. Algebra 42 (1986), no. 3, 287–295. 8.4.4

[P00a]   Teimuraz Pirashvili, *Hodge decomposition for higher order Hochschild homology*, Ann. Sci. École Norm. Sup. (4) 33 (2000), 151–179. 15

[P00b]   Teimuraz Pirashvili, *Dold-Kan type theorem for Γ-groups*, Math. Ann. 318 (2000), no. 2, 277–298.

[P02]   Teimuraz Pirashvili, *On the PROP corresponding to bialgebras*, Cah. Topol. Géom. Différ. Catég. 43 (2002), no. 3, 221–239. 14.1.6

[PR00]   Teimuraz Pirashvili, Birgit Richter, *Robinson-Whitehouse complex and stable homotopy*, Topology 39 (2000), no. 3, 525–530. (4), 15, 15.3, 15.3.3, 15.5

[PR02]   Teimuraz Pirashvili, Birgit Richter, *Hochschild and cyclic homology via functor homology*, K-Theory 25 (2002), no. 1, 39–49. 15.3, 15.3.3, 15.3.9, 15.4.1, 15.4.3, 15.4

[Q67]   Daniel G. Quillen, *Homotopical algebra*, Lecture Notes in Mathematics, No. 43 Springer-Verlag, Berlin-New York (1967), iv+156 pp. (document), 10

[Q69]   Daniel G. Quillen, *Rational homotopy theory*, Ann. of Math. (2) 90 (1969), 205–295. 10.4.1

[Q70]   Daniel G. Quillen, *On the (co-)homology of commutative rings*, Applications of Categorical Algebra (Proc. Sympos. Pure Math., Vol. XVII, New York, 1968), Amer. Math. Soc., Providence, R.I. (1970), pp. 65–87. 5.1

[Q73]   Daniel G. Quillen, *Higher algebraic K-theory: I*, in: Algebraic K-theory, I: Higher K-theories (Proc. Conf., Battelle Memorial Inst., Seattle, Wash., 1972), Lecture Notes in Math., Vol. 341, Springer, Berlin (1973), 85–147. (document), 10.8, 11.2, 11.5, 11.5, 11.5.6, 11.6, 11.7.2, 11.7, 11.7.7, 11.7.8, 16.2, 16.2.3

[Q94]   Daniel G. Quillen, *On the group completion of a simplicial monoid*, Appendix Q in: Eric Friedlander, Barry Mazur, *Filtrations on the homology of algebraic varieties*, Mem. Amer. Math. Soc. 110 (1994), no. 529, x+110 pp. 13.4.2

[Ree∞]   Christopher Leonard Reedy, *Homotopy theory of model categories*, unpublished preprint, available at www-math.mit.edu/~psh/ 10.8.6

[Re93]   Christophe Reutenauer, *Free Lie algebras*, London Mathematical Society Monographs. New Series, 7. Oxford Science Publications. The Clarendon Press, Oxford University Press, New York (1993), xviii+269 pp. 2.4.2

[Rez10]   Charles Rezk, *A Cartesian presentation of weak n-categories*, Geom. Topol. 14 (2010), no. 1, 521–571. See also: Charles Rezk, *Correction to 'A Cartesian presentation of weak n-categories'*, Geom. Topol. 14 (2010), no. 4, 2301–2304. 14.8, 14.8.3

[Rez∞]   Charles Rezk, *When are homotopy colimits compatible with homotopy pullbacks*, note available at www.faculty.math.illinois.edu/~rezk/i-hate-the-pi-star-kan-condition.pdf 11.4.10

[Ri03]   Birgit Richter, *Symmetry properties of the Dold-Kan correspondence*, Math. Proc. Cambridge Philos. Soc. 134 (2003), no. 1, 95–102. 9.3, 12.4, 12.4.2, 12.4.3, 12.4

[RiRo04]   Birgit Richter, Alan Robinson, *Gamma homology of group algebras and of polynomial algebras*, Homotopy theory: relations with algebraic geometry, group cohomology, and algebraic K-theory, Contemp. Math., 346, Amer. Math. Soc., Providence, RI (2004), 453–461. 15.5, 15.5.3

[RiSa∞]   Birgit Richter, Steffen Sagave, *A strictly commutative model for the cochain algebra of a space*, preprint, arXiv:1801.01060. 11.4.7

[RiS17]   Birgit Richter, Brooke Shipley, *An algebraic model for commutative $H\mathbb{Z}$-algebras*, Algebraic and Geometric Topology 17 (2017), 2013–2038. 9.8.9, 10.15.3

[Rie14]   Emily Riehl, *Categorical Homotopy Theory*, New Mathematical Monographs, 24. Cambridge University Press, Cambridge (2014), xviii+352 pp. 4, 8.5, 9, 11.4.7

[Rie16]   Emily Riehl, *Category theory in context*, Aurora Dover Modern Math Originals, Mineola, NY: Dover Publications (2016), xvii, 234 p. (document), 6.3

[Ro03]   Alan Robinson, *Gamma homology, Lie representations and E_∞ multiplications*, Invent. Math. 152 (2003), no. 2, 331–348. 15.5

[RoWh02]   Alan Robinson, Sarah Whitehouse, *Operads and Γ-homology of commutative rings*, Math. Proc. Cambridge Philos. Soc. 132 (2002), 197–234. (4), 15.5

[RG12]   Beatriz Rodríguez González, *Simplicial descent categories*, J. Pure Appl. Algebra 216 (2012), no. 4, 775–788. 11.4.7

[RG14]   Beatriz Rodríguez González, *Realizable homotopy colimits*, Theory Appl. Categ. 29 (2014), No. 22, 609–634. 11.4.7

[R07]   Jiří Rosický, *On homotopy varieties*, Adv. Math. 214 (2007), no. 2, 525–550. 11.4.1, 11.4.9

[SaSc12]   Steffen Sagave, Christian Schlichtkrull, *Diagram spaces and symmetric spectra*, Adv. Math. 231 (2012), no. 3-4, 2116–2193. 9.7.1, 9.8.9, 13.3.8, 14.5

[Sc08]   Christian Schlichtkrull, *The homotopy infinite symmetric product represents stable homotopy*, Algebr. Geom. Topol. 7 (2007), 1963–1977. 14.5.3

[Sc09]   Christian Schlichtkrull, *Thom spectra that are symmetric spectra*, Doc. Math. 14 (2009), 699–748. 14.5

[ScSo16]   Christian Schlichtkrull, Mirjam Solberg, *Braided injections and double loop spaces*, Trans. Amer. Math. Soc. 368 (2016), no. 10, 7305–7338. 8.6.8, 14.6, 14.6, 14.6.3, 14.6, 14.6.6

[Sch70]   Horst Schubert, Kategorien. I, II, Heidelberger Taschenbücher, Bände 65, 66, Springer (1970). (document), 1.4.5, 2.3.3

[Schw99]   Stefan Schwede, *Stable homotopical algebra and Γ-spaces*, Math. Proc. Cambridge Philos. Soc. 126 (1999), no. 2, 329–356. 14.3.15

[Schw∞]   Stefan Schwede, *An untitled book project about symmtric spectra*, file available at www.math.uni-bonn.de/people/schwede/SymSpec-v3.pdf 10.15

[Se68]   Graeme Segal, *Classifying spaces and spectral sequences*, Inst. Hautes Études Sci. Publ. Math. No. 34 (1968), 105–112.

[Se73]   Graeme Segal, *Configuration-spaces and iterated loop-spaces*, Invent. Math. 21 (1973), 213–221. 13.4.5

[Se74]   Graeme Segal, *Categories and cohomology theories*, Topology 13 (1974), 293–312. 9.7.1, 10.6.10, 10.9.1, 10.9.2, 10.9.4, 10.9, 11.2, 14.2.3, 14.2, 14.3, (2), 14.3.13, 14.4

[Sh96]   Brooke E. Shipley, *Convergence of the homology spectral sequence of a cosimplicial space*, Amer. J. Math. 118 (1996), no. 1, 179–207. 10.4.1

[Sł03]   Jolanta Słomińska, *Decompositions of the category of noncommutative sets and Hochschild and cyclic homology*, Cent. Eur. J. Math. 1 (2003), no. 3, 327–331. 15.3.10

[Sm70]   Larry Smith, *Lectures on the Eilenberg-Moore spectral sequence*, Lecture Notes in Mathematics, Vol. 134 Springer-Verlag, Berlin-New York (1970), vii+142 pp. 10.4.1

[Smi89]   Jeffrey Henderson Smith, *Simplicial group models for $\Omega^n S^n X$*, Israel J. Math. 66 (1989), no. 1-3, 330–350. 12.3.9

[Sr96]   Vasudevan Srinivas, *Algebraic K-theory*, Second edition. Progress in Mathematics, 90. Birkhäuser Boston, Inc., Boston, MA, 1996. xviii+341 pp. 11.7

[St99]   Richard P. Stanley, *Enumerative combinatorics. Vol. 2*, Cambridge Studies in Advanced Mathematics, 62. Cambridge University Press, Cambridge (1999), xii+581 pp. 12.3.4

[Sta63]   James D. Stasheff, *Homotopy associativity of H-spaces. I, II*, Trans. Amer. Math. Soc. 108 (1963), 275–292 and 293–312. 12.3.4, 12.3.4

[Sta92]   James D. Stasheff, *Differential graded Lie algebras, quasi-Hopf algebras and higher homotopy algebras*, Quantum groups (Leningrad, 1990), Lecture Notes in Math., 1510, Springer, Berlin (1992), 120–137. 14.6

[Ste43]   Norman E. Steenrod, *Homology with local coefficients*, Ann. of Math. (2) 44, (1943), 610–627. 1.4.6

[Ste67]   Norman E. Steenrod, *A convenient category of topological spaces*, Michigan Math. J. 14 (1967), 133–152. 8.5

[Sto93]   Christopher R. Stover, *The equivalence of certain categories of twisted Lie and Hopf algebras over a commutative ring*, J. Pure Appl. Algebra 86 (1993), no. 3, 289–326. 9.7.1

[Su57]   Masahiro Sugawara, *A condition that a space is group-like*, Math. J. Okayama Univ. 7 (1957), 123–149. 12.3.4

[Sw75]   Robert M. Switzer, *Algebraic topology – homotopy and homology*, Reprint of the 1975 original; Classics in Mathematics. Springer-Verlag, Berlin (2002), xiv+526 pp. 2.2

[T77]   Robert W. Thomason, *Homotopy colimits in CAT, with applications to algebraic K-theory and loop space theory*, Thesis (Ph.D.) Princeton University. ProQuest LLC, Ann Arbor, MI (1977), 132 pp. 5.5.5

[T79]   Robert W. Thomason, *Homotopy colimits in the category of small categories*, Math. Proc. Cambridge Philos. Soc. 85 (1979), no. 1, 91–109. 5.5, 9.6.5

[T82]   Robert W. Thomason, *First quadrant spectral sequences in algebraic K-theory via homotopy colimits*, Comm. Algebra 10 (1982), no. 15, 1589–1668. 11.4

[tD08]   Tammo tom Dieck, *Algebraic topology*, EMS Textbooks in Mathematics. European Mathematical Society (EMS), Zürich (2008), xii+567 pp. 8.5, 8.5, 8.5.8, 8.5.10, 8.5.17

[V71]   Rainer M. Vogt, *Convenient categories of topological spaces for homotopy theory*, Arch. Math. (Basel) 22 (1971), 545–555. 8.5

[V77]   Rainer M. Vogt, *Commuting homotopy limits*, Math. Z. 153 (1977), no. 1, 59–82. 11.4.10

[W94]   Charles A. Weibel, *An introduction to homological algebra*, Cambridge Studies in Advanced Mathematics, 38. Cambridge University Press, Cambridge (1994), xiv+450 pp. (document), 10.4.1, 10.11, 16.3, 16.5, 16.5

[W13]   Charles A. Weibel, *The K-book. An introduction to algebraic K-theory*, Graduate Studies in Mathematics, 145. American Mathematical Society, Providence, RI (2013), xii+618 pp. (document), 11.3.4

[We05]   Michael Weiss, *What does the classifying space of a category classify?*, Homology Homotopy Appl. 7 (2005), no. 1, 185–195. 10.9.2

[Wh78] George W. Whitehead, *Elements of homotopy theory*, Graduate Texts in Mathematics, 61. Springer-Verlag, New York-Berlin (1978), xxi+744 pp. 1.4.6, 16.2

[Z16]   Stephanie Ziegenhagen, E_n-*cohomology with coefficients as functor cohomology*, Algebr. Geom. Topol. 16 (2016), no. 5, 2981–3004. 15.3

Index

Printed in the United States
by Baker & Taylor Publisher Services